D0026594

Advances in EXERCISE IMMUNOLOGY

Laurel T. Mackinnon, PhD
The University of Queensland, Australia

Human Kinetics

Library of Congress Cataloging-in-Publication Data

Mackinnon, Laurel T., 1953-
Advances in exercise immunology / Laurel T. Mackinnon.
p. cm.
Includes bibliographical references and index.
ISBN 0-88011-562-9
1. Exercise–Immunological aspects. I. Title.
[DNLM: 1. Exercise–physiology. 2. Immunity. QT 260M158a 1999]
QP301.M159 1999
616.07'9–dc21
DNLM/DLC
for Library of Congress 98-38797
CIP

ISBN: 0-88011-562-9

Aquisition Editor: Michael Bahrke, PhD; **Developmental Editor:** Christine Drews; **Assistant Editor:** John Wentworth; **Copyeditor:** Joyce Sexton; **Proofreader:** Myla Smith; **Indexer:** Barbara E. Cohen; **Graphic Designer:** Nancy Rasmus; **Graphic Artist:** Roberta Edwards; **Photo Editor:** Boyd LaFoon; **Cover Designer:** Jack Davis; **Printer:** Edwards Brothers

Printed in the United States of America 10 9 8 7 6 5 4 3 2 1

Human Kinetics
Web site: http://www.humankinetics.com/

United States: Human Kinetics
P.O. Box 5076
Champaign, IL 61825-5076
1-800-747-4457
e-mail: humank@hkusa.com

Canada: Human Kinetics
475 Devonshire Road Unit 100
Windsor, ON N8Y 2L5
1-800-465-7301 (in Canada only)
e-mail: humank@hcanada.com

Europe: Human Kinetics, P.O. Box IW14
Leeds LS16 6TR, United Kingdom
(44) 1132 781708
e-mail: humank@hkeurope.com

Australia: Human Kinetics
57A Price Avenue
Lower Mitcham, South Australia 5062
(088) 277 1555
e-mail: humank@hkaustralia.com

New Zealand: Human Kinetics
P.O. Box 105-231, Auckland 1
(09) 523 3462
e-mail: humank@hknewz.com

To Ian, Scott, and Cameron

In memory of Lawrence Traeger

Contents

Preface

The field of exercise immunology has expanded rapidly in the past six years, attracting interest from researchers in diverse fields such as exercise science, medicine, immunology, physiology, and behavioral science. Interest in the immune response to exercise has arisen for many reasons. Athletes, coaches, and team physicians want to keep athletes healthy during training and competition. They have long believed that athletes are susceptible to illness, in particular upper respiratory tract infection, during periods of intense training and after major competition. Some illnesses adversely affect the athlete's ability to train and compete, and continued training or competition during illness may be detrimental to the athlete's health. In addition, in elite athletes, frequent illness is associated with overtraining syndrome, a neuroendocrine disorder resulting from excessive training.

The attention to exercise immunology is also shaped by community interest in health promotion and preventive medicine. Physical inactivity is now accepted as a major risk factor for several diseases, including heart disease, obesity, non-insulin-dependent diabetes, hypertension, and osteoporosis, which are major causes of death and disability in developed countries; regular moderate exercise appears to be an important strategy in prevention of these diseases. Researchers are now focusing on other diseases with significant lifestyle-associated risk factors such as cancer, and there is evidence to suggest that physical activity also helps lower the risk of certain cancers.

Exercise has come to be prescribed as adjunct therapy for certain diseases, including cancer, arthritis, and human immune deficiency (HIV) infection. Exercise was first introduced to counteract the physically debilitating effects of illness and treatment or to improve the patient's psychological state. Because the immune system is intimately involved in some of these diseases (e.g., certain cancers, HIV infection, rheumatoid arthritis), there is interest in studying the immune response to exercise in order to determine its effects on disease progress.

Clinical applications of exercise and immunology also extend to understanding and preventing compromised immune function associated with other conditions, such as aging, immobility, and spaceflight. Although at present there is only a limited body of research literature, attention is increasingly directed to the question whether exercise may reverse or prevent such adverse changes, for example, whether exercise training in astronauts may provide a "countermeasure" against immune suppression occurring during long-term spaceflight, or whether a lifetime of moderate exercise may prevent age-related changes in immune function.

Finally, there is continued interest in exercise and immunology in terms of "psychoneuroimmunology" or "behavioral immunology." It has been known for some time that there is significant interaction between the immune and neuroendocrine systems; indeed, these two systems share many messenger molecules (e.g., cytokines). Stress has long been identified as a modulator of immune function. Exercise can be considered a form of physical stress, because many of the hormones capable of immunomodulation also increase during exercise. Exercise may provide a unique model for studying adaptation to stress because exercise may be easily quantified and is highly reproducible; moreover, repeated exposure (i.e., exercise training) induces hormonal and physiological adaptation.

In this book, I begin by considering whether athletes are more susceptible to illness than the general public, and then review the literature on the effects of exercise on resistance to infectious disease and on the question whether infection adversely influences exercise performance. Chapter 2 provides a very brief and cursory overview of the immune system. Because the immune system is exceedingly complex, and appearing more so with continuing advances in technology, I have deliberately limited the scope of this section, intending only to introduce the nonimmunologist to relevant terminology and concepts. The interested reader may wish to consult any number of excellent recent immunology texts for more detail.

Chapters 3 through 7 address the effects of exercise on different aspects of immunity, focusing on the responses to exercise of immune cell function and other mediators of immunity; separate chapters are devoted to discussing the effects of exercise on specific aspects of immune function. Chapter 8 focuses on recent interest in the potential clinical applications of exercise immunology; I use the word "potential" with purpose, since, at present, research has just begun to empirically extend knowledge in this field to clinical applications. Finally, chapter 9 concludes with an overview of the current state of knowledge and some predictions about future directions of this exciting field of study.

The book is designed to be read from different perspectives, depending on the reader's background and interest. Those wishing to gain a comprehensive overview of the topic may wish to read it in sequential order. Alternatively, each chapter has been written as a separate review, and selected chapters may be of value to those seeking information about the effects of exercise on particular aspects of immune function. Throughout the book I have intentionally focused on the effects of exercise on the human immune system, bringing in examples from research on animal models only when human data are lacking or when a particular animal study provides additional important information. Although animal models can yield vital data on regulatory mechanisms that may not be obtainable in humans, there are major differences in terminology and function between species, and it is beyond the scope of this book to address all these issues.

As some readers may be aware, in 1992 Human Kinetics published my monograph, *Exercise and Immunology,* the first book written on this topic. Although the current book maintains a similar overall structure (e.g., chapter titles and parts of the first three chapters), it is very much an entirely new book, in both its breadth and depth of inquiry as well as its discussion of relevant issues. In the preface to the 1992 monograph, I noted that the study of exercise immunology was in its infancy. Although those of us working in this area at that time expected an increase in interest in the coming years, I think we have been pleasantly surprised by the tremendous growth in the number of active research groups in this field. Indeed, there has been far more published in the seven years since the monograph than in all the time before. I think it is fair to claim, then, that the field of exercise immunology has entered a mature phase of inquiry directed at more complete understanding of underlying mechanisms and broad clinical applications.

Acknowledgments

There are many people to thank for their support in the writing of this book: Doctors Edith Peters-Futre (Republic of South Africa), Hoger Gabriel (Germany), David Rowbottom (Australia), David Nieman, Gerald Sonnenfeld, and Jeff Woods (USA), who either pointed me in the direction of relevant references or provided me access to their most recent data; Jane McCosker, Vicki Gedge, and Natalie Hiscock for their assistance with library searches; Rainer Martens, other staff, and especially Christine Drews at Human Kinetics for their support, patience, and editorial expertise; my colleagues in the Department of Human Movement Studies who, through their cooperative spirit, have created a working environment that encourages completion of such projects; and finally, my husband and children for their constant support and tolerance of my occupation of our home study.

Chapter 1

Exercise and Resistance to Infectious Illness

© Human Kinetics

Physical fatigue, whether caused by exercise or manual work, has long been considered a factor affecting susceptibility to illness. Work early in this century linked physical fatigue with increased susceptibility to, and severity of illness resulting from, serious infections such as poliomyelitis. More recently, evidence has been mounting to support the long-held perception that athletes are susceptible to certain illnesses during intense training and major competition. At the same time, there is a common belief that regular moderate exercise reduces susceptibility to infectious illness, such as the common cold. Although not yet extensively studied, recent epidemiological evidence on athletes (mainly distance runners) is consistent with such a dual effect of exercise on immunity to infectious illness; that is, prolonged periods of intense exercise training increase, and regular moderate exercise reduces, susceptibility to infectious illness.

Historical Perspectives

There is a long history of a perceived association between physical activity and health in many cultures. In ancient Greece, Aristotle succinctly noted:

> "A man falls into ill health as a result of not caring for exercise."

In 17th-century England, John Dryden observed in verse:

> Better to hunt in fields for health unbought,
>
> Than fee the doctor for naseous draught.
>
> The wise, for cure, on exercise depend;
>
> God never made his work for man to mend.
>
> Epistle to John Driden of Chesterton

Dryden clearly recognized the economic benefits of preventive medicine.

Work early in this century was concerned with the major health risks of that time—issues related to hygiene, manual labor in industrial settings, and common infectious diseases such as influenza and polio. In the 1940s, Russell published retrospective studies showing that the severity of paralysis resulting from polio was directly related to the intensity of physical activity at the onset of infection (Russell 1947, 1949). He classified patients by the amount of physical activity performed during the early stage of infection and by the severity of paralysis resulting from infection (figure 1.1). Of patients who did little or mild exercise (open bars in figure 1.1), more than 80% recovered completely with no long-term ill effects. In contrast, of those who performed intense physical activity after the onset of symptoms (gray bars in figure 1.1), more than 60% became severely paralyzed. These data suggest that intense physical activity during the early stages of viral infection adversely affects the body's response to infection. Bed rest was generally recommended during the early stages of infection with the polio virus. Advances in community health, such as modern hygiene and mass immunization against diseases like polio, have altered our need to follow such advice in many countries. However, the emergence of newly identified viruses (e.g., HIV), and their rapid spread through much of the less developed world where manual labor is still an essential economic activity, pose new challenges to under-

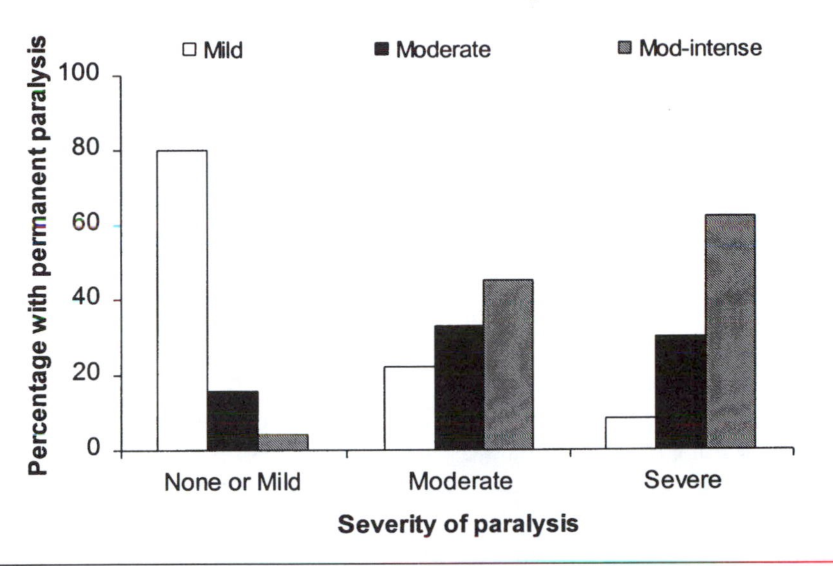

Figure 1.1 Physical activity and risk of developing paralysis during polio infection. Percentage of patients (y-axis) developing no or mild, moderate, or severe paralysis (x-axis) related to the amount and intensity of physical activity after the onset of pre-paralytic symptoms. Open bars = little or no physical activity after the onset of symptoms; black bars = moderate activity for less than 24 hours after the onset of symptoms; gray bars = moderate to intense activity for more than 24 hours after the onset of symptoms. Those who engaged in little or no activity were far more likely to fully recover compared with those who performed intense activity, who were more likely to experience severe permanent paralysis.

Adapted from Russell, W.R. 1949. Paralytic poliomyelitis: The early symptoms and the effect of physical activity on the course of the disease. *British Medical Journal*: March 19, 1949: 465-471.

standing the relationship between infectious illness and physical activity.

Are Athletes Susceptible to Infectious Illness?

A relationship between intense exercise and susceptibility to illness was noted early in this century. Cowles (1918) reported that virtually all cases of pneumonia at a boys' school occurred in athletes and that respiratory infections seemed to progress toward pneumonia after intense exercise and competitive sport. Some athletes experience high rates of certain illnesses, such as infectious mononucleosis

(Foster et al. 1982) and upper respiratory tract infection (URTI; e.g., common cold, sore throat, middle ear infection) (Berglund and Hemmingsson 1990; Douglas and Hanson 1978; Heath et al. 1991; Linde 1987; Mackinnon et al. 1993b; Mackinnon and Hooper 1996; Nieman et al. 1990a; Peters and Bateman 1983; Peters et al. 1993, 1996). Frequent illness has also been associated with "overtraining syndrome" in athletes (Fitzgerald 1991; Fry et al. 1991), a neuroendocrine disorder in athletes characterized by persistent fatigue and poor performance and due primarily to excessive training (Lehmann et al. 1993).

Although not extensively documented, URTI appears to be the most prevalent infectious illness among athletes (Berglund and Hemmingsson 1990; Hanley 1976; Linde 1987; Mackinnon et al. 1993b). Symptoms of URTI are generally obtained through self-report, especially in large epidemiological studies in which it is difficult to obtain medical diagnoses for hundreds to thousands of subjects. Self-report may introduce errors of diagnosis, for example by not distinguishing between allergic (noninfectious) and infectious symptoms. However, in one study on a small group (N = 33) of elite athletes that included both independent physician diagnosis and athletes' self-report of illness, athletes correctly identified URTI in all cases (Mackinnon et al. 1993b), suggesting that athletes' self-report of symptoms may indeed be accurate.

In the past, much of the evidence has been anecdotal or indirect, but recent evidence is consistent in supporting a high incidence of infectious illness among endurance athletes (e.g., distance runners, competitive swimmers) (reviewed in Heath et al. 1992; Nieman and Nehlsen-Cannarella 1992; Weidner 1994a). URTI appears to be the most common illness experienced by athletes (Berglund and Hemmingsson 1990; Hanley 1976; Linde 1987; Mackinnon et al. 1993b). Although the incidence of URTI varies by sport, type of athlete, and duration of study period, illness rates as high as 68% have been noted after an endurance competition (ultramarathon run) (Peters et al. 1993). Published rates of URTI among endurance athletes include up to 47% during the two weeks after a 56-km ultramarathon (Peters and Bateman 1983); 40% of marathon runners in the two months before and 12% of runners in the week after a marathon (Nieman et al. 1990a); 68% of runners after a 90-km ultramarathon (Peters et al. 1993); 50% of runners in the week after a marathon and in rowers after competition (Castell et al. 1996); and 40% of elite swimmers during four weeks of intensified training (Mackinnon and Hooper 1996).

Athletes also appear more likely to perceive illness and to seek medical care for it than nonathletes (Douglas and Hanson 1978), possibly because even minor illness may impair normal function (i.e., the ability to train and compete) in athletes. While URTI may seem a minor inconvenience to most people, it is of great importance to the elite athlete who must be in top physical and mental form every day; for the athlete, even minor illness at a critical point in training or competition may mean the difference between success and failure. In a study of collegiate athletes followed over a season, Weidner (1994b) reported that 18% of athletes missed practice and 5% missed competition due to URTI. Moreover, although a relatively minor illness, URTI is a major reason for consulting a medical practitioner and for absence from work in many developed countries; the annual associated cost has been estimated in the billions of dollars in the United States (Nieman and Nehlsen-Cannarella 1992).

However, although there is a general perception that athletes are more susceptible to infectious mononucleosis, nearly 90% of young adults in developed countries exhibit antibodies to its causal agent, the Epstein-Barr virus, by age 30 years (Eichner 1987), indicating a high incidence of infection among all young people. Again, it is possible that infection is more readily noticed or considered significant in athletes who are more affected by temporary functional impairment than the average young nonathlete.

Infectious Illness Rates After Competition

While the topic has not been studied extensively, several recent investigations suggest a dual effect of endurance competition, such as marathon running, on the incidence of URTI. Most studies have focused on distance runners, probably because of the large numbers of participants and frequent competitions. Although the data are far from conclusive at this point, it does appear that the incidence of URTI is little affected, if at all, by relatively short, moderate races (e.g., 21-km half-marathons or less), whereas longer races are associated with an increased incidence of URTI. The intensity or pace of running also appears to influence susceptibility to URTI. The following discussion will first focus on the incidence of URTI after longer, more intense races and then on illness rates after shorter, milder efforts.

Table 1.1 provides a summary of studies on physical activity and the incidence of URTI.

Illness Rates After Intense Endurance Exercise

The first study to quantify illness in an athletic population was the elegant and often quoted study by Peters and Bateman (1983) in which 140 runners were surveyed before and after a 56-km ultramarathon about symptoms of URTI (e.g., sore throat, runny nose, cough). Each

Table 1.1 Summary: Physical Activity and Incidence of URTI

Subjects	Activity	Time period	Effects of incidence of URTI	Reference
141 runners, 124 nonrunners	56 km ultra-marathon	2 wk post-race	Higher in runners; related to race pace and pre-race training	Peters and Bateman 1983
44 elite orienteers and 44 controls	Normal training season	12 mo	Higher in orienteers (2.5 vs. 1.7 episodes per year)	Linde 1987
137 children 11-12 yr	Normal activity	12 mo prospective	No relationship with sport participation or activity level	Osterback and Qvarnberg 1987
200 young adults	Normal activity	3 mo retrospective	No relationship with activity level or $\dot{V}O_{2max}$	Schouten et al. 1988
294 runners	5, 10, 21 km races	2 mo before and 1 wk after races	No effect of race	Nieman et al. 1989b
> 2000 runners	42 km marathon	2 mo before and 1 wk after race	6 × higher in participants; higher risk > 97 vs. < 32 km/wk	Nieman et al. 1990a
36 overweight, sedentary women	45 min walking, 5 × wk	15 wk	Fewer symptom days in walkers vs. controls (5.1 vs. 10.8)	Nieman et al. 1990b

174 elite and nonelite cross-country skiers	Normal training	12 mo	No difference between groups; no relationship with training distance	Berglund and Hemmingsson 1990
530 runners	Normal training	12 mo	Positive correlation with yearly distance	Heath et al. 1991
36 runners, 34 nonrunners	90 km ultra-marathon	2 wk post-race	Higher in runners compared with sedentary; incidence lowest in moderately trained runners; no effect of vitamin A supplementation	Peters et al. 1992
32 sedentary, 12 trained elderly women	37 min walking, 5 × wk	23 wk	Lowest in trained (8%), intermediate in walkers (21%), highest in nonexercisers (50%)	Nieman et al. 1993a
84 runners, 73 nonrunners	90 km ultra-marathon	2 wk post-race	67% lower in runners on vitamin C vs. placebo; no effect of vitamin C on nonrunners	Peters et al. 1993
178 runners, 162 nonrunners	90 km ultra-marathon	2 wk post-race	50% lower in runners on vitamin C with or without vitamin E or β-carotene compared with placebo and controls	Peters et al. 1996
24 elite swimmers	Intensified training	4 wk	Lower in overtrained (12.5%) compared with well trained (56%)	Mackinnon and Hooper 1996

runner nominated as a control subject an age-matched nonrunner living in the same household. During the two weeks following the race, 33% of all runners exhibited symptoms of URTI, compared with only 15% of control subjects. Moreover, there was a highly significant (R = 0.995) graded response—the incidence of illness increased as race time decreased, with nearly half (47%) of the fastest runners experiencing illness after the race (figure 1.2). Post-race incidence of illness was also correlated with pre-race training distance, although less strongly than with race time (R = 0.470). The most prevalent symptoms of URTI were sore throat and then nasal symptoms (runny nose, sneezing). Of the symptoms reported, nearly half (47%) lasted more than seven days and 33% were of four to seven days duration. Importantly, data from athletes with history of allergic rhinitis (e.g., hay fever) were excluded from the analysis. Subsequent studies from the same laboratory confirm high rates of URTI after long races (e.g., 90 km ultramarathon) (Peters et al. 1992, 1993, 1996). These studies have consistently shown a twofold risk of symptoms in the two weeks after the race in runners compared with nonrunners. Moreover, the lowest incidence of URTI symptoms appears to occur in runners who

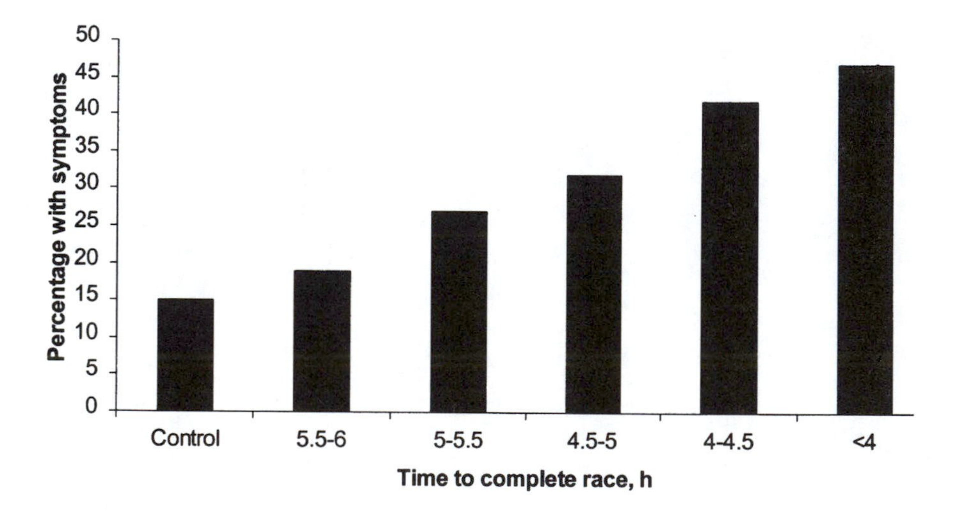

Figure 1.2 Incidence of URTI after an ultramarathon. Percentage of runners and matched nonrunner controls (y-axis) reporting symptoms of URTI in the two weeks after a 56 km ultramarathon. Runners are categorized by time to complete the race (x-axis). The incidence of URTI increased directly with race pace.

Adapted from Peters, E.M., and E.D. Bateman. 1983. Ultramarathon running and upper respiratory tract infections. *South African Medical Journal* 64: 582-584.

trained moderately in the weeks before the race compared with those completing too little or excessive training (Peters et al. 1992). Interestingly, vitamin C supplementation (600 mg per day) for three weeks before a race greatly reduced the incidence of URTI among runners but not among controls (further discussed later in this chapter and in chapter 8).

These data are consistent with an epidemiological study of over 2000 randomly selected participants in the 1987 Los Angeles Marathon (Nieman et al. 1990a). Participants were nearly six times more likely to exhibit URTI (self-reported) during the week after the race compared with similarly trained runners who entered but did not participate in the marathon, for reasons other than illness. Taken together, these data suggest that the incidence of symptoms of URTI is higher in the two weeks after competition involving prolonged exercise, such as marathon or ultramarathon running, and that the incidence increases with pre-race training distance and race pace. Risk may be elevated between two and six times compared with that for matched noncompetitors.

Illness Rates After Moderate Exercise

In contrast to the high rates of URTI after long-duration events such as a marathon or ultramarathon, susceptibility to infection does not appear to be altered by participation in shorter and less competitive events (Nieman et al. 1989b). For example, in a random survey of 273 runners in "fun runs" and 5, 10, and 21-km races, 30% reported infectious episodes during the two months before the races. The incidence of infection was similar during the week before and after the races, suggesting no short-term effect on illness. Incidence of illness was more related to illness in the home than to training distance. Moreover, there was a nonsignificant trend toward lower rates of illness before the race among runners in 21 km races than among those in the shorter runs (23% vs. 31%, respectively). In a recent report on runners, the incidence of URTI was about 50% lower during the week after competition in middle distance compared with long distance runners; URTI rates were 25% in middle distance runners who ran 10 km and 47% in marathon and ultramarathon runners who ran ≥42 km (Castell et al. 1996).

Exercise Training and URTI

There are also few studies focusing on the incidence of URTI associated with regular exercise training, that is, a chronic (long-term) effect as opposed to the acute (short-term) effect of a single bout of exercise. Compared with the acute studies, these studies have included subjects from more diverse activities (e.g., running, swimming, skiing, orienteering) and differing levels of expertise (from previously sedentary to elite athletes); it is noteworthy, however, that only endurance sports have been studied. Although far from conclusive, these data suggest a dose-response relationship between training volume and incidence of URTI, similar to that observed acutely after a single competition.

Table 1.1 also summarizes data from studies focusing on the effects of exercise training on URTI incidence.

Illness Rates and High-Volume Exercise Training

Illness rates also appear to be elevated by high-volume exercise training, such as distance running, in a dose-dependent manner. As discussed later, however, risk of illness is elevated only by high volume, not lower volume or more moderate-intensity training. Frequency of illness rather than duration appears to be most affected by high-volume training. For example, Heath et al. (1991) reported a dose-response relationship between training volume and the incidence of URTI over a one-year period in more than 500 runners who experienced an average of 1.2 episodes of URTI during the study. Compared with runners training < 778 km per year, backward stepwise logistic regression showed that the odds ratio of an URTI was twice as high in those running 778-1384 km, and 3.5 times higher for those running 1384-2222K per year (figure 1.3). There was no significant difference in annual incidence of URTI between runners in the latter category compared with > 2222 km per year. Living alone was also a risk factor for URTI, suggesting the possible influence of social factors.

In a study on more than 2000 marathon runners, those who trained more than 97 km per week were twice as likely to exhibit infectious illness (e.g., cold, flu, sore throat) during the two months before a marathon compared with runners who averaged less than 32 km per week; data were adjusted for confounding variables such as age, perceived stress levels, and illness in the home (Nieman et al. 1990a). In another study on runners entering shorter races (5, 10, and 21-km runs), there was no significant relationship between training distance and incidence of URTI during the two months before or one week after the races (Nieman et al. 1989a). If anything, in the latter study, there was a nonsignificant trend for runners entering the longer races (21 km) to experience slightly fewer episodes of URTI compared with those entering the shorter races (5 and 10 km). Furthermore, URTI in runners appeared to be more related to the presence of illness in the home than to training-related factors.

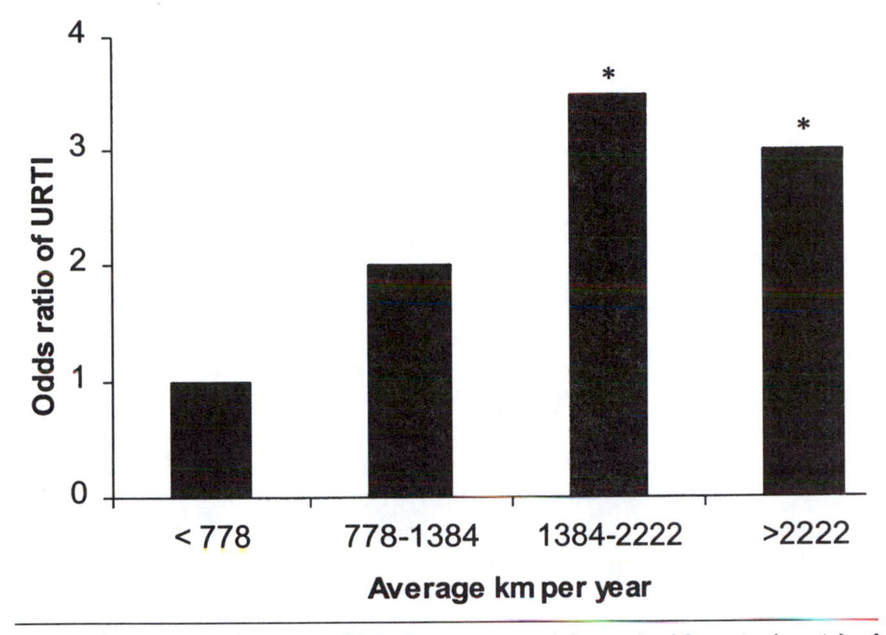

Figure 1.3 Yearly incidence of URTI in runners. Adjusted odds ratio (y-axis) of developing an URTI related to average yearly training distance (x-axis) during a one-year study.

Adapted from Heath, G.W., E.S. Ford, T.E. Craven, C.A. Macera, K.L. Jackson, and R.R. Pate. 1991. Exercise and the incidence of upper respiratory tract infection. *Medicine and Science in Sports and Exercise* 23: 152-157.

In a one-year study of 44 elite orienteers, the orienteers were found to have significantly more episodes of URTI compared with age-, sex-, and occupation-matched nonathletes (2.5 vs. 1.7 episodes, respectively) (Linde et al. 1987). About one third of the controls (32%) experienced no illness, whereas only 10% of the orienteers were symptom free during the one-year study. The common cold accounted for 74% of illness overall (70% in orienteers and 80% in controls). Despite the more frequent illness among athletes, there was no significant difference between groups for length of illness (7.9 vs. 6.4 days for orienteers and controls, respectively). These data suggest that exercise training increases the frequency but not duration of illness.

Another prospective study documented URTI over a one-year period in 174 elite Finnish cross-country skiers (Berglund and Hemmingsson 1990). The frequency of URTI was 1.5 infections per year in national team members and 1.88 per year in younger collegiate skiers, although this difference was not significant. Although a nonathlete control group was not included for comparison, no differences were noted for training hours between athletes who exhibited illness and those who remained healthy, suggesting that illness was not related to training volume.

Moderate Exercise Training and URTI

In contrast to the high incidence of URTI associated with high-volume endurance training, regular moderate activity does not appear to increase, and may possibly decrease, susceptibility to illness. In a randomized prospective study on 36 previously sedentary overweight women, a 15-week period of moderate exercise training was associated with a decrease in the number of symptom days (Nieman et al. 1990b). Training consisted of 45 minutes walking, five times per week. Women who exercised exhibited 5.1 days with URTI compared with 10.8 days of symptoms in a control group that did not exercise. In another study from the same laboratory (Nieman et al. 1993a), 32 inactive elderly women were randomly assigned to either a walking or nonexercising control group, and the incidence of URTI was compared with that for age-matched highly trained women runners. Exercise consisted of 23 weeks of 37 minutes walking five days per week. URTI incidence was lowest in the highly trained women (8%), intermediate in the walkers (21%), and highest in the sedentary control group (50%). These data suggest that moderate exercise may reduce the incidence or duration of URTI, at least over the short term. The lower rate of URTI among the conditioned runners who trained an average of 1.5 hours each day of moderate exercise seems contrary to data from younger distance runners, who seem to exhibit a higher rate of infection, and may reflect a more moderate approach to exercise training in older individuals.

In contrast to these studies suggesting a relationship between exercise and incidence of URTI, physical activity was not associated with URTI in a study of nearly 200 young Dutch adults (Schouten et al. 1988). Subjects were retrospectively surveyed by questionnaire about their activity patterns over the previous three months and URTI symptoms over the previous six months. Except for a low, but significant, correlation between sport activity and URTI incidence in women ($R = -0.18$), URTI incidence was not related to physical activity or $\dot{V}O_{2max}$. Moreover, activity level and $\dot{V}O_{2max}$ were not related to length of URTI symptoms. The retrospective survey of both symptoms and activity patterns may not have given clear and consistent definitions of these variables.

In a 12-month prospective study of 137 Finnish children (age 12.7 years at the end of the study), no relationship was observed between sport participation (gymnastics, ice hockey, swimming) and infectious illness (Osterback and Qvarnberg 1987). Athletes were matched for age and sex with children who did not participate in sports other than usual school activities. Information about training was obtained from the children at three examinations during the year, and data regarding respiratory infections were obtained by a nurse who contacted the children's parents every two months, thus enhancing the accuracy of reporting of illness. The overall infection rate was 2.8 (3.3 in girls and 2.5 in boys), which did not differ between sports or between athletes and nonathletes; sex differences were due to a higher incidence of the common cold among girls. The incidences of common cold and more serious respiratory infections (e.g., otitis media, tonsillitis, sinusitis) were similar in athletes and control subjects. These data suggest that sport participation outside of school does not influence susceptibility to URTI among children. However, since even the children in the control group were physically active during normal school activities and these activities were not quantified, it is not clear whether the groups differed sufficiently to enable discrimination based on exercise levels.

Are Overtrained Athletes More Susceptible to URTI?

Overtraining syndrome is a generalized stress response occurring in elite athletes and is characterized by persistent fatigue, poor performance despite continued training, and changes in mood state and other psychological variables (Lehmann et al. 1993). Recent terminology distinguishes between overtraining in which symptoms are present for weeks to months, requiring long-term recovery, and a shorter-term "overreaching" in which symptoms are similar but less persistent (days to weeks), requiring less time for recovery (Kreider et al. 1997). Recent evidence strongly suggests that overtraining is a neuroendocrine disorder, as evidenced by, for example, catecholamine (norepinephrine) depletion in overtrained runners (Lehmann et al. 1992) and swimmers (Mackinnon et al., 1997). Although overtraining is associated with frequent URTI (Fitzgerald 1991; Fry et al. 1991), data in support of such a perception are indirect (Fitzgerald 1991) and there are few studies to specifically address this issue (Mackinnon and Hooper 1996).

In a recent study on URTI and intensified training in elite swimmers, illness rates were compared between swimmers showing symptoms of overreaching and those considered well trained (Mackinnon and Hooper 1996). Training was intensified over four weeks in 24 elite swimmers (16 female, 8 male) by increasing both swim and dryland (resistance) training volume progressively by 10% in each of the four weeks. Overreaching was identified in 8 (33%) of the swimmers (6 female, 2 male), based on performance decrements in a criterion swim, persistent high fatigue ratings recorded in daily logbooks, and supporting comments from the swimmers in the logbooks indicating poor adaptation to the training loads. Of the 24 swimmers, 10 or 42% exhibited URTI during the four weeks. Surprisingly, the incidence of URTI over the four weeks was higher in the well-trained (9 of 16, 56%) compared with overreached (1 of 8, 12.5%) swimmers. It was suggested that increased risk of URTI may not necessarily be associated with overreaching/overtraining itself (i.e., presence of symptoms), but may occur as a consequence of intense training in all athletes. At present, this paper appears to be the only study to specifically assess the incidence of URTI among athletes showing symptoms of overreaching or overtraining. Obviously, larger prospective studies are needed to follow different types of athletes over extended periods to determine whether increased risk of URTI is in actuality a consequence of overtraining.

Conclusions About the Susceptibility of Athletes to URTI

Taken together, then, these studies on URTI among athletes and recreationally active individuals suggest that top competitive athletes, and those who engage in very long or intense exercise (e.g., marathon, ultramarathon), are more susceptible to URTI. In contrast, less competitive athletes and those engaging in less strenuous exercise are not more susceptible to URTI than the general population, and resistance to infection may actually be enhanced by regular moderate exercise.

This purported relationship, although not extensively documented at present, has been described by a model using a "J" curve (Nieman and Nehlsen-Cannarella 1992). In this model, an individual participating in regular moderate activity such as that prescribed for general health would have a lower risk of URTI compared with a sedentary person. On the other hand, athletes undertaking strenuous training would exhibit increased incidence of infection. Although this model is consistent with the epidemiological studies discussed earlier, more work is needed to confirm the model in various types of athletes, to more clearly define the quantity of physical activity within each category (sedentary, moderate, intense), to identify minimal and optimal levels of activity to promote resistance to infectious illness, and to explain the mechanisms involved.

Mechanisms Responsible for URTI in Athletes

The mechanisms responsible for the apparently high incidence of URTI among athletes are not currently known, and several possible explanations have been proposed (table 1.2). High ventilatory flow rates during intense prolonged exercise may adversely affect mucosal surfaces of the upper respiratory tract that are exposed to the environment (Peters and Bateman 1983; Tomasi et al. 1982). Impaired nasal mucociliary clearance was significantly reduced for several days after a marathon (Muns et al. 1995). Intense exercise may adversely influence immune function important to host defense against pathogens causing URTI. Intense exercise training may deplete certain factors necessary for optimal functioning of the immune system, such as glutamine (Keast et al. 1995; Parry-Billings et al. 1992; Rowbottom et al. 1995, 1996) or vitamin C (Peters et al. 1993, 1996). Psychological stress associated with training and competition may also contribute to altered immune function among elite athletes, presumably by altering circulating levels of immunomodulatory stress hormones (Lehmann et al. 1993). These mechanisms are not mutually exclusive, and frequent URTI in athletes may result from any combination of these factors.

Can URTI Be Prevented in Athletes?

Athletes and coaches perceive that intense exercise training and competition increase risk of URTI among athletes, and these perceptions are generally supported by the recent scientific literature. Thus, athletes, their coaches, and team physicians are interested in ways to prevent or limit URTI among athletes. Since prolonged intense exercise, especially during competition, appears to be most associated

Table 1.2 Possible Mechanisms Related to URTI in Athletes

High ventilatory flow rates altering mucosal surfaces

Infiltration of inflammatory cells into mucosa

Suppression of one or more immune system functions

Depletion of factor(s) needed by immune system

Psychological stress

with increased risk of URTI, limiting the volume of high-intensity training and frequency of competition would seem obvious preventive measures to reduce susceptibility in athletes. Although these restrictions may apply to the general public or noncompetitive athletes who exercise for health or other reasons, they are impractical for high-performance athletes whose training and competition schedules require extensive periods of high-intensity work and frequent competition. Attention has focused on ways to alter susceptibility to URTI among athletes during training and competition, in particular a potential role for easily obtained, legal (i.e., nonbanned) supplements. Two such substances that have been studied to date include vitamin C (ascorbic acid) and the amino acid glutamine. The potential role of these substances in preventing URTI in athletes is discussed in detail in chapter 8, and only a brief summary is presented here.

Two studies have suggested that vitamin C supplementation may reduce the incidence of URTI after endurance competition (Peters et al. 1993, 1996). In a double-blind, placebo-controlled study, vitamin C supplements (600 mg per day) or placebo was given to distance runners and matched control subjects for three weeks before a 90 km ultramarathon run (Peters et al. 1993). The percentage of runners reporting symptoms of URTI in the two weeks after the race was significantly lower in runners taking vitamin C compared with placebo (68% vs. 33% for placebo and vitamin C, respectively). In contrast, vitamin C supplementation had no effect on the incidence of URTI among nonrunners (49% for placebo and supplemented controls each), although this incidence appears rather high for nonathlete groups. In the second study (Peters et al. 1996), vitamin C alone was as effective as vitamin C in combination with vitamin E and β-carotene in reducing the incidence of URTI after a 90 km race. Although controversial and not consistently documented, vitamin C has long been thought to provide protection against URTI (reviewed in Hemila 1992, 1996). Peters et al. (1993, 1996) concluded that intense exercise training may deplete the body's stores of vitamin C, increasing susceptibility to URTI. Nonathletes, presumably, would not suffer such depletion and thus would not benefit from supplementation. The study is intriguing, although these data await further verification in other studies on other types of athletes.

A recent report suggests that oral supplementation with the amino acid glutamine may also reduce frequency of URTI among endurance athletes (Castell et al. 1996). As explained in chapter 8, glutamine is used as both a metabolic substrate and precursor for ribonucleotide synthesis in lymphocytes. Plasma glutamine concentration may decrease during prolonged or intense exercise (Castell et al. 1996; Keast et al. 1995; Parry-Billings et al. 1992; Rowbottom et al. 1995, 1996) and may also be lower in athletes exhibiting overtraining/overreaching (Mackinnon and Hooper 1996; Parry-Billings et al. 1992; Rowbottom et al. 1995). It has been speculated that low plasma glutamine concentration resulting from periods of intense exercise training may compromise lymphocyte function and the ability to mount an immune response (Newsholme 1994; Rowbottom et al. 1996). In a recent study, ultramarathon and marathon runners were fed a drink containing either 5 g L-glutamine or placebo (n = 72 and 79, respectively) immediately and then again 2 hours after a full marathon or ultramarathon race (Castell et al. 1996). During the week after the race, athletes reported symptoms of URTI (e.g., cold, cough, sore throat, influenza) via questionnaire. Glutamine supplementation was associated with significantly fewer athletes reporting symptoms after the race: 19% of athletes ingesting glutamine reported symptoms compared with 51% of runners consuming placebo. Whether glutamine supplementation is effective in altering resistance to URTI among other athletes or in other conditions (e.g., during periods of intense training) awaits confirmation by further study.

Resistance to Experimentally Induced Infectious Illness

Epidemiological data on the incidence of illness among athletes are not easily obtained and do not always provide information on resistance to specific illnesses. It is also difficult to quantify or control the level of exercise and other possibly confounding factors such as diet, travel, and competition stress. Hence, some studies have taken an experimental approach to the question of whether exercise influences resistance to infectious illness. Experimental work on immune reactivity in humans has obvious ethical limitations, and although some work has focused on humans, most experimental studies have been performed on laboratory animals. However, many studies using animal models of exercise include forced activity, either swimming or treadmill running, sometimes to exhaustion, which may induce a stress response that affects immune function.

The earliest work on exercise and immune function focused on the effects of physical fatigue on resistance to infectious illnesses, such as pneumonia and influenza, that were major killers before modern antibiotics (Bailey 1925; Nicholls and Spaeth 1922; Oppenheimer and Spaeth 1922). These studies suggested that exercise training prior to infection enhanced resistance, whereas exercise at the time of infection appeared to reduce resistance. Later studies on the effects of exercise on experimentally induced poliomyelitis followed clinical observations that paralysis was often preceded by intense physical activity at the time of infection (discussed earlier in this chapter; Horstman 1950; Russell 1947, 1949). For example, monkeys (Levinson et al. 1945) and mice (Rosenbaum and Harford 1953) forced to exercise to exhaustion at the time of poliomyelitis infection exhibited reduced survival rates and increased severity of illness.

Intense exercise also appears to influence resistance to other experimentally induced viral illnesses, such as coxsackievirus and influenza in mice (Cabinian et al. 1990; Ilback et al. 1984, 1989; Kiel et al. 1989; Reyes and Lerner 1976; Woodruff 1980). For example, forced swimming during the early stages of viral infection has been shown to increased mortality (Cabinian et al. 1990; Ilback et al. 1984; Kiel et al. 1989), the number of viruses in serum and the heart, and the extent of myocardial lesions in exercised compared with nonexercised animals (Gatmatin et al. 1970; Ilback et al. 1989; Kiel et al. 1989; Reyes and Lerner 1976). Exercised animals also had fewer antibodies to the virus in the blood and a delayed response in the appearance of interferon (a naturally occurring cytokine with antiviral activity) in the blood (Reyes and Lerner 1976). These data suggest that host resistance to viral infection is severely compromised by enforced exercise in a mouse model. Whether these data are applicable to humans is not known.

The effects of exercise on resistance to viral infection depend on when exercise is introduced in relation to the infection (Ilback et al. 1984). In mice, for example, exhaustive swim training for six weeks prior to influenza infection increased survival by 25% compared with that in nonexercised, infected control animals. In contrast, exhaustive swimming at the time of and for six days after infection decreased survival by 33%. Prior exercise training also partially protected against myocardial protein degradation, but exercise at the time of infection did not. These data suggest that prior exercise training may protect against viral infection, whereas exercise during the early stages of infection may reduce resistance; this is the pattern observed in humans during early stages of infection by polio virus, as already discussed. It should be noted, however, that in the study on mice, the exercised mice ceased training one day before infection and were permitted to rest during the infection. It is not clear whether the protective effect of prior training is maintained when training is continued through the early stages of infection.

The effects of exercise on resistance to bacterial infection have received relatively less attention. From the limited data available, it appears that exercise during bacterial infection is not as detrimental as exercise during viral infection, although few studies have compared the effects of exercise on the immune response to viral and bacterial infections (Ilback et al. 1984). Survival may be unchanged (Ilback et al. 1984, 1991) or enhanced (Cannon and Kluger 1984) by exercise training before bacterial infection. Ilback et al. (1984) reported that in mice, exercise training (swimming) for six weeks before or during bacterial infection *(tularemia)* did not influence mortality or bacterial counts in the spleen. However, decreases in myocardial oxidative enzyme activity normally seen with infection were prevented by exercise training before and during infection, suggesting a possible beneficial effect of exercise during bacterial infection.

In rats infected with the bacterium *Streptococcus pneumoniae,* four weeks prior exercise training (swimming) did not alter lethality of infection compared with that in untrained animals (48% in both groups) (Friman et al. 1991; Ilback et al. 1991). Rats were swum to exhaustion 72 hours after infection, and trained rats had been rested for 96 hours prior to infection. Compared with that in uninfected rats, exercise capacity, as measured by time to exhaustion, declined significantly in both trained and untrained infected animals, although trained infected rats were still able to swim for longer than untrained infected animals. Prior exercise training appeared to improve metabolic status during exercise in infected rats, as suggested by less pronounced changes during exercise in concentrations of blood insulin and glucagon as well as skeletal muscle and total body glycogen. The authors concluded that in trained rats, improved metabolic status, as evidenced by increased reliance on fatty acid oxidation during exercise, is at least partially preserved during acute bacterial infection.

Prior exercise training may also assist in recovery from parasitic infection, although there are few studies on this topic (Chao et al. 1992). To study the immune response to exercise training during and after parasitic infection, a relatively avirulent strain of the intracellular parasite *Toxoplasma gondii* was injected into mice on the first day of a daily swim training program (45 minutes per day). Survival of mice and swim performance were unaffected by infection, which was not surprising given the relatively mild form of infectious agent. Food intake and body weight decreased for several days after infection, but these variables returned to normal levels sooner in swim-trained compared with nontrained infected animals. Activation of macrophages (an immune cell active in the early stages of infection) was unaffected by swim training in infected animals, which may account for the lack of effect of exercise on survival. Exercise training appeared to blunt some aspects of immune function during infection, as evidenced by attenuated increases in splenic size and tumor necrosis factor (TNF, a cytokine or soluble factor produced in response to infection; see chapters 2 and 6) compared with values in infected nonexercised animals. TNF is associated with loss of appetite and body mass during acute infection, and the faster recovery of body mass and food intake in trained mice was attributed to the lower TNF response. The authors concluded that exercise training during acute mild parasitic infection had no deleterious effects on the mice, and that recovery from infection appeared to be enhanced by simultaneous exercise training. Although this study is an interesting observation in terms of documenting the immune response to mild parasitic infection in an animal model, its relevance and application to understanding the human response to exercise during infection are uncertain at present.

Mechanisms by Which Resistance to Infectious Illness May Be Altered in Experimental Models

The mechanisms by which intense exercise may alter resistance to infectious illness are not known at present and are speculative based on relatively few data. As discussed in the next chapter, the immune response to viral and bacterial infections includes antibodies to viral and bacterial proteins, killing activity by certain immune cells, and direct action on foreign agents by soluble factors such as interferon.

In viral infections, a delayed interferon response or a decrease in specific antibodies to viral proteins may permit viral replication (Reyes and Lerner 1976). Other mechanisms that have been suggested include altered T lymphocyte function, enhanced spread of virus by the killing of virally infected cells, or increased uptake of virus into cells (Ilback et al. 1989). For example, a threefold increase in the number of cytotoxic/suppressor T lymphocytes localizing to the heart has been observed in mice exercised 48 hours after injection with coxsackievirus, and this change was correlated with myocardial structural damage (Ilback et al. 1989). An increase in the number of these T cells may suppress normal immune responsiveness or may contribute to structural damage.

There is some evidence that soluble serum factors such as hormones or cytokines may be involved in exercise-induced immunosuppression during viral infection. For example, in vitro replication of bovine rhinotracheitis virus increased when serum from exercised steers was added to the incubation system (Blecha and Minocha

1983). Serum cortisol levels increased after exercise in these animals, and addition of hydrocortisone to the assay system (without serum from exercised animals) partially mimicked immunosuppression observed with exercise. It is well known that corticosteroid release increases dramatically during and after intense exercise (Brooks and Fahey 1995) and that corticosteroids suppress a variety of immune parameters. As discussed throughout this book, it is likely that exercise-induced changes in host defense against infectious illness involve a variety of factors.

Does Acute Illness Affect Exercise Performance?

Most research on exercise and immunity focuses on the effects of exercise on the body's capacity to mount an immune response. The converse question, whether acute infection and illness influence exercise capacity, has been given relatively less attention. However, there are several reasons and some evidence to suggest that exercise performance may indeed be impaired during infection or illness. Impairment of exercise capacity has obvious practical importance to athletes who may wish to continue training or to compete during minor illness such as URTI; prolonged recovery after illness may also interfere with an athlete's ability to resume training at a critical point in the competitive season.

In athletes, decrements in performance have been associated with subclinical viral infection and prolonged recovery after viral infection (Maffulli et al. 1993; Roberts 1985, 1986). Aerobic exercise capacity during prolonged submaximal work may be compromised during febrile illness (Daniels et al. 1985); moreover, blood volume and muscular strength also appear to be adversely affected by viral illness (Daniels et al. 1985; Friman et al. 1977; Friman and Ilback 1992; Hedin and Friman 1982). For example, during brief viral infection caused by sandfly fever virus, isometric and dynamic strength and muscular endurance declined by 10-30% during febrile illness (Friman et al. 1985) (figure 1.4). Impairment of strength could not be attributed to changes in muscle enzymes or structural damage, but correlated significantly with perceived symptoms, especially myalgia. Strength returned to pre-illness levels after the fever subsided, suggesting only transitory effects on muscle function. In an earlier study from the same laboratory, single-fiber EMG recordings suggested impaired neuromuscular transmission in subjects exhibiting myalgia due to acute viral infection (Friman et al. 1977).

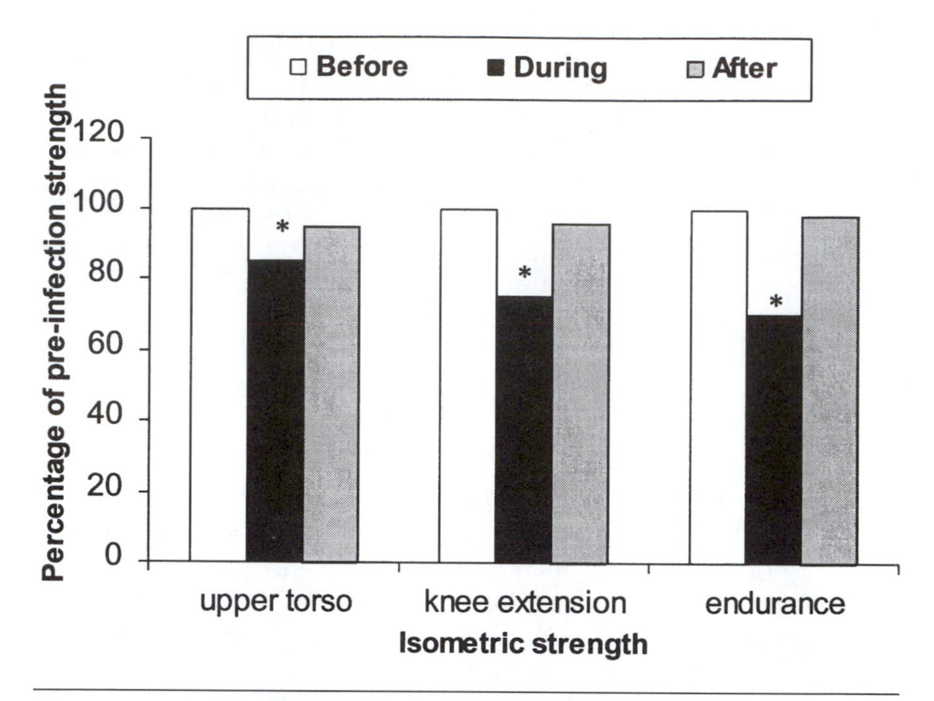

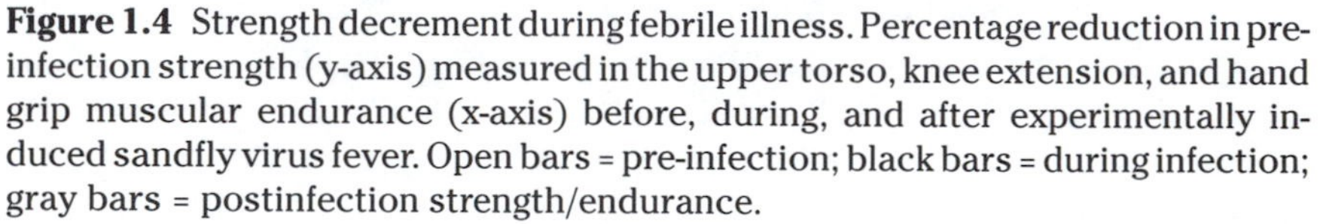

Figure 1.4 Strength decrement during febrile illness. Percentage reduction in pre-infection strength (y-axis) measured in the upper torso, knee extension, and hand grip muscular endurance (x-axis) before, during, and after experimentally induced sandfly virus fever. Open bars = pre-infection; black bars = during infection; gray bars = postinfection strength/endurance.

Adapted from Friman, G., J.E. Wright, N.G. Ilback, W.R. Beisel, J.D. White, D.S. Sharp, E.L. Stephen, W.L. Daniels, and J.A. Vogel. 1985. Does fever or myalgia indicate reduced physical performance capacity in viral infections? *Acta Medica Scandinavica* 217: 353-361.

Aerobic exercise capacity also appears to be altered during febrile illness (Daniels et al. 1985). For example, Daniels et al. (1985) experimentally induced sandfly fever in seven subjects. Isometric and isokinetic strength, and submaximal exercise capacity, were significantly reduced during the febrile stage; three subjects could not complete a submaximal exercise test. Hedin and Friman (1982) reported that, during febrile illness, moderate physical activity decreased adverse hemodynamic changes normally associated with bed rest (e.g., reduced plasma and red cell volumes). Thus, although physical performance may be impaired and intense exercise should be avoided during febrile illness, some level of light to moderate physical activity may be beneficial in recovering physical capacity after resolution of illness. Mental performance may also be adversely affected by minor illness (Beisel et al. 1974); performance in many skill-based sports such as archery, fencing, and shooting may thus be influenced by the presence of illness despite the relatively low level of physical activity required.

Certain infections cause disturbances of cellular structure or energy metabolism in a variety of tissues, notably myocardial and skeletal muscle (Cabinian et al 1990; Friman et al. 1982, 1991; Ilback et al. 1984, 1991; Woodruff 1980). Protein degradation and impairments of protein synthesis or energy metabolism may limit exercise capacity as well as long-term adaptations to exercise training (e.g., synthesis of oxidative enzymes; ability of skeletal muscle to hypertrophy). There is evidence that acute febrile illness causes temporary negative nitrogen balance (Friman and Ilback 1992).

Infection also causes release of factors that may impair or interfere with normal physiological responses to exercise. For example, the cytokine interleukin-1 (IL-1) is released early in the immune response. IL-1 is pyrogenic (raises core temperature), which may tax thermoregulatory mechanisms and thus limit the capacity for sustained exercise, especially in the heat. IL-1 also increases muscle protein degradation (Janeway and Travers 1996; Scales 1992), possibly limiting muscle adaptations to exercise training (see chapter 6).

Postviral fatigue syndrome (PVFS), or chronic fatigue syndrome (CFS), is associated with prolonged fatigue, muscle weakness, and delayed recovery of exercise capacity (Maffulli et al. 1993; McCully et al. 1996; Parker et al. 1996). For example, in eight varsity endurance runners, prolonged impairment in a variety of parameters including $\dot{V}O_{2max}$, work rate at anaerobic threshold, muscular endurance and strength, and uphill running ability was reported for up to one year after resolution of viral infection (Maffulli et al. 1993). Four athletes showed serological evidence of Epstein-Barr virus infection, three of cytomegalovirus infection, and one of coxsackievirus infection. CFS has also been associated with prior intense exercise training in athletes (Parker et al. 1996). In a case study, Arnold et al. (1984) reported a 30-year-old male with postviral fatigue syndrome who exhibited extreme fatigue and muscular pain following even mild activity. Nuclear magnetic resonance spectroscopy indicated an abnormal early acidosis during an exercise test involving the forearm muscles, suggestive of excessive glycolytic activity and disturbed energy metabolism. Such disturbance of energy metabolism accompanied by, presumably, excessive production of lactic acid and acidosis would certainly limit exercise capacity and the ability of an athlete to train for competitive sport. However, such impairment has not been consistently observed in CFS patients (McCully et al. 1996). Rowbottom et al. (1998b) suggested that impaired exercise capacity in CFS patients may be related more to detraining imposed by illness rather than infection-related physiological impairment of metabolic capacity. For example, a cross-sectional comparison of 16 CFS patients and 16 matched sedentary subjects showed similar responses to submaximal exercise, but lower peak $\dot{V}O_2$ and heart rate in CFS compared with sedentary subjects (Rowbottom et al. 1998b). Rating of perceived exertion (RPE) was significantly higher at all work rates for CFS patients. Measurements on these patients taken six months later showed increases in peak $\dot{V}O_2$ and heart rate and a decrease in RPE at the same work rates in CFS patients who had recovered from the syndrome (i.e., no symptoms), but evidence for further declines in CFS patients who had not recovered. The authors concluded that limited exercise capacity in CFS may result from deconditioning due to the long-term rest required for recovery coupled with an increased perception of exertion during exercise. Resumption of normal activity upon symptomatic recovery appears to restore exercise capacity in CFS.

Practical Considerations for the Athlete

It has been suggested that certain illnesses, such as influenza or other systemic bacterial or viral infections, require at least two to four weeks recovery (Roberts 1986; Nieman and Nehlsen-Cannarella 1992). Serious illness such as infectious mononucleosis may require extended recovery, up to six months, before an athlete resumes intense exercise training (Roberts 1986). Other viral illnesses such as the common cold without systemic involvement may not require cessation of training (Roberts 1986; Simon 1987). However, impaired strength and exercise capacity during viral infection (discussed above) may lead to musculoskeletal injury or overtraining in athletes who attempt to continue training at high intensity during illness (Simon 1987). Moreover, systemic viral infections may cause structural and functional changes in cardiac and skeletal muscle. Intense exercise during infection may exacerbate illness, leading to complications and, possibly but rarely, death. Roberts (1986) suggested that viral illness with systemic involvement (e.g., fatigue, fever, muscle aches, enlarged lymph nodes) requires one month for complete recovery before resumption of intense training. The general rule is that athletes may continue training with a head cold only ("above-the-neck" rule), that is, with nasal symptoms only. Athletes should not train intensely during any other type of illness unless it is first diagnosed by a physician who can recommend the proper course of treatment and training.

Summary and Conclusions

Competitive endurance athletes appear to experience elevated rates of infectious illness, especially URTI, during periods of intense exercise training and after competition. In contrast, "recreational" athletes do not appear to have elevated risk of infection, and may in fact lower their risk by participation in regular moderate exercise. These data suggest that prolonged intense exercise may compromise some aspect(s) of immune function, increasing susceptibility to minor infectious illness such as the common cold. The physiological and psychological stress of training and competing at the elite level most likely exert a combined effect on susceptibility to illness in athletes.

Animal models suggest that moderate exercise training prior to experimentally induced viral or bacterial infection enhances resistance, whereas intense exercise at the time of infection reduces resistance. Acute infectious illness in athletes may adversely affect exercise capacity, and intense exercise during systemic infection may exacerbate illness. Prolonged recovery after some systemic viral infections (e.g., influenza) or associated with postviral fatigue (chronic fatigue) syndrome may limit the ability of an athlete to regain competitive form. Athletes should not train intensely during acute systemic infection involving more than a head cold until a medical diagnosis is obtained and the infection is resolved.

Research Findings

- Compared with nonathletes, competitive endurance athletes such as distance runners experience high rates of symptoms of URTI during the two weeks after major competition.
- In distance runners, the incidence of URTI symptoms increases with training volume and intensity.
- The risk of URTI is not elevated, and may possibly be reduced, in moderately active individuals or "recreational" athletes.
- Exercise capacity is reduced during febrile viral illness.

Possible Applications

- Athletes should not be expected to perform to usual standards during febrile illness, and attempting to do so may be detrimental to their health.
- Athletes and their coaches should consider ways to modify training volume and intensity to avoid excessive exercise that may increase susceptibility to URTI.

- If regular moderate exercise can be shown to reduce risk of URTI, the health of inactive individuals may be enhanced by a physically active lifestyle.

Yet to Be Explored

- Is risk of URTI elevated by exercise competition and training in nonendurance athletes? If so, is there a dose-response relationship? If not, what is different about their training (compared to endurance training) that does not increase susceptibility?
- What specific components of endurance training (e.g., volume, intensity, frequency, number of rest days) contribute to increased risk of URTI?
- What specific mechanisms underlie the apparent increased risk of URTI among athletes?
- To what extent does illness influence an athlete's ability to train and compete?
- What preventive measures may athletes use to reduce the risk of URTI during intense training and major competition?
- Does regular moderate exercise reduce the risk of URTI? If so, what are the minimal and optimal "doses" for good health?

Chapter 2

Overview of the Immune System

© Human Kinetics

The immune system probably developed as a means of self-identification (i.e., distinguishing the body's own cells from those originating outside the body) and of maintaining homeostasis. As such, it is exquisitely complex, capable of recognizing and defending the body against, theoretically, an infinite number of challenges in the environment. The immune system covers the body's responses to foreign or novel molecules, usually proteins called immunogens; microorganisms including viruses, bacteria, fungi, and parasites; tumor growth; cell and tissue transplantation; and allergens. The immune system response to any challenge requires complex communication and coordination among diverse cells, tissues, and messenger molecules throughout the body.

Although immunology is a relatively new science, having originated about 200 years ago (Janeway and Travers 1996), it is now one of the fastest moving fields of study, especially in the past 20 years with the advent of modern biotechnology. For example, advancement in the field occurs with such regularity that textbooks are rewritten every two to three years. The literature on exercise and immunology focuses on diverse aspects of immune function (described separately in subsequent chapters). Because of the complexity of the immune system and the rate at which the field is advancing, this chapter provides only a very simplified introduction to those aspects of immune function specifically addressed in the exercise immunology literature.

General Scheme of the Immune System

A simplified scheme of the immune response to an infectious agent is presented in figure 2.1. The immune response begins when an invading foreign agent, usually a microorganism, penetrates the chemical and physical barriers protecting the body. The pathogen meets and is engulfed by phagocytes, which kill the microbe and degrade its proteins. The foreign proteins are processed by the phagocyte, appearing on the cell surface in combination with the phagocyte's own cell surface proteins. Specialized immune cells, called T helper lymphocytes (CD4, see figure 2.1), recognize and are activated by presentation of the foreign protein on the phagocyte's surface. Upon activation, these helper T cells then stimulate other immune cells to proliferate and secrete substances that combat the microorganism. Antibodies are produced against the foreign protein by mature B cells, another type of immune cell; antibodies neutralize some microbes and stimulate killing by other immune cells. During the initial encounter with a microbe, "memory" T and B cells are generated that respond quickly to subsequent infection by the same agent, conferring immunity in many instances. The combination of these processes is usually sufficient to eliminate most pathogenic microorganisms within days to weeks, but in some situations the host's defenses are ineffective or inappropriate and infection may persist.

Innate and Adaptive Immunity

The immune response can be divided into two broad functions: innate (natural or nonadaptive) and adaptive (acquired) immunity (table 2.1). As their names imply, innate immunity occurs naturally and immediately, while adaptive immunity occurs as an adaptive response

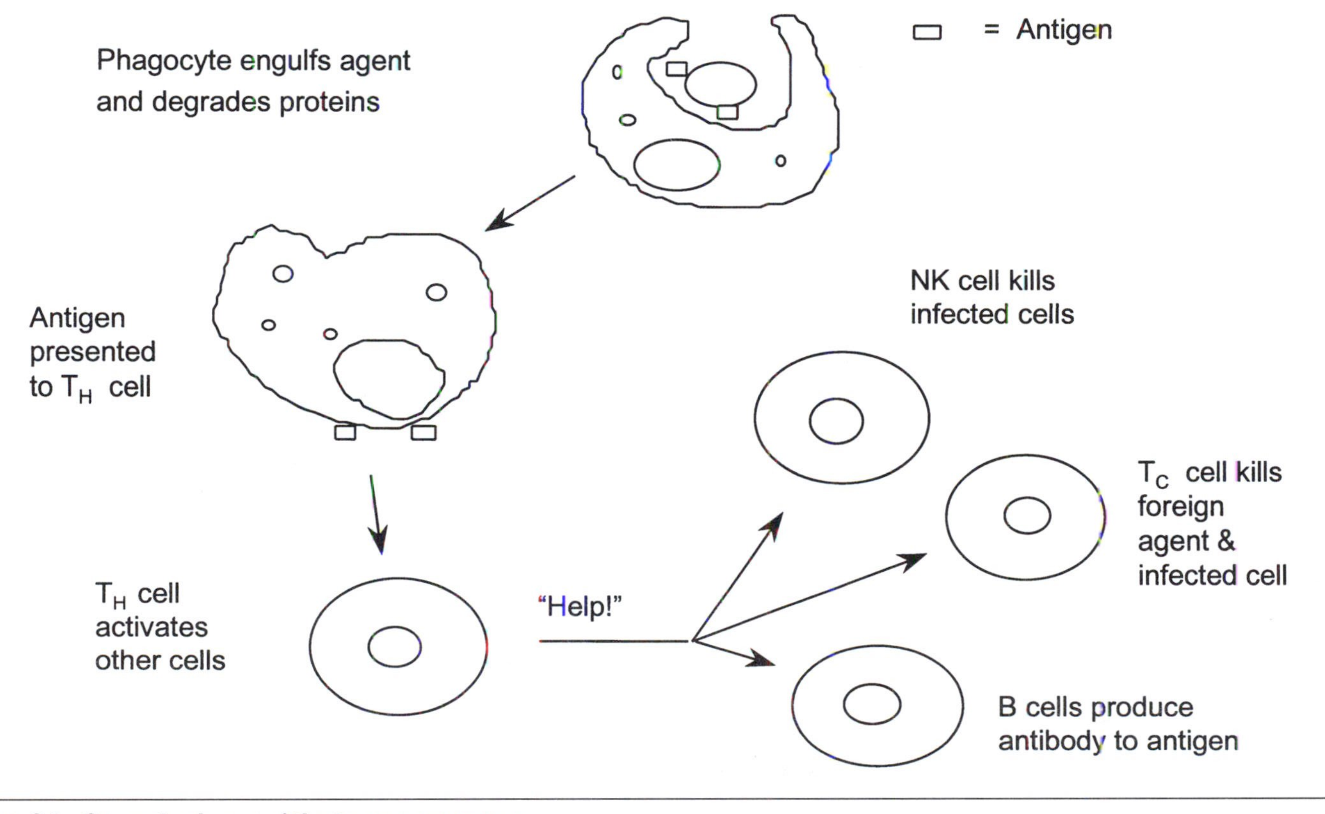

Figure 2.1 General scheme of the immune response.

Abbreviations: T_H = helper (CD4) T cell; T_C = cytotoxic (CD8) T cell; NK = natural killer cell.

Table 2.1 Innate and Adaptive Immunity

Innate immunity	Adaptive immunity
Physical barriers	Humoral
Skin, epithelial cell barrier	Antibody
Mucus	Memory
Chemical barriers	Cell mediated
Complement	T cells
Lysozyme	
pH of body fluids	
Acute phase proteins	
Other secretions	
Cells	
Monocytes/macrophages	
Granulocytes (e.g., neutrophils)	
Natural killer (NK) cells	

to a pathogen. Innate immunity is the first aspect of the immune system encountered by a foreign agent and includes soluble mediators as well as physical and chemical barriers on the body's surface. Cells involved in innate immunity can recognize and act against "nonself" (cells from another organism) without prior exposure. Innate immunity does not improve with repeated exposure. In contrast, adaptive immunity is characterized by specificity to the infectious agent and a short lag period (a few days) to become fully activated. The adaptive response generates memory of prior exposure that provides a faster and more effective level of protection with subsequent exposure to the same agent; this is the basis for immunization to prevent disease. Because of its specificity and memory, adaptive immunity is more complex and versatile. However, both systems work synergistically in a coordinated manner and are essential to optimal functioning of the immune system. For example, cells of the innate immune system (phagocytes) play an integral role in the initiation of the adaptive response by presenting foreign proteins (antigens) and secreting factors (cytokines) that stimulate cells of the adaptive system. Importantly, since the adaptive system requires a few days to become fully activated, the innate system provides an essential first line of defense in the early stages of infection.

Innate immunity involves three general mechanisms to prevent infection:

- Structural barriers preventing entry of pathogens into the body
- Chemical means (pH and soluble factors) that create an inhospitable environment for microbes
- Phagocytic cells that recognize and kill microorganisms

Adaptive immunity involves action of specialized immune cells, called lymphocytes, that inactivate and destroy microorganisms by several mechanisms including secretion of antibodies, direct killing of microorganisms or infected cells, and secretion of substances called cytokines that can activate other cells or participate in killing of pathogens. Memory B cells are generated by the first exposure to the foreign agent, and subsequent exposure produces a faster and more effective antibody response. The adaptive immune response can be broadly divided into responses mediated by humoral agents such as antibodies or by immune cells that activate other immune cells or directly kill foreign or infected cells.

Cells of the Immune System

Leukocytes (or "white blood cells" when in the blood) are immune cells found in several lymphoid organs and tissues throughout the body and in the blood and lymph circulation. Leukocytes originate from stem cells in the bone marrow and then undergo further maturation and differentiation in primary lymphoid tissues such as the thymus (T cells) and bone marrow (B cells). These cells interact with other cells and foreign agents in secondary lymphoid tissue in the lymph nodes, spleen, and gut. In lymph nodes located throughout the body and in the spleen, antigens filtered from lymph are taken up by antigen-presenting cells and presented to lymphocytes to initiate the immune response.

Leukocytes migrate between different lymphoid tissues via the circulation and the lymphatic system. Lymphocytes exit the circulation into various lymphoid tissues via specialized venules, and phagocytes (neutrophils and macrophages) can also exit the circulation into various tissues, aided by selective binding to special adhesion molecules on the endothelial cells (discussed later). At any given time only 1-2% of the body's total lymphocyte pool are in the circulation, with most remaining within lymphoid tissues; as we shall see in chapter 3, this number may increase dramatically during exercise.

Immune cells circulate continuously between tissues and blood, and it has been estimated that 1-2% of all lymphocytes recirculate through the blood every hour (Roitt et al. 1993).

Types of Leukocytes

Table 2.2 presents information on the major types of leukocytes found in the circulation—granulocytes, monocytes, and lymphocytes. All cells involved in the immune response, as well as other blood cells such as red blood cells and platelets, arise from a common ancestor, the hematopoietic stem cell found in bone marrow. This pluripotent progenitor cell gives rise to two specialized progenitor cells, the lymphoid and myeloid progenitors. Lymphocytes (T and B cells) derive from the lymphoid progenitor, and monocytes/macrophages, granulocytes, and other blood cells from the myeloid progenitor. The exact lineage of the large granular lymphocyte or natural killer cell is still uncertain, and this cell may derive from a "third population" progenitor (Roitt et al. 1993). Leukocytes (and many other types of cells) display unique cell surface proteins that can be used to identify, classify, and study the origins and functions of these cells, primarily by using monoclonal antibodies to the specific cell surface proteins. Many of these cell surface proteins have specific functions, such as receptors or adhesion molecules. By international agreement, many of these proteins (and the cells they identify) are now designated by the prefix CD (clusters of differentiation) (Janeway and Travis 1996).

Clusters of Differentiation (CD)

Leukocytes display an array of cell surface proteins (antigens). Functional studies of lymphocytes and other cells revealed that particular combinations of cell surface proteins appeared during stages of differentiation and activation, suggesting a role for these molecules in immune and other cellular functions. Monoclonal antibodies raised to specific proteins, now standardized and commercially available, are used to identify and quantify these cell surface proteins and the cells they identify. There are now over 100 CD antigens identified and characterized for the study of immune cells. For the most part, the CD numbers have been arbitrarily assigned, sometimes in order of discovery or description, and consecutive numbers often bear no relationship to each other. Table 2.3 presents a very simplified overview of some of the commonly used antigens in the exercise immunology literature.

Table 2.2 Circulating Leukocytes and Lymphocytes

a. Leukocytes

Cell	% of circulating leukocytes	Normal cell number per L blood	Primary functions
Granulocyte	60-70		
Neutrophil	90% of granulocytes	3.00-5.55	Phagocytosis
Eosinophil	2.5% of granulocytes	0.05-0.25	Phagocytosis of parasites
Basophil	0.2% of granulocytes	0.02	Chemotactic factor production Allergic reactions
Monocyte	10-15	0.15-0.60	Phagocytosis Antigen presentation Cytokine production Cytotoxicity
Lymphocyte	20-25	1.0-2.5	Lymphocyte activation Cytokine production Antigen recognition Antibody production Memory Cytotoxicity

(continued)

Table 2.2 ***(continued)***

b. Lymphocytes

Subset	**% of lymphocytes**	**Normal number per L blood**	**Major CD antigens**	**Primary functions**
T cell	60-75	1.0-2.5	CD2, CD4, CD8	Lymphocyte regulation
Helper T and inflammatory T	60-70% of T cells	0.5-1.6	CD4	Cytokine secretion Antigen recognition B cell activation
Cytotoxic T	30-40% of T cells	0.3-0.9	CD8	Cytotoxicity Lymphocyte regulation
B cell	5-15	0.3	CD19-23	Antibody production Memory
LGL/NK cell	10-20	0.1-0.5	CD16, CD56	Cytotoxicity Cytokine production

Compiled from Dacie and Lewis 1984; Janeway and Travers 1996; Roitt et al. 1993.

Eosinophils compose a small percentage of circulating leukocytes and are capable of phagocytosis of some microorganisms; these cells are most active in resistance to parasitic infection. Basophils and mast cells constitute a very small percentage of circulating leukocytes and are primarily involved in allergic and inflammatory reactions.

Immune Cells of the Lymphoid Cell Line

Lymphocytes (T, B, and NK cells) are leukocytes derived from the bone marrow lymphoid progenitor, and are the major effectors of adaptive immunity (B and T cells) and some aspects of natural immunity (NK cells). Lymphocytes compose about 20-25% of all leukocytes and represent a heterogeneous group of cells that can be further classified according to function and cell surface markers into subsets, as listed in table 2.2b. Lymphocytes are essential to a wide variety of immune functions, including initiation of the immune response (T cells); production of cytokines (T and NK cells) and antibody (B cells); cytotoxicity (killing foreign, tumor, or virally infected cells; T and NK cells); and importantly, memory of previous infection (memory B and T cells).

T cells

T lymphocytes are small (6-10 μm) lymphoid cells distinguished by the T cell receptor (TCR, CD3) displayed on the cell surface. T cells are intimately involved in initiating and regulating virtually all aspects of the adaptive immune response because of their ability to modulate the activity of several immune cells. Examples of this modulation include activation of B cells to proliferate and differentiate into antibody-producing cells, killing of tumor and virally infected cells by cytotoxic T cells, and secretion of myriad cytokines (see below) that modulate the activity and other immune cells. Memory T cells are also generated in response to initial exposure to a novel antigen.

T Cell Subsets

There are two distinct subpopulations of T cells—cytotoxic T cells, which can recognize and kill virally infected and some tumor cells, and helper and inflammatory T cells, which activate other immune cell functions (Janeway and Travers 1996). These cells are distinguished by their functions and by cell surface antigens, notably CD8 on cytotoxic T cells and CD4 on helper and inflammatory T cells. All T cells recognize targets (e.g., infected cells, tumor cells) via foreign antigen fragments bound to specialized cell surface receptors on the infected cell called major histocompatibility complex (MHC) molecules (Janeway and

Travers 1996). All T cell subsets also release various soluble factors (cytokines) that activate and regulate other immune cell functions.

CD8 cells are capable of recognizing body cells infected with viruses via the viral proteins displayed on the infected cell surface along with MHC class I molecules. CD8 cells help control infection by releasing molecules that penetrate infected cells, causing degradation of DNA and cell death, before the virus can replicate and spread farther. The CD8 cell remains functional and continues to find and kill more infected target cells.

CD4 cells can be further subdivided into helper and inflammatory T cells, which recognize antigen fragments bound to MHC class II molecules. Helper T cells participate in killing of extracellular pathogens (primarily bacteria) by activating B cells to produce antibody to the foreign antigens. Inflammatory T cells help destroy intracellular pathogens by stimulating monocyte/macrophage phagocytic and antibacterial activities. CD4 cells also produce several cytokines. The importance of CD4 cells to immune function is highlighted by the severe immune deficiency resulting from infection with the human immunodeficiency virus (HIV, the cause of AIDS), which gains entry into and destroys CD4 cells.

B cells

Composing about 5-15% of circulating immune cells, B cells are defined as lymphocytes bearing surface antibody (antigen receptor). Resting B cells are small (6-10 μm) and do not secrete antibody. B cells are identified mainly by the markers CD19 and 20. Upon stimulation by helper T cells in response to infection, B cells are activated and differentiate into plasma cells that produce and secrete large amounts of antibody. Each B cell produces antibody that recognizes only a single antigen (see below). B cells carry "memory" of earlier encounters with antigen, and subsequent exposure results in faster and greater production of antigen-specific antibody.

Large Granular Lymphocytes or Natural Killer Cells

The exact lineage of larger granular lymphocytes (LGL) or natural killer (NK) cells is uncertain, but may represent a third population of lymphocyte arising from the common bone marrow progenitor cell (Roitt et al. 1993). These cells constitute about 15% of circulating lymphocytes, and are histologically characterized as large (15 μm) lymphocytes with many cytoplasmic granules, hence the name LGL. NK cells lack the T cell marker CD3 and are identified by their surface antigens CD16 and CD56 (i.e., NK cells are $CD3^-$, $CD16^+$, $CD56^+$ cells). NK cells are capable of recognizing and killing certain tumor and virally infected cells, and also mediate antibody-dependent cell-mediated cytotoxicity (ADCC) against antibody-coated cells or particles. NK cells play an important role in innate immunity and are believed to function early in host defense against some types of tumor growth and viral infection, in particular by helping to control viral replication (Janeway and Travers 1996). NK cells kill target cells by binding to and releasing toxic substances similar to those released by cytotoxic T cells. NK cells also release some cytokines in response to stimulation by T cells.

Immunoglobulin and Antibody

Representing a class of glycoproteins produced and secreted by B and plasma cells, immunoglobulin (Ig) is found in serum and other body fluids such as tears and saliva. An antibody is an Ig molecule that reacts specifically with an antigen; all antibodies are Igs, but not all Igs exhibit antibody activity. Antibody is important to antigen recognition and memory of earlier exposure to a specific antigen. Antibody on the B cell surface acts as a receptor for antigen, and antigen-antibody binding is an important step in initiating the adaptive immune response.

Antibody serves many functions (table 2.4). Antibody-antigen binding initiates a variety of responses that act both directly and indirectly to combat the foreign agent carrying the antigen. Direct antibody actions include binding to antigens on microorganisms inhibiting their access to the host's cells. More prevalent, however, are indirect actions that include stimulating recognition and killing by phagocytes and other

Table 2.4 Functions of Antibody
Antigen binding
Complement binding and activation of complement cascade
Binding to and neutralizing bacterial toxins
Inhibiting nutrient uptake by bacteria
Inhibiting bacterial movement and access to host cells
Inhibiting viral entry into host cells
Facilitating cytotoxicity by cytotoxic T and NK cells
Enhancing phagocytosis and cytotoxicity against parasites
Passive immunity in the newborn

cytotoxic cells. Antibody binds complement, a complex series of more than 20 proteins involved in the immune and inflammatory responses (see table 2.4); antibody-bound complement can directly kill certain bacteria and virally infected cells. Other functions of antibody include direct binding to bacterial toxins, thus neutralizing their deleterious effects; interfering with movement and binding of bacteria to the host's cells; and inhibiting the uptake of essential nutrients by bacteria. Antibody plays a central role in host defense against viral infection by binding directly to extracellular viruses, inhibiting their entry into the body, and by facilitating recognition and killing by cytotoxic lymphocytes (T and NK cells). In addition to its antibacterial and antiviral activity, antibody is active in host defense against some parasitic infections by enhancing phagocytosis and antibody-dependent cytotoxicity against parasites. Antibody is also important to passive immunity in special conditions. For example, Ig or specific antibodies may be administered to immunodeficient or immunocompromised patients or in certain infections. Passive immunity conferred by antibodies that cross the placenta or are present in breast milk is essential to the newborn, whose immune system requires several months after birth to fully develop.

Ig Structure and Recognition of Antigen

All Igs share a common basic structure but vary in the specific structure of their antigen-binding sites. The basic Ig structure consists of four polypeptide chains composed of two identical light chains and two identical heavy chains (figure 2.2). Ig is a bifunctional molecule, with different functions at the two ends. On one end are two identical Fab (fragment antibody binding) segments, which, as the name implies, are involved in antibody binding. The opposite end consists of the Fc (fragment crystallizable) portion. The Fab end has a unique three-dimensional structure complementary to the antibody that it binds; this unique structure of the antigen binding site confers specificity of antibody-antigen binding. The Fc portion binds selectively to complement and to Fc receptors on the surface of various immune cells; Fc receptor binding to the antigen-antibody complex stimulates various aspects of the immune response.

Each B or plasma cell produces a single type of antibody to a specific antigen (in fact, to only a particular part of a specific antigen). Upon exposure to a specific antigen, a particular B cell proliferates and differentiates into myriad mature plasma cells secreting antibody against the antigen. Because each pathogenic organism may express several different antigens, and several antibodies may be directed against each

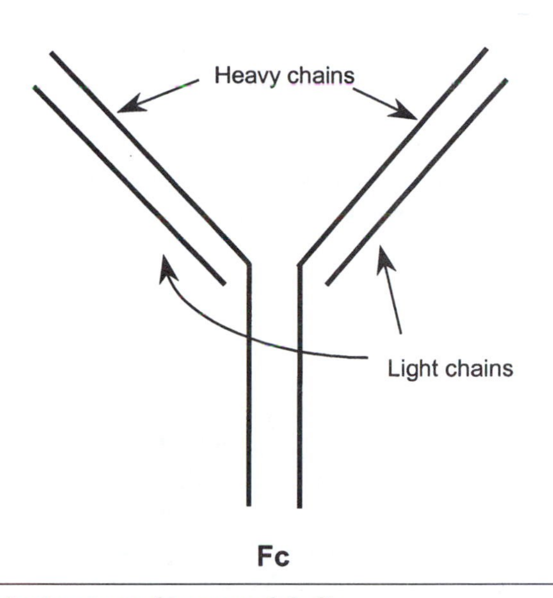

Figure 2.2 Basic structure of immunoglobulin.

antigen, production of antibody is usually a highly effective immune response.

There are five classes of Ig, each related to the basic Ig structure but with distinct properties. IgG is the major Ig in serum, and IgA is the predominant Ig in mucosal secretions found in saliva, tears, breast milk, and fluids of the respiratory, genitourinary, and gastrointestinal tracts. IgA contained in mucosal secretions is a major effector of host defense against microorganisms that gain entry through the mucosal surfaces. IgM constitutes approximately 10% of total serum Ig and is found in small amounts in mucosal secretions. IgD constitutes less than 1% of total serum Ig and is present on the B lymphocyte cell surface; its function is not well understood. IgE is involved in the allergic response by stimulating release of vasoactive and other soluble factors. The exercise literature has focused primarily on three classes of Ig—IgA, IgG, and IgM.

Phagocytosis

Phagocytosis is an important early step in the innate immune response and also to initiation of the later adaptive response. As mentioned above, the major effectors of phagocytosis are the monocytes, resident macrophages, and neutrophils. Phagocytosis involves a complex series of steps including migration of cells to the site of infection, binding to the organism, ingestion and killing of the organism, and degradation and processing of antigens for presentation on the cell surface. Phagocytes are attracted to sites of infection, injury, or inflammation by local release of chemotactic factors such as cytokines and small peptide fragments called chemokines (Janeway and Travers 1996). Monocytes and neutrophils are directed from the circulation to sites of inflammation by interaction of complementary adhesion molecules on these cells and on the vascular endothelium (see below); expression of these adhesion molecules increases in response to inflammation, enabling these cells to leave the circulation to enter the site of infection or inflammation. As described above, phagocytic cells release several types of toxic substances that kill and degrade pathogens.

Once initiated, phagocytosis results in three important actions that generally ensure an effective immune response (Janeway and Travers 1996). First, as described above, phagocytes kill and degrade pathogens, by itself an effective response to many types of microorganisms. Second, phagocytes such as macrophages release cytokines that stimulate killing by cytotoxic and NK cells. Third, antigen processing and presentation further stimulates the adaptive immune response with activation of helper and inflammatory T and B cells and subsequent production of antibody.

Soluble Factors in the Immune Response

Soluble factors are found in blood and other body fluids and represent a diverse range of mediators of immune function. Soluble mediators may act in several ways:

- To activate immune cells
- As chemical messengers between different immune cells
- As mediators of leukocyte trafficking throughout the body
- As agents able to neutralize or kill pathogens or tumor cells
- To regulate the immune response

The major soluble mediators of immune function include cytokines (a heterogeneous group of regulatory and growth factors), complement, acute phase proteins, and antibody. Except for antibody, which was discussed previously, each of these major mediators will be discussed separately below.

Cytokines

Cytokines are polypeptides that are involved in communication between lymphoid and other cells (Janeway and Travers 1996; Kunkel and Remick 1992; Liles and Van Voorhis 1995). The term cytokines refers to the general class of regulatory factors, which includes lymphokines made by lymphocytes and monokines made by monocytes. However, the more generic term cytokine is now used, reflecting the fact that different cell types (and not only immune cells) may be capable of producing these substances. Cytokines are primarily regulatory growth factors although they may have other functions, for example as mediators of the inflammatory response. A particular cytokine may act on many different target cells, and the cellular response varies by target cell. Cytokines are involved in virtually all aspects of immune function and may act on nonimmune cells as well. There are also naturally occurring factors that interact with or inhibit the action of several cytokines.

There are several classes of cytokines, each with diverse functions. Table 2.5 lists the major cytokines studied in the exercise literature; bear in mind that this is only a partial listing of the various cytokines and their actions. Molecules with the same prefix (e.g., interleukins) are not necessarily (and usually are not) related in structure or function.

Interleukins

Interleukins (IL) represent a generic class of 13 (in humans) generally unrelated cytokines produced by leukocytes, mainly by T cells but also by monocytes/macrophages and NK cells. IL share some functional properties (e.g., stimulation of immune cell function), but are generally not structurally related.

IL-1 exists in two forms, IL-1α and IL-1β, is produced by monocytes/macrophages, and exerts a wide range of actions mainly as a mediator of inflammation (Scales 1992). Immune functions include T and B cell activation via inducing IL-2 receptor expression; enhancement of macrophage and neutrophil activation, adhesion, and migration; production of other cytokines, in particular IL-6 and TNFα; and release of acute phase reactants. IL-1 is pyrogenic (i.e., induces fever) and also exerts wide-ranging actions such as prostaglandin release and proteolysis in many other types of cells.

IL-2 is released by activated inflammatory CD4 T cells and stimulates many immune cell functions including expression of high-affinity IL-2 receptors on B and T cells (Anderson 1992); further stimulation of T and B cell proliferation and differentiation into inflammatory T cells and antibody-producing plasma cells, respectively; release of other cytokines such as IFNγ; and stimulation of NK cytotoxic activity. Expression of the high-affinity IL-2 receptor is an essential early step in initiation of the immune response. Unstimulated lymphocytes usually express low-affinity IL-2 receptor subunits; but upon activation by IL-1 and IL-2, expression of high-affinity receptor subunits increases, greatly enhancing sensitivity of these cells to IL-2.

Table 2.5 Major Cytokines Studied in Exercise Literature

Cytokine	Primary producer(s)	Major immune actions
IL-1α, IL-1β	Macrophages	Fever T cell activation Macrophage activation
IL-1ra	Macrophages	IL-1 receptor antagonist
IL-2	T cells	T cell activation, proliferation IL-2 receptor expression
IL-3	Activated T cells	Granulocyte, monocyte differentiation
IL-4	Activated T cells	B cell activation, IgE expression
IL-5	Activated T cells	Eosinophil differentiation, maturation
IL-6	Activated T cells, macrophages	T and B cell growth, differentiation Stimulate acute phase response
IFNα	Leukocytes	Antiviral activity MHCI expression Stimulate cytotoxicity
IFNβ	Fibroblasts	Antiviral activity MHCI expression Stimulate cytotoxicity
IFNγ	T, NK cells	Stimulate cytotoxicity Macrophage activation MHC expression
TNFα (cachectin)	Macrophages, NK cells	Stimulate cytotoxicity against tumor cells Local inflammation

Compiled from Janeway and Travers 1996; Kunkel and Remick 1992.

IL-3 is produced by activated inflammatory and helper T cells and stimulates production of monocytes/macrophages and neutrophils from precursors in bone marrow (Jordan 1992). IL-4, IL-5, and IL-10 are produced by helper T cells and act as B cell growth factors, stimulating B cell differentiation and Ig production; IL-4 also stimulates differentiation of helper CD4 T cells. IL-6 is produced by helper T cells and activates B cell growth and differentiation into antibody-forming plasma cells. In addition, IL-6 acts synergistically with other cytokines such as IL-1 and TNFα to initiate the acute phase response and fever. IL-12, secreted by activated macrophages and neutrophils in response to infection, activates NK cytotoxic activity and production of IFN-γ, as well as differentiation of inflammatory CD4 T cells; IL-12 also appears to exert direct antitumor activity (Cox and Gauldie 1992; Janeway and Travers 1996; Lamont and Adorini 1996).

Interferons

Interferons (IFN) are factors capable of interfering with viral replication both in vitro and in vivo (Jordan 1992). IFN are believed to play an important role in preventing or limiting the spread of virus between infected and noninfected cells. IFN-α is produced primarily by leukocytes and IFNβ by fibroblasts, although many other cells also produce these cytokines in response to viral infection. IFNγ is produced by CD4 and CD8 T cells and NK cells. IFNα and IFNβ are structurally related while IFN-γ is quite dissimilar.

IFNα and IFNβ are released in response to viral infection and exert antiviral activity in several ways. Viral replication is inhibited by activation of enzymes that degrade viral mRNA and inhibit viral protein translation (Janeway and Travers 1996). These IFN also stimulate expression of MHC class I molecules on several cell types, which enhances presentation of viral proteins to CD8 (cytotoxic) T cells. In addition, both cytokines activate NK cytotoxicity against virally infected cells.

IFN-γ produced by CD8 cytotoxic T cells, CD4 inflammatory T cells, and NK cells is also active against viral and bacterial infection. This cytokine inhibits viral replication; increases expression of both MHC class I and II molecules, which subsequently enhances antigen recognition by cytotoxic T and NK cells; and also stimulates macrophage cytotoxic, bactericidal, and antigen-presenting activities.

Tumor Necrosis Factor

Tumor necrosis factor-α (TNFα) produced by macrophages and cytotoxic CD8 and inflammatory CD4 T cells is an important mediator of host defense against viral and bacterial infection as well as inflammation (Tsuji and Torti 1992). TNFα and IFNγ act synergistically in direct killing of some target cells; they also act to stimulate cytotoxic activity of CD8 T cells, NK cells, and macrophages and to enhance macrophage antibacterial activity. This cytokine acts both locally (within tissues) to increase vascular permeability and migration of leukocytes to sites of infection or inflammation, and systematically, in combination with IL-1 and IL-6, to induce the acute phase response and fever. The latter effects can become detrimental, possibly leading to sepsis and shock if uncontrolled.

TNFβ (lymphotoxin) released by inflammatory CD4 T cells acts synergistically with IFNγ to increase direct killing of some infected cells. This cytokine is also chemotactic for macrophages, helping to recruit macrophages to sites of infection or inflammation.

Complement

Complement is a complex system of more than 20 plasma proteins that act in a coordinated manner to kill extracellular pathogens. One of the first soluble factors to be identified, the complement system is a central feature of resistance to bacterial infection and the inflammatory process. Complement activation initiates three main consequences—recruitment of inflammatory cells to the site of infection or inflammation, opsonization (coating) of pathogens stimulating subsequent detection and killing by phagocytes, and direct killing of pathogens, especially bacteria.

The complement system involves activation of a sequence or cascade of reactions, which in turn further activate other complement components. There are three pathways of complement activation, all of which result ultimately in the same action (Janeway and Travers 1996):

- Classical pathway activated by antibody-antigen binding on surface of pathogens
- Lectin-binding pathway via a serum protein binding directly to molecules on the pathogen's surface
- Alternate pathway activated spontaneously by binding of certain complement components directly to the surface of a pathogen

Regardless of the activation pathway, all eventually result in activation of specific complement components (e.g., C3a, C3b, C4a, C5a, C5b, C6-9; the numbers refer to order of discovery and not the sequence of activation). Ultimately, the terminal complement components assemble into a porelike "membrane-attack complex" (Janeway and Travers 1996) that, once inserted into the pathogen's outer membrane, kills the pathogen by disrupting the ionic gradient across this membrane. Other functions of activated complement components include opsonization (coating) of pathogens, which stimulates phagocytosis; direct stimulation of phagocytic adherence to blood vessel walls and migration to sites of infection or inflammation; and enhanced sensitivity of B cells for certain antigens.

The Acute Phase Response

Release of pro-inflammatory cytokines, IL-1, IL-6, and TNFα in particular, induces production and secretion of several unrelated glycoproteins from the liver, collectively called acute phase proteins or reactants (APP or APR) (Baumann and Gauldie 1994; Steel and Whitehead 1994). These proteins exert myriad functions, some of which enhance the immune response to a pathogen. Serum concentrations of APP may increase 100-1000 times during acute infection or inflammation. Since many stimuli may cause release of the cytokines IL-1, IL-6, and TNFα, release of APPs may also occur in response to any number of insults to the body and not just infection.

C-reactive protein, the predominant APP, is released during bacterial infection and other traumas such as surgery, myocardial infarction, tissue injury, and prolonged intense exercise (see chapter 4). C-reactive protein exhibits antibody-like activity by binding to bacterial cell surface proteins and opsonizing bacteria, which stimulates phagocytosis and activates complement. Other APP can also activate the complement cascade and phagocytosis, and some are chemotactic for phagocytic cells. Other APPs, the proteinase inhibitors, limit proteolysis of muscle and other tissue proteins, whereas metal-binding APPs bind iron and copper-containing compounds, which may inhibit bacterial growth by limiting availability of these metals. Some APPs are also free radical scavengers that may help neutralize free radicals produced during exercise.

Adhesion Molecules

Adhesion molecules mediate binding between cells or between cells and extracellular matrix proteins; adhesion molecules are often displayed as a membrane or transmembrane protein on the cell surface. Adhesion molecules are important to directing leukocyte trafficking within the circulation and from the circulation to lymphoid and other tissues, especially to sites of inflammation or infection. There are four general classes of adhesion molecules: selectins and vascular addressins, which are important for leukocyte homing to specific sites; integrins, which mediate binding between cells and with extracellular proteins; and some members of the Ig superfamily that are important to antigen recognition and T cell activation (Janeway and Travers 1996).

Selectins are found on both leukocytes and vascular endothelial cells. Under resting conditions, selectins mediate only weak binding to complementary adhesion molecules (vascular addressins) on the endothelium, allowing leukocytes to "roll along" the endothelial surface (Janeway and Travers 1996). During infection or inflammation, dilation of blood vessels in the affected site slows blood flow and leukocyte movement, increasing the frequency of interaction between selectins and addressins. Moreover, upregulation by cytokines increases expression of other adhesion molecules (e.g., integrins and adhesion molecules of the Ig superfamily) on both leukocytes and endothelial cells; these latter adhesion molecules provide stronger binding between leukocytes and the endothelium. In this way, leukocytes adhere to the endothelium, allowing "captured" leukocytes to squeeze between endothelial cells in the vessel and to migrate into the lymph nodes or into infected tissues (a process called diapedesis). As discussed in chapter 3, selective recruitment of certain leukocytes into and exit from the circulation after exercise is mediated to some extent by adhesion molecules. Adhesion molecules are also important in cell-cell interaction required for antigen presentation to and subsequent activation of T cells. Activation of T cells is accompanied by changes in expression of various cell adhesion molecules, allowing selective homing of T cells at different stages of differentiation and activation.

Overview of Methods to Assess Immune Function

As would be expected in any large and complex scientific field, there are myriad methods to assess immune function, and recent advances in biotechnology have allowed for greatly increased sensitivity and specificity of these assays. Because of the complexity and number of these assays, only a few methods that have been specifically applied in the context of exercise in humans will be briefly described here. Immunological assays applied to exercise research can be roughly divided into those quantifying the concentration of the variable of interest (e.g., cell number, antibody or cytokine levels) and those assessing functional status of immune cells (e.g., cell proliferation, cytotoxic or phagocytic activity).

Quantifying Immune Cell Number

Peripheral blood leukocyte (total white blood cell) number and differential counts are routinely assessed via electronic counter in which a given volume of whole blood passes through an orifice one cell at a time. Leukocyte differential counts give the proportion of leukocyte subsets (e.g., neutrophils, lymphocytes, monocytes) and can be performed using an electronic counter or histologically on a whole blood smear on a glass slide. A given number of cells, usually > 100-200, is counted and the proportion and absolute number of subsets are calculated.

Lymphocyte subsets (e.g., T, B, and NK cells) and other cells can be identified and quantified using the powerful tool of flow cytometry. As described above, immune cell subsets can be identified by their cell surface proteins (antigens, classified by CD nomenclature); fluorescently labeled monoclonal antibodies that react specifically with these cell surface antigens are now commercially available. Whole blood or isolated lymphocytes are incubated with these antibodies, and a small sample of cells (usually > 10^6) is then passed cell by cell in front of a special laser that excites the fluorescent dye; different dyes can be used that are excited at different wavelengths. Highly sensitive photomultiplier detectors then measure light scatter (related to cell size) and light emission when excited at different wavelengths (related to the amount of monoclonal antibody bound). Data are generated and analyzed by computer, yielding quantitative information about the number of cells expressing these cell surface proteins. Cells may be assessed for more than one surface antigen at a time, allowing complex study and identification of lymphocyte subsets; for example, NK cells are identified as $CD3^-$, $CD16^+$, $CD56^+$, requiring labeling with three monoclonal antibodies. Gabriel and Kindermann (1995) have published a comprehensive discussion of the applications of flow cytometry to exercise-related research.

Quantifying Ig and Antibody Concentration

Igs and antibodies in serum or mucosal fluids may be quantified by several methods that may differ in their sensitivities, such as antibody precipitation, isoelectric focusing, immunoelectrophoresis, and the more recently developed immunoassays. Immunoassays are commonly used, highly specific and sensitive assays that rely on the fact that Igs are also antigens; that is, they are proteins to which antibodies can be made. In the enzyme-linked immunoassay (ELISA), an enzyme-linked antibody against the antigen (Ig) of interest is bound to a solid support (e.g., test tube or microtiter plate well). The sample containing antigen is then incubated in the tube or well, allowing the enzyme-linked antibody to bind antigen. After a given incubation period, excess (unbound) antigen is washed away, and a substrate for the enzyme is added to the tube or well. Bound antibody is detected by a color change induced by the enzyme attached to the antibody; the amount of color is proportional to the amount of antibody bound to antigen and hence to the concentration of antigen in the sample. A similar principle is used for the radioimmunoassay (RIA), only instead of an enzyme-linked antibody, an antibody linked to a radioactive compound, usually ^{125}I, is used. As with the ELISA, after incubation, unbound antigen is washed away and the radioactivity measured is proportional to the concentration of antigen in the sample.

Quantifying Cytokine Concentration

Until recently, cytokine concentration was assessed in bioassays using cell lines in which growth is dependent on the presence (and was thus proportional to the concentration of) specific cytokines. In such bioassays, a pure cell line is incubated with the sample (e.g., plasma), which presumably contains the cytokine of interest. Growth of the cells is quantified and compared with growth under identical conditions with the addition of known quantities of the purified cytokine. Although this type of assay measures the biological activity of cytokines, which is an important variable, its sensitivity is limited by other factors in the sample that may mimic or inhibit the action of the cytokine. For example, there are naturally occurring antagonists to some cytokines, and several cytokines have overlapping functions (discussed further in chapter 6). More recently, cytokines have been measured in blood using more sensitive and highly specific commercially available immunoassays, which are similar in principle to immunoassays as described above.

Assessing Lymphocyte Proliferation

To be effective, adaptive immunity depends upon proliferation of lymphocytes in response to antigenic challenge. Such response is difficult to measure in vivo and in response to naturally occurring challenges. Consequently, in vitro assays have been developed that use polyclonal mitogens, so called because they stimulate mitosis of lymphocytes without specificity to a particular antigen. Standard mitogens used in the study of human lymphocyte proliferation include phytohemagglutinin (PHA) and concanavalin A (ConA), which stimulate T cell proliferation, and pokeweed mitogen (PWM), which stimulates T cell-dependent B cell proliferation; in mice, ConA and lipopolysaccharide (LPS) stimulate T and B cell proliferation, respectively, and have been used in experimental animal models. Although these substances do not appear to act directly via antigen receptors, they do stimulate a response similar to that elicited by antigen, and are useful in assessing lymphocyte growth and proliferative capacity.

Whole blood containing lymphocytes, or isolated lymphocytes, are cultured in sterile conditions in the presence of optimal amounts of the relevant mitogen and other factors required for growth. Once stimulated, these lymphocytes enter the cell cycle, synthesizing new DNA in order to divide and proliferate. Cell proliferation is assessed by measuring incorporation into DNA of ^{3}H-thymidine added to the culture medium. These data give a general idea of the ability of lymphocytes to respond to antigenic challenge but yield little information regarding the specificity of that response.

Antigenic challenge also initiates cytokine production by T cells and antibody production by B cells. Cytokine or antibody concentrations can be measured (as described above) in supernatants obtained from cell cultures of mitogen-stimulated lymphocytes. The presence of antibody and cytokines in these supernatants also gives a general idea of the degree of lymphocyte activation in response to mitogenic challenge.

Assessing Phagocytic Antimicrobial Activity

As discussed above, monocytes/macrophages and neutrophils are important effectors of the innate immune response, exhibiting phagocytic activity especially against bacterial pathogens. Phagocytosis is a complex process involving several steps from recognition and ingestion to degradation and killing of the pathogen; thus, phagocytic activity can be estimated at various steps in this process. The most commonly used assays include measurement of oxidative burst activity (to assess priming or killing) and particle ingestion (to assess ingestion of pathogens). To measure in vitro particle ingestion, fluorescently labeled latex beads are incubated with isolated cells for a given time during which the phagocytes ingest the latex beads. The number of ingested beads per given number of cells can be quantified histologically or with flow cytometry. Oxidative burst activity can be assessed several ways; one commonly used method is to measure production of reactive oxygen species in the form of hydrogen peroxide. Isolated cells are incubated with known stimulants of phagocytic activity (e.g., opsonized zymosan particles). Upon stimulation of phagocytosis, these cells produce hydrogen peroxide, which can be quantified by flow cytometry or chemiluminescence.

Assessing Cytotoxic Activity

NK cells and cytotoxic T cells are capable of cytotoxicity (killing) against certain tumor and infected cells. Cytotoxic activity is often quantified in an in vitro chromium release assay against target cells specifically sensitive to NK or cytotoxic T cells. Target cells are incubated with ^{51}Cr, which is internalized and released only upon target cell death. The sample (e.g., whole blood or isolated lymphocytes) containing cytotoxic cells is then incubated with the labeled target cells for a designated time, usually 4 hours. Cytotoxic lymphocytes bind to and kill the targets, releasing ^{51}Cr into the medium upon target cell death. The amount of radioactivity released is proportional to the number of target cells killed.

Many of these techniques are discussed in more detail in the context of the exercise literature in subsequent chapters.

Immune Response to Tumor Cells

Cancer arises from the uncontrolled growth of cells descended from a single transformed (abnormal) cell; transformation of a cell may occur in response to a number of stimuli relating to genetic or environmental factors. T cells appear to be central to the immune response to some tumor cells that display specific cell surface antigens, and may provide some type of "immunosurveillance" function, especially against virally induced tumors (Janeway and Travers 1996). Histological analysis of human tumors shows infiltration of inflammatory cells, primarily lymphocytes and macrophages. Both CD4 and CD8 T cells localize to some human tumors, and some of these T cells appear to react specifically against certain tumors (Roitt et al. 1993). However, the degree of infiltration of these cells does not appear to be related to patient prognosis, and the exact role of these cells in the body's defense against tumor growth is at present unclear (Roitt et al. 1993).

Some tumor cells express specific antigens, called tumor-specific transplantation antigens (TSTA), which can be recognized by T cells and macrophages in much the same way as other foreign antigens are recognized by these cells (Herberman 1991; Janeway and Travers 1996). Cytotoxic NK cells may also recognize and act against certain types of tumor cells. TSTA can arise via several mechanisms, including mutations in naturally occurring cellular peptides altering their structure and thus antigenicity; increased expression of some normal proteins usually found only in small quantities; or activation of embryonic genes and production of their proteins in mature cells, in which the embryonic proteins are antigenic. However, despite the appearance of TSTA on the tumor cell surface, some tumor cells are also capable of escaping detection via producing suppressive cytokines that suppress T cell function, or via selective loss of factors needed for detection by T cells (e.g., loss of MHC molecules) (Janeway and Travers 1996).

Since the immune system is involved in at least some aspects of defense against tumor growth, it has long been postulated that immunological interventions to enhance immune responsiveness may have some merit in the treatment of certain cancers. Although few such interventions have been successful to date, some recent advances in immunotherapy that focus on stimulating the immune system to fight tumor cells in the body include the theoretical possibility of developing immunization using TSTA (but there are many practical difficulties); genetic manipulation of tumor cells to make them more immunogenic and easily recognized by cytotoxic cells; introduction of genes coding for cytokines into tumor cells so that the tumor produces cytokines that locally stimulate immune cells to reject tumor cells; extraction of tumor-infiltrating T cells and in vitro stimulation with IL-2 followed by reinfusion into the body; and in vitro stimulation with IL-2 of the patient's mononuclear cells with subsequent reinfusion of these cells, presumably to stimulate cytotoxic activity of NK and cytotoxic T cells (called lymphokine-activated killer or LAK cells) (Janeway and Travers 1996; Roitt et al. 1993).

Summary and Conclusions

The immune system is extremely complex, capable of distinguishing "self" from "non-self" and of generating "memory" of prior exposure to foreign or novel agents. The immune response to a challenge involves coordinated activity of many cell types, soluble factors, and messenger molecules located throughout the body. Innate immunity provides nonspecific and immediate defense against microorganisms, while the more versatile and specific adaptive immune system acts against pathogens that are not completely eradicated by the innate system. Three basic types of immune cells (leukocytes) are involved in the immune response: neutrophils, monocytes/macrophages, and other antigen-presenting cells; lymphocytes; and NK cells. Together these cells activate several pathways of host defense, including direct killing of pathogens or infected cells; production of antibody specific to a foreign antigen; and secretion of cytokines, which activate and regulate various immune cell functions. Other soluble factors such as complement or APP act to enhance phagocytosis and direct killing of pathogens. Although certain immune cells (cytotoxic T cells, NK cells, monocytes/macrophages) are capable of recognizing and killing tumor cells, their role in "immunosurveillance" against neoplastic growth is not clearly proven in humans.

Chapter 3

Exercise and Leukocytes: Number, Distribution, and Proliferation

Leukocytes are involved in virtually all aspects of immune function, either directly via their cellular activities or indirectly via release of soluble factors. Exercise causes profound changes in the number and subset distribution of circulating leukocytes, and may also induce changes in lymphocyte proliferation. Redistribution of leukocytes has been attributed to release of hormones, such as catecholamines and corticosteroids, and some cytokines, such as IL-1. Together these hormones and cytokines regulate the distribution of leukocytes in the blood via influencing cardiac output and blood flow, release of cells from the spleen, and adhesion of leukocytes to the vascular endothelium. Although even brief exercise (e.g., 1 min) may increase leukocyte number in the circulation, exercise-induced changes in leukocyte number, subset distribution, and lymphocyte proliferation are generally transitory. Leukocyte number usually returns to baseline levels within hours of most types of exercise, with the possible exception of intense, prolonged exercise, which may induce long-lasting (e.g., 24 h or beyond) changes in cell number. Whether immune function is influenced by these changes is still subject to much debate.

Leukocyte Number and Subset Distribution

Leukocytosis (increase in circulating leukocyte number) is one of the most striking and consistent changes observed during exercise.[1] Circulating leukocyte number may increase up to four times resting values (see table 2.2 for normal cell counts), may continue to rise after cessation of exercise, and may remain elevated for several hours after some types of exercise. In general, the magnitude of leukocytosis appears to be directly related to exercise intensity and duration and inversely related to fitness level; exercise duration may be the most important factor (McCarthy and Dale 1988; Galun et al. 1987; Oshida et al. 1988; Shek et al. 1995; Tvede et al. 1993). The degree of leukocytosis may also be influenced by factors that modify the hormonal response to exercise, in particular release of corticosteroids, supporting a central role of these hormones in exercise-induced distribution of immune cells (discussed below). The increase in leukocyte number during and immediately after exercise is due predominantly to increases in neutrophil, and to a lesser extent, lymphocyte counts, although monocyte number may also increase.

Resting Leukocyte Number in Athletes

Most studies of moderately trained and well-trained (i.e., not overtrained) athletes indicate normal resting circulating leukocyte counts when compared with clinical norms (4 to $11 \times 10^9 \cdot L^{-1}$) or values in nonathletes (Gleeson et al. 1995; Gray et al. 1993b; Hooper et al. 1995; Nieman et al. 1995b, 1995f; Tvede et al. 1989b). Moreover, several studies have shown no changes in resting leukocyte number after moderate training in previously untrained individuals (Nehlsen-Cannarella et al. 1991a; Soppi et al. 1982) and in athletes observed over several months of training and competition (Baum et al. 1994; Gleeson et al. 1995; Hooper et al. 1995; Janssen et al. 1989).

[1]There are many published papers documenting exercise-induced changes in leukocyte and subset numbers, and it is impossible to fully cite all of them throughout this chapter. Consequently, only representative references will be cited in the text in this chapter.

In contrast, low resting leukocyte counts have been reported in some endurance athletes such as distance runners or cyclists (Blannin et al. 1996; Green et al. 1981; Keen et al. 1995; Lehmann et al. 1992, 1996). In some of these studies, although values were reported to be within the clinically normal range for most athletes, mean levels were at the low end of the range ($4\text{-}5 \times 10^9 \cdot L^{-1}$), and counts in several athletes were below the lower limit. For example, Green et al. (1981) reported leukocyte counts of less than $4.3 \times 10^9 \cdot L^{-1}$ in 4 of 20 male distance runners, with three of these also having low lymphocyte counts. These low leukocyte counts may suggest long-term suppression of circulating leukocyte number in endurance athletes (discussed further below).

Acute Changes in Leukocyte Number

A large body of literature shows increases in leukocyte number after a variety of exercises, ranging in duration from a few seconds (100-yard run) to hours (marathon running, marching) and in intensity from moderate to intense. Table 3.1 presents a summary of exercise-induced changes in leukocyte number. The magnitude of increase in cell counts varies and is determined by a combination of duration and intensity effects; moreover, factors that modify release of stress hormones also influence the magnitude of increase in leukocyte count during exercise (discussed below). The time course of increase in circulating cell counts is complex, and often shows a biphasic response, depending on the intensity and duration of exercise (figure 3.1) (Gabriel et al. 1992a, 1992c; Gray et al. 1993b; Hansen et al. 1991; Shinkai et al. 1992). In general, the longer and more intense the exercise, the higher and more persistent the increase in leukocyte number. The pattern of leukocytosis appears to be similar in males and females, at least during prolonged exercise (Wells et al. 1982).

Brief Moderate and Intense Exercise

Leukocyte number increases dramatically after even brief (e.g., 30-60 seconds) maximal exercise (Gabriel et al. 1992c; Nieman et al. 1992). For example, Gabriel et al. (1992c) reported a 50% increase in leukocyte number immediately and 15 minutes after 60-second all-out cycle ergometry (figure 3.1). Leukocyte number declined toward resting values between 30 and 60 minutes postexercise, but showed a second rise to 50% above resting levels between 2 and 4 hours postexercise. The increase in cell number during and soon after exercise was attributed to rises in neutrophil, monocyte, and lymphocyte counts, whereas the delayed increase 2-4 hours postexercise was due only to increasing neutrophil number; lymphocyte number declined to below baseline

Table 3.1 Acute Effects of Exercise on Leukocyte Number

Exercise	During or immediately after*	Recovery*	Representative references
Brief (< 45 min)			
Moderate	↑ 0-40%	ND	Fairbarn et al. 1993; Iversen et al. 1994; Nieman et al. 1994
Intense	↑ 50%	↑ 50-100% 2 h post	Deuster et al. 1988; Lewicki et al. 1987; Ndon et al. 1992
Intense interval	↑ 65-80%	↑ 75% 2-6 h post	Fry et al. 1992b; Gray et al. 1993b; Nieman et al. 1995e
Prolonged (1-3 h)			
Moderate	↑ 25-50%	↑ 25-65% 2 h post	Shinkai et al. 1992; Tvede et al. 1993
Intense	↑ 200-300%	↑ 200-300% 2-6 h post	Nieman et al. 1995f; Shek et al. 1995; Tvede et al. 1989
Very long (> 3 h)	↑ 300%	↑ 300% 3 h post	Gabriel et al. 1994a

* = Percentage change compared with resting or preexercise values; ↑ = increase; ↓ = decrease; ND = no available data.

levels 2 hours postexercise. Leukocyte number increases up to 50% over resting levels during moderate or intense exercise of 30 minutes or less in duration (Deuster et al. 1988; Fairbarn et al. 1993; Fry et al. 1992b; Iversen et al. 1994; Lewicki et al. 1987; Ndon et al. 1992). Cell count may remain elevated up to 2 hours postexercise (Deuster et al. 1988; Ndon et al. 1992).

Intense Interval Exercise

Circulating leukocyte number also increases during and after intense interval exercise in a similar pattern to that observed with brief but continuous exercise (Gray et al. 1993b; Nieman et al. 1995e). For

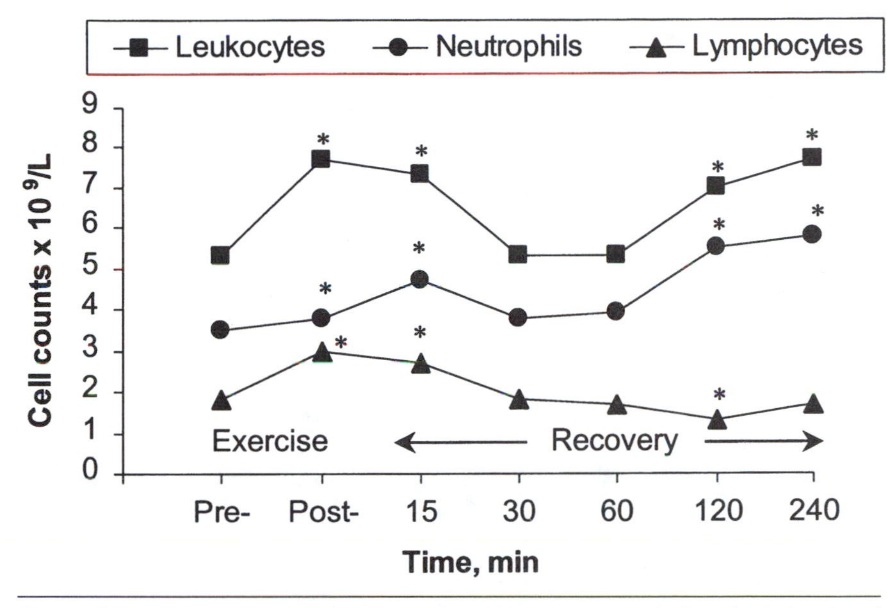

Figure 3.1 Leukocyte, neutrophil, and lymphocyte counts before and after brief intense exercise. Peripheral blood was sampled before (pre-) and immediately post-, and then at various times up to 4 hours after a 1-minute maximal cycle ergometry test. * = Significantly different from preexercise values.

Adapted from Gabriel, H., A. Urhausen, and W. Kindermann. 1992. Mobilization of circulating leucocyte and lymphocyte subpopulations during and after short, anaerobic exercise. *European Journal of Applied Physiology* 65: 164-170.

example, in eight male triathletes, repeated 1-minute treadmill sprints at 100% $\dot{V}O_{2max}$ (average of 16 total) resulted in a 65-75% increase in leukocyte number that persisted for 6 hours after exercise (Gray et al. 1993b). In another study, repeated weight lifting using the lower body (squats) to exhaustion (average time 37 minutes) was associated with an 80% increase in leukocyte number for at least 2 hours after exercise (Nieman et al. 1995e). Although plasma stress hormone levels changed during exercise in these two studies, increased cell counts could not be attributed to changes in these hormones (see discussion below).

Endurance Exercise

Leukocyte number increases shortly after the onset of endurance exercise, continues to increase throughout, and may remain elevated for at least 6 hours after prolonged exercise (figure 3.2) (Gabriel et al. 1994a; Nieman et al. 1989a, 1995f; Shek et al. 1995). In 10 male distance runners, leukocyte count increased by 250% during 3 hours treadmill running at marathon pace, and remained nearly three times resting

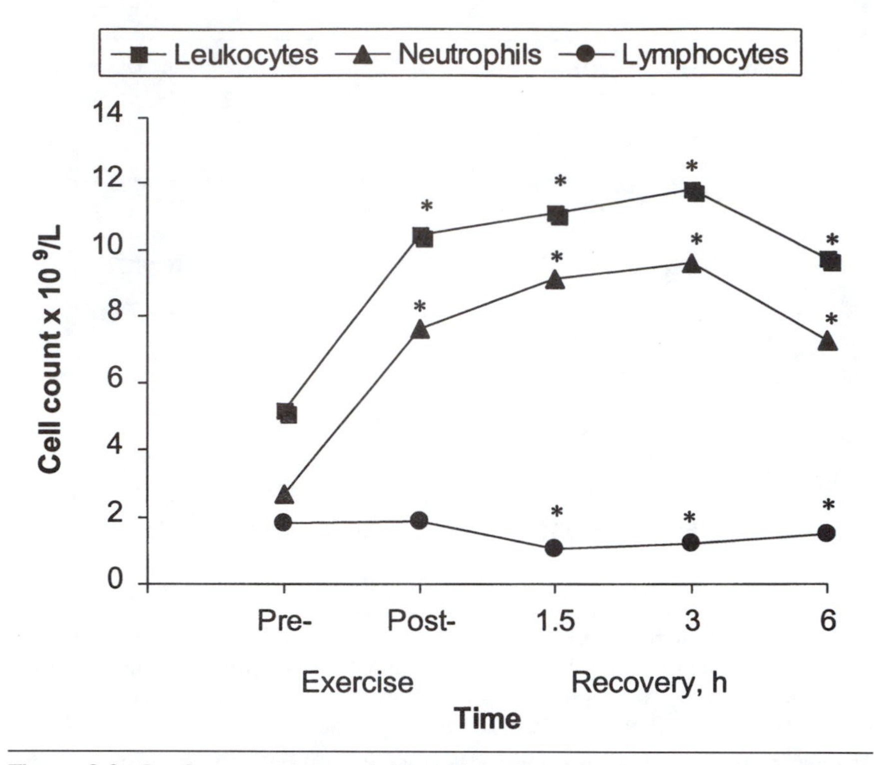

Figure 3.2 Leukocyte, neutrophil, and lymphocyte counts before and after a 2.5-hour treadmill run in distance runners. Peripheral blood was sampled before (pre-) and immediately post-, and then up to 6 hours during recovery postexercise. * = Significantly different from preexercise values.

Data from Nieman, D.C., S. Simandle, D.A. Henson, B.J. Warren, J. Suttles, J.M. Davis, K.S. Buckley, J.C. Ahle, D.E. Butterworth, O.R. Fagoaga, and S.L. Nehlsen-Cannarella. 1995f. Lymphocyte proliferative response to 2.5 hours of running. *International Journal of Sports Medicine* 16:404-408.

values at 1.5 and 2 times resting levels at 6 hours postexercise (Nieman 1989a). In another study on active males, leukocyte concentration rose progressively during 2-hour treadmill running at 65% $\dot{V}O_{2max}$ (blood samples taken at 30-minute intervals), reaching values three times resting level by the end of exercise; these high values were maintained for at least 2 hours postexercise (Shek et al. 1995).

During rigorous exercise lasting about 1 hour, leukocytes are rapidly mobilized during the first 10 minutes. For example, in trained and untrained subjects who cycled to exhaustion (> 60 minutes) at 100% individual anaerobic threshold (IAT, usually about 60-80% $\dot{V}O_{2max}$), leukocyte and other subset numbers achieved 50-80% of maximum values within the first 10 minutes, after which more gradual increases occurred; a similar pattern was observed for the increase in heart rate (Gabriel et al. 1994a). In contrast, blood catecholamine concentrations increased gradually throughout the entire exercise. It was suggested that hemodynamic changes such as increased cardiac output at the onset of prolonged exercise, as well as changes in adhesion molecules (see below), are largely responsible for the rapid mobilization of cells into the circulation at the onset of exercise.

The magnitude of change in leukocyte number during prolonged exercise is a function of exercise intensity. For example, prolonged exercise (mean time 87 minutes) at 85% IAT produced significantly smaller increases in leukocyte and neutrophil counts during and up to 4 hours after exercise compared with the same-duration exercise at 100% IAT. The largest differences between conditions were observed 1-2 hours postexercise.

Very Prolonged Exercise

During very long exercise lasting more than several hours (e.g., marching in soldiers, ultramarathons, multistage events), leukocyte number may initially increase for several hours and then decline close to or below resting values. For example, during a 160 km march lasting 24 hours, leukocyte number increased progressively to 33% above resting values during the first 16 hours, then decreased and remained about 17% below resting values for 50 hours after the march (Galun et al. 1987). The prolonged decrease in leukocyte count observed toward the end of exercise and throughout several days of recovery was attributed to migration of cells to damaged skeletal muscle and possibly to altered responsiveness of these cells to catecholamines (discussed further below). Gabriel et al. (1994a) reported a 300% increase in leukocyte count during a 100 km (8 hours) ultramarathon run; cell counts remained at that level for 3 hours postexercise. Thus, leukocytes are recruited into the circulation early in very prolonged exercise, and as for shorter exercise, the magnitude of recruitment varies with exercise intensity. However, this recruitment does not appear to be maintained during excessively long activity (e.g., 24-hour marching). This may reflect opposing stimuli on leukocyte distribution; that is, during the early stage of very prolonged exercise leukocyte demargination may be driven by release of stress hormones, while in the later stage cytokines or other chemoattractants may stimulate migration of leukocytes to damaged skeletal muscle (discussed further below and in chapter 6).

Biphasic Response of Exercise-Induced Leukocytosis

A biphasic increase in leukocyte number in response to exercise has been noted after maximal (Gabriel et al. 1992a, 1992c; Gray et al. 1993b;

Hansen et al. 1991), but not moderate (Nehlsen-Cannarella et al. 1991b), exercise. This biphasic response is characterized by a large increase in lymphocyte number early in exercise, followed by a large increase in neutrophil number and decline in lymphocyte number toward the end or after cessation of exercise (see figure 3.1). Hansen et al. (1991) studied the leukocyte response to maximal distance running over three distances (1.7, 4.8, and 10.5 km) in seven healthy males. During exercise, lymphocyte number doubled whereas neutrophil number increased only slightly, about 20%. Leukocyte, lymphocyte, and neutrophil counts returned toward resting values by 30 minutes postexercise. However, lymphocyte number continued to decline to 32-39% below resting values between 30 minutes and 4 hours postexercise. In contrast, neutrophil number exhibited a delayed increase between 2 and 4 hours postexercise, reaching levels 68-170% above resting. The magnitude of changes was directly related to running distance both during and after exercise. Gabriel et al. (1992c) reported a 30% increase in leukocyte number due to an equivalent rise in lymphocyte and a smaller (10%) increase in neutrophil number immediately after 1-minute maximal cycle ergometry at supramaximal power output (equivalent to 150% power output at $\dot{V}O_{2max}$). As noted for longer exercise, cell numbers returned to baseline values within 30 minutes postexercise, after which lymphocyte number declined and neutrophil count increased about 30%. Resting lymphocyte number was restored by 2 hours postexercise, although neutrophil number remained elevated 4 hours after exercise. These data suggest that even very brief intense exercise may significantly alter circulating leukocyte number and distribution for several hours after exercise. In contrast, moderate exercise does not appear to induce such a response. For example, in moderately trained women, there was only a slight increase (about 20%) in leukocyte count during and then 1.5 hours after moderate exercise (45 minutes walking at 60% $\dot{V}O_{2max}$) (Nehlsen-Cannarella et al. 1991b). Although lymphocyte count increased significantly during exercise, cell number returned to but did not decline below baseline levels during the 5 hours postexercise.

Exercise Leukocytosis in Athletes Versus Nonathletes

The magnitude of exercise-induced leukocytosis during and after various types of exercise is similar in athletes and nonatheletes during brief maximal exercise (Lewicki et al. 1987). For example, in young adult male cyclists and nonathletes, leukocyte number increased to a similar extent (about 50%) immediately after progressive cycling exercise to volitional fatigue and returned to similar baseline levels within 2 hours after exercise (Lewicki et al. 1987). In contrast, athletes may exhibit less of an increase in leukocyte number when exercising at the same absolute or relative submaximal work rate (Gimenez et al. 1987; Oshida et al. 1988). For example, although 2 hours of cycling at 60% $\dot{V}O_{2max}$ resulted in a 100% elevation of leukocyte number immediately after exercise in five nonathletes and six athletes, cell counts were significantly higher in the untrained subjects (Oshida et al. 1988).

Persistence of Leukocytosis After Exercise

Leukocyte count may remain elevated for some time after exercise, and this effect may persist far longer than the duration of the exercise itself (Gabriel et al. 1992a, 1992c; Gray et al. 1993b; Hansen et al. 1991; Nieman et al. 1989a, 1995e; Shek et al. 1995). For example, a 30% elevation in leukocyte number was observed for 4 hours after only 1-minute maximal exercise (power output at 150% of $\dot{V}O_{2max}$) (Gabriel et al. 1992c). Leukocyte count may also remain elevated for several hours and even longer after prolonged exercise (Nieman et al. 1989a; Shek et al. 1995). Nieman et al. (1989a) reported leukocyte number to be more than twice resting values for 6 hours after a 3-hour simulated marathon run on a treadmill in trained runners. Shek et al. (1995) noted a threefold increase in leukocyte concentration that persisted for at least 2 hours after 2-hour treadmill running at 65% $\dot{V}O_{2max}$. This persistent elevation of leukocyte count is due almost entirely to a concomitant increase in neutrophil number, since lymphocyte number may decline below resting values between 1 and 6 hours postexercise (discussed further below) (Nieman et al. 1989a; Shek et al. 1995).

Leukocyte Response to Exercise Training

Short-term exercise training in previously untrained individuals does not appear to alter the magnitude of leukocytosis during maximal exercise (Busse et al. 1980; Soppi et al. 1982). For example, in naval recruits, resting and postexercise leukocyte numbers were unchanged after a six-week training period in which $\dot{V}O_{2max}$ and work capacity increased significantly (Soppi et al. 1982).

As mentioned above, resting leukocyte number is generally clinically normal in athletes and does not change appreciably during normal periods of training. Normal leukocyte counts were reported in elite swimmers sampled at five times during a six-month season (early, mid, and late season, during taper, and postcompetition) (Hooper et al. 1995). Gleeson et al. (1995) and Tvede et al. (1989b) also reported no differences in resting leukocyte counts between low- and high-intensity training periods in elite swimmers and cyclists, respectively.

Ndon et al. (1992) found no significant changes in resting and post-exercise leukocyte counts after four weeks increased training in 10 male triathletes and cyclists who appeared to adapt well to the increased training (50% increase in volume) and did not show signs of overtraining. Baum et al. (1994) observed no changes in leukocyte number in male track athletes during various phases of a training season (e.g., endurance training, speed and strength training, competition). Janssen et al. (1989) found no changes in circulating leukocyte or subset numbers in 114 male and female previously untrained subjects who participated in an 18- to 20-month progressive training program in preparation for marathon running.

In contrast, recent studies suggest that, in some athletes, prolonged periods of intense training may result in lower than normal cell counts. Lehmann et al. (1996) reported a progressive decrease in leukocyte counts over four weeks intensified training in male distance runners. Training was intensified by either increasing distance (volume) by 100% or pace (intensity) by 150% over the four weeks. Subjects also rated "complaints" such as severe muscular stiffness or fatigue on a 1 to 4 scale, daily throughout the study. By the end of the four weeks, subjects in the increased-volume group showed clear signs of overtraining, such as a significant increase in the complaints index after day 20 and performance and hormonal changes. Leukocyte counts declined significantly from a mean $5.4 \times 10^9 \cdot L^{-1}$ before the study to 4.9 after two weeks and 4.2 after four weeks intensified training. In contrast, leukocyte counts remained unchanged at $5.3 \times 10^9 \cdot L^{-1}$ throughout the four weeks in the increased-intensity group, and these athletes showed no signs of overtraining syndrome. For ethical reasons, the study was not extended past four weeks, and it is unclear whether leukocyte number would have continued to decline in the increased-volume group, possibly below clinically normal levels, with continued training at that level. These data suggest that long periods of high-volume training leading to symptoms of overtraining may be associated with low leukocyte counts.

Decreased resting leukocyte counts were also reported in male competitive cyclists after a five-month training cycle compared with before (Ferry et al. 1990). However, mean counts were well within the normal range (7.6 and $6.8 \times 10^9 \cdot L^{-1}$ before and after training, respectively), and there were no significant differences between posttraining values in athletes and those obtained at the same time from control subjects. Keen et al. (1995) reported resting leukocyte numbers in well-trained cyclists that were in the clinically low end of the normal range; leukocyte counts were 5.04 and $7.5 \times 10^9 \cdot L^{-1}$ in cyclists and nonathletes, respectively. However, when measured over 12 days of intense exercise during competition (a multistage cycling race over 117-185 km · d^{-1}), neither early morning preexercise nor evening (5-7 hours post-race) leukocyte counts varied significantly in the cyclists. Similarly, Fry et al. (1992a) reported no significant changes in leukocyte number during 10 days of intense interval training in five well-trained male endurance athletes who completed two sessions per day of repeated 1-minute maximal treadmill runs (25 intervals per day); leukocyte counts remained above $5.7 \times 10^9 \cdot L^{-1}$ throughout the 10 days. Mujika et al. (1996) reported no significant changes in leukocyte count during 12 weeks normal training followed by 4 weeks of tapering (reduced training) in competitive swimmers. Taken together, these data suggest that daily intense exercise over shorter periods (less than two weeks) may not significantly alter circulating leukocyte number and that cell counts change only after prolonged periods (i.e., weeks to months) of intense, high-volume exercise associated with symptoms of maladaptation to training (i.e., the overtraining syndrome). The mode of training may also influence resting leukocyte number, in that low cell counts have been observed in distance runners (Green et al. 1981; Lehmann et al. 1996) but not swimmers (Gleeson et al. 1995; Hooper et al. 1995; Mujika et al. 1996), cyclists (Keen et al. 1995; Tvede et al. 1989b), or triathletes (Gray et al. 1993b; Ndon et al. 1992).

Although leukocyte count may increase acutely after exercise, very long exercise (e.g., hours) may cause a long-lasting suppression of circulating cell number. For example, compared with resting levels, leukocyte counts were lower for up to 40 hours after a 24-hour, 120 km march in endurance-trained males (Galun et al. 1987), and this was attributed to migration of leukocytes out of the circulation, possibly to damaged skeletal muscle. Elite endurance athletes often train several hours each day, and it is possible that low resting leukocyte counts may reflect a long-lasting effect of prior exercise sessions. Low leukocyte counts may also arise from migration of cells out of the circulation to sites of tissue injury; it is well known that skeletal muscle damage frequently occurs during very prolonged or intense exercise (see below).

To summarize, resting circulating leukocyte number is generally clinically normal in athletes, although values at the low end of the normal range have been reported during high-volume training in some endurance athletes. Leukocyte number increases dramatically during exercise, reflecting increases in all the main subsets, neutrophils and lymphocytes in particular. The magnitude of increase is related to exercise intensity and duration, and increases may persist for several hours after cessation of exercise. There is often a biphasic response of leukocyte number, characterized by a large increase in lymphocyte and relatively smaller increase in neutrophil number early in exercise, followed by a continued rise in neutrophil number and plateau or decline in lymphocyte number toward the end of exercise. Lymphocyte number may decline below baseline levels 1 to 2 hours postexercise, whereas neutrophil number may continue to rise during this time. Changes in leukocyte number and relative proportion of subsets result from recruitment of cells into the circulation, mediated at least in part by stress hormones such as corticosteroids and catecholamines (discussed further below).

Granulocyte Number

In general, exercise-induced changes in neutrophil number parallel changes in total leukocyte count, since neutrophils account for about 70% of all leukocytes. Neutrophil number increases during exercise and remains elevated for several hours after cessation of exercise. Recruitment of neutrophils into the circulation appears to be related to cortisol release during prolonged, but not necessarily brief (< 30 minutes), exercise. Because neutrophils constitute the vast majority of granulocytes, and the responses of other types of granulocytes are infrequently reported, this section will focus only on the neutrophil response to exercise.

Resting Neutrophil Number

Resting neutrophil number is usually within the clinically normal range in athletes (Fry et al. 1992a; Gabriel et al. 1994b; Hooper et al. 1995; Pyne et al. 1995; Tvede et al. 1991). For example, neutrophil number was within the clinically normal range in elite male and female swimmers studied throughout 12 weeks of intense training (Pyne et al. 1995). Moreover, a significant increase in neutrophil number was reported from the start to end of the 12 weeks. In another study on elite swimmers followed over a six-month season, neutrophil counts were clinically normal, and no differences were observed between overtrained and non-overtrained swimmers, except in overtrained swimmers during the taper before major competition (Hooper et al. 1995). During the taper, neutrophil counts were significantly elevated by 80% in swimmers diagnosed as overtrained (as evidenced by persistent fatigue, performance decrements, subjective comments from swimmers in daily logbooks) (Hooper et al. 1995). The high neutrophil counts in overtrained swimmers were attributed to a long-lasting effect of exercise prior to blood sampling and to maintenance of high-volume, high-intensity work by these swimmers, who, in fact, did not reduce training during the taper period.

Gabriel et al. (1994b) reported no changes in resting or postexercise neutrophil counts in cyclists tested in the week before and after a strenuous 240 km cycle race. Fry et al. (1992a) found no changes in resting neutrophil number in male endurance athletes over a 10-day period of twice-daily intense interval training. Suzuki et al. (1996) reported no changes in resting neutrophil number in previously untrained males who performed seven consecutive days of intense exercise (1.5-hour cycle ergometry at 70% $\dot{V}O_{2max}$). These data suggest that resting neutrophil count is relatively impervious to short-term intense training lasting up to 12 weeks.

In contrast, lower neutrophil counts have been reported in male distance runners during intense compared with moderate training periods and with control subjects (Hack et al. 1994). Neutrophil number was compared between matched nonathletes and seven male distance runners during moderate and intense training, consisting of 89 $km \cdot wk^{-1}$ of continuous running versus 102 $km \cdot wk^{-1}$ of harder continuous and interval running, respectively, over a three-month period. Resting cell counts were similar in nonathletes and athletes during moderate training, but were 30% lower during intense training. In another study involving middle-aged male cyclists, resting neutrophil concentration was significantly lower than in matched nonathletes, and about 15% lower than the clinically normal lower limit of $3.0 \times 10^9 \cdot L^{-1}$ (Blannin et al. 1996). Morphological evidence of immature neutrophils (e.g., less lobulated and indented nuclei) was observed in cyclists exhibiting low neutrophil counts (Keen et al. 1995). The presence of immature neutrophils may suggest increased turnover of cells, which may account for the lower neutrophil number. Thus, although most evidence indicates that athletes generally exhibit clinically normal neutrophil concentration, recent data suggest that neutrophil counts may be low in some endurance athletes during prolonged periods of very intense training.

Acute Exercise

Neutrophil number increases markedly during and after virtually all forms of exercise, but the magnitude and persistence of these changes are related to exercise intensity and duration (Bury and Pirnay 1995; Hansen et al. 1991; Shek et al. 1995; Suzuki et al. 1996). Table 3.2 summarizes the current literature on the effects of exercise on neutrophil number.

Table 3.2 Acute Effects of Exercise on Neutrophil Number

Exercise	During or immediately after*	Recovery*	Representative references
Brief (< 45 min)			
Moderate	↑ 30-50%	ND	Camus et al. 1992; Fairbarn et al. 1993
Intense	↑ 30-150%	Normal 30 min-2 h post; ↑ 25-100% 2-4 h post	Deuster et al. 1988; Gabriel et al. 1992c; Hack et al. 1992; Nielsen et al. 1996a
Intense interval	↑ 25%	↑ 60-100% 2-6 h post	Fry et al. 1992b; Gray et al. 1993b; Nieman et al. 1995e
Prolonged (1-3 h)			
Moderate	↑ 20-50%	↑ 50-150% 2 h post	Nieman et al. 1993b; Shinkai et al. 1992; Tvede et al. 1993
Intense	↑ 300%	↑ 300-400% 2-6 h post	Gabriel et al. 1992a; Nieman et al. 1995f; Shek et al. 1995; Suzuki et al. 1996
Very long (> 3 h)	↑ 300%	↑ 300% 3 h post	Gabriel et al. 1994a

* = Percentage change compared with resting or preexercise values; ↑ = increase; ↓ = decrease; ND = no available data.

Brief Moderate and Intense Exercise

Neutrophil count appears to be transiently increased during and immediately after brief (< 30 minutes) intense exercise. For example, neutrophil count increased by about 30% immediately after a treadmill test to volitional fatigue in males of varying fitness level, but returned to preexercise values by 1 hour postexercise (Deuster et al. 1988). In a study on rowers, neutrophil count increased by 50% during and 150% immediately after 6-minute all-out rowing but returned to resting concentration by 2 hours postexercise (Nielsen et al. 1996a). In untrained subjects, 6 minutes of cycle ergometry to volitional fatigue produced a brief 25% increase in neutrophil count, with resting levels restored within 30 minutes (McCarthy et al. 1992).

Although these data suggest only transient disturbance of neutrophil number after brief exercise, most studies have sampled blood only a few times after exercise (e.g., immediately and at one other time postexercise). Other studies including more frequent and later blood sampling suggest that neutrophil number may exhibit a biphasic response with an initial increase during or immediately after brief exercise, followed by a return to resting values between 30 and 60 minutes postexercise and a second increase in cell number beginning 1-2 hours postexercise (Hansen et al. 1991). In active males, runs of varying duration (1.7, 4.8, and 10.5 km, about 5 to 40 minutes) produced slight increases (about 15%) in neutrophil number immediately after exercise. Cell counts remained at this level 30 minutes after the longer run, but returned to resting values within 30 minutes after the two shorter runs. However, at the next sample time (2 hours postexercise), neutrophil number was again elevated by 25% to 100% and remained at this level for the next 2 hours. The delayed elevation of neutrophil number was directly related to exercise duration.

Intense Interval Exercise

Intense discontinuous exercise also induces a prolonged increase in neutrophil number (Gray et al. 1993b; Nieman et al. 1995e). In male endurance athletes, neutrophil count increased by 25% immediately after repeated 1-minute treadmill sprints to exhaustion, and continued to increase a further 50% up to 6 hours postexercise (Gray et al. 1993b). Neutrophil number was also reported to double 2 hours after repetitive weight lifting to exhaustion in experienced weight lifters (Nieman et al. 1995e).

Endurance Exercise

During longer-duration exercise, neutrophil number rises early (Gabriel et al. 1993) and may continue to increase throughout and for some time after prolonged exercise (Gabriel et al. 1994a; Nieman et al. 1995f; Shek et al. 1995). For example, in male runners, neutrophil counts tripled immediately after a 2.5-hour treadmill run at 75% $\dot{V}O_{2max}$, increased to four times resting values by 1.5 hours postexercise, and remained at three to four times resting concentration for 6 hours after exercise (Nieman et al. 1995f). During and after exercise, neutrophil count increases to a much greater extent than lymphocyte number (which may decrease after exercise, discussed below), resulting in an increase in the neutrophil-to-lymphocyte ratio (Ne/Ly ratio) (Nieman et al. 1995f).

The extent to which neutrophil concentration increases during and after exercise depends on an interaction between exercise mode, intensity, and duration. In recreationally active males, neutrophil number increased to a similar extent (doubled) after three different endurance exercise sessions consisting of (mode unspecified) 2 hours at 75% $\dot{V}O_{2max}$, 3 hours at 60% $\dot{V}O_{2max}$, and 4 hours at 45% $\dot{V}O_{2max}$ (Bury and Pirnay 1995). In a study comparing the effects of downhill versus level treadmill running (60 minutes at 70% $\dot{V}O_{2max}$), circulating neutrophil counts were higher 1.5 and 12 hours postexercise in the downhill protocol, suggesting that exercise with an eccentric bias elicits a greater and more persistent mobilization of neutrophils (Pizza et al. 1995) (discussed further below).

Training Effects

There are contradictory data regarding the effect of training or fitness level on the neutrophil response to exercise; some studies show similar responses in trained and untrained individuals, while other studies have noted differences. Discrepancies between studies may be explained by different exercise protocols. For example, in untrained controls, distance runners, and triathletes, neutrophil number increased to a similar extent, by about 30% to 40%, immediately after a treadmill test to volitional fatigue; resting values were restored by 30 minutes postexercise in the three groups of subjects (Hack et al. 1992). In another study from the same laboratory, the relative increase in neutrophil concentration was similar after exercise to volitional fatigue in nonathletes and distance runners assessed during both moderate and intense training periods (Hack et al. 1994). In another study, a similar 50-60% increase in neutrophil number was observed immediately and 2 hours after a progressive cycle ergometry test to volitional fatigue in trained cyclists and matched control subjects (Lewicki et al. 1987).

In contrast, Blannin et al. (1996) reported less of an increase in neutrophil count after 15-minute cycling to 150 W in middle-aged cyclists compared with matched nonathletes. However, it is apparent from the heart rate data that the athletes were working at a far lower level relative to maximum (127 vs. 161 beats per minute in athletes and nonathletes, respectively). Oshida et al. (1988) noted a larger increase in neutrophil number after 2-hour cycle ergometry at 60% $\dot{V}O_{2max}$ in untrained compared with trained subjects. In another study, neutrophil concentration increased more after 15-minute cycling in active (training 4-5 hours per week) compared with inactive subjects (50% vs. 20% increases, respectively) (Benoni et al. 1995b). However, because of hemodynamic training effects that lower submaximal heart rate at any given work rate, the active subjects would have performed more work during the 15 minutes.

Intense training in previously untrained individuals may attenuate the acute increase in neutrophil number following prolonged exercise. For example, untrained males performed seven consecutive days of exercise consisting of 1.5-hour cycle ergometry at 70% $\dot{V}O_{2max}$ (which could be considered vigorous training in previously untrained individuals) (Suzuki et al. 1996). Compared with resting values obtained on days 1, 4, and 7, neutrophil count doubled immediately and 1 hour after exercise on the first day, but increased only 20-60% after exercise on days 4 and 7 of training (figure 3.3). On each day of exercise, an increase in immature "band" neutrophils was observed immediately and 1 hour after exercise, suggesting release of cells from bone marrow during exercise. However, the attenuated postexercise increase in total neutrophil count was paralleled by a similar decline in postexercise band neutrophils, suggesting that fewer immature cells were recruited into the circulation after training. Taken together, these data suggest that the neutrophil response to brief maximal exercise is similar in trained and untrained individuals and that the response to brief submaximal exercise is related to the total amount of work performed. Intense endurance training appears to reduce the extent of mobilization of neutrophils, possibly via attenuating the hormonal response to exercise (discussed below).

In summary, acute changes in total leukocyte count during and after exercise are paralleled by changes in neutrophil number. Neutrophil concentration increases during and after exercise and may remain elevated for several hours after the end of exercise, and the magnitude of these changes varies with exercise intensity and duration. Neutrophil count may exhibit a biphasic response, characterized by an initial small increase during, followed by a decline to resting values 30-60 minutes after, exercise and a delayed larger (twofold) increase in cell number 2 to 4 hours postexercise. These increases reflect recruitment into the circulation of cells, possibly immature cells with lower functional activity. Neutrophil count is generally normal at rest in athletes, although training may attenuate the acute response of cell number to exercise. The effects of exercise on neutrophil function are addressed in chapter 4.

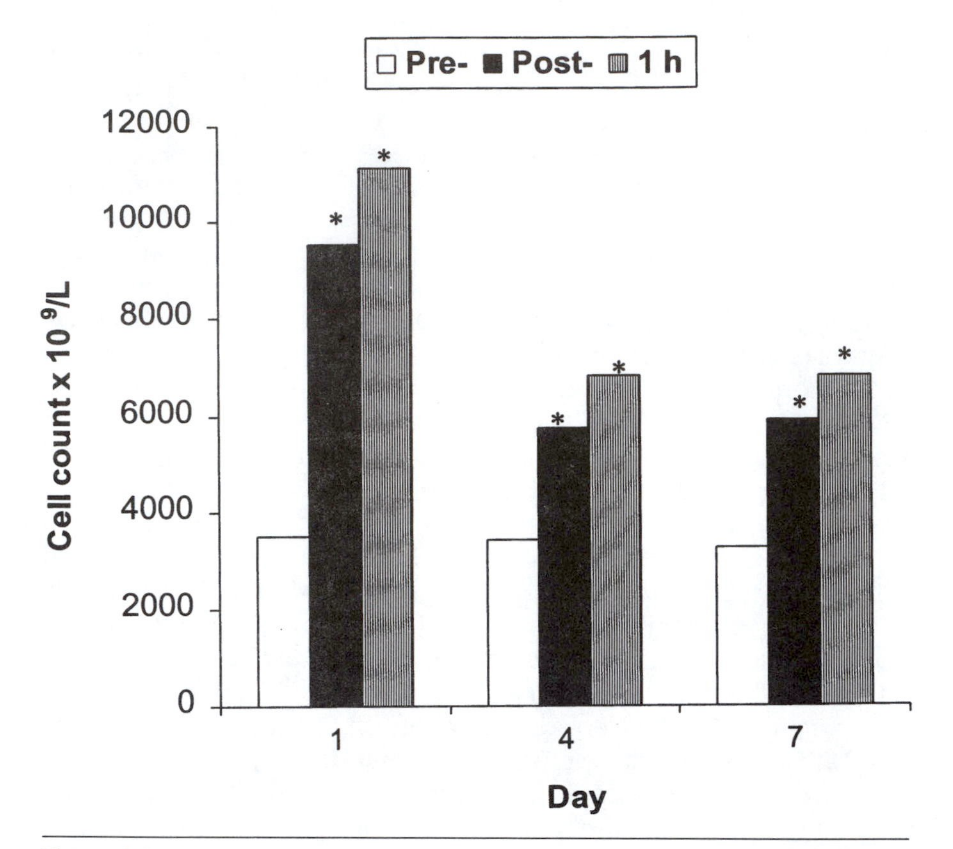

Figure 3.3 Effects of exercise training on the acute response of neutrophil count to exercise. Untrained men performed intense exercise (1.5 hours at 70% $\dot{V}O_{2max}$) on seven consecutive days. Peripheral blood was sampled pre-, immediately post-, and 1 hour postexercise on days 1, 4, and 7 of the seven days training. * = Significantly different compared with preexercise values on the same day.

Data from Suzuki, K., S. Naganuma, M. Totsuka, K.-J. Suzuki, M. Mochizuki, M. Shiraishi, S. Nakaji, K. Sugawara. 1996. Effects of exhaustive endurance exercise and its one-week daily repetition on neutrophil count and functional status in untrained men. *International Journal of Sports Medicine* 17: 205-212.

Lymphocyte Number and Subset Distribution

Lymphocytosis (increase in lymphocyte number) occurs during and immediately after exercise under a variety of conditions, from very brief exercise such as sprints or stair climbing to extended exercise such as marathon running. The effects of exercise on lymphocyte count are summarized in table 3.3. As noted above, lymphocyte number increases to a far lesser extent than neutrophil number, resulting in a higher Ne/Ly ratio both during and after exercise (Nieman et al. 1995f). Lymphocyte number generally returns to baseline levels early in recovery after exercise, but may decline below resting

Table 3.3 Acute Effects of Exercise on Lymphocyte Number

Exercise	During or immediately after*	Recovery*	Representative references
Brief (< 45 min)			
Moderate	↑ 50-100% ≈ work rate	Normal	Fairbarn et al. 1993; Foster et al. 1986; Nehlsen-Cannarella et al. 1991
Intense	↑ 100-200%	↓ 40% 1-2 h post; normal by 6 h post	Deuster et al. 1988; Ndon et al. 1992; Nielsen et al. 1996; Nieman et al. 1992
Intense interval	↑ 100%	↓ 40% 2 h post; normal by 6 h post	Fry et al. 1992b; Gray et al. 1993b; Nieman et al. 1995e
Prolonged (1-3 h)			
Moderate	↑ 0-25%	↓ 25-40% 1-2 h post	Shinkai et al. 1992; Tvede et al. 1993
Intense	↑ 30-100%	↓ 15-40% 1-6 h post	Gabriel et al. 1992a; Nieman et al. 1995f; Pizza et al. 1995; Shek et al. 1995; Tvede et al. 1993
Very long (> 3 h)	↑ 50%	↓ 40% 3 h post	Gabriel et al. 1994a

* = Percentage change compared with resting or preexercise values; ↑ = increase; ↓ = decrease; ND = no available data.

values before returning to preexercise concentration (discussed further below).

Resting Lymphocyte Number in Athletes

Resting lymphocyte number is usually normal in athletes from a variety of endurance and power sports (Baum et al. 1994; Fry et al. 1992a; Gleeson et al. 1995; Gray et al. 1993b; Mujika et al. 1996; Nieman et al. 1995b, 1995c, 1995f; Nielsen et al. 1996a). There is, however, one report of low lymphocyte counts in marathon runners, in which 10 of 20 marathon runners had resting lymphocyte counts of less than 1.5 $\times 10^9 \cdot L^{-1}$ (Green et al. 1981). However, it was also noted that at least 5 athletes had completed marathon or other long runs within three days of blood sampling. It is possible that the low lymphocyte counts reflected a long-lasting effect of the last run; lymphocyte count has been shown to decline below resting levels after distance running (discussed below). Since the vast majority of papers on different types of athletes, including distance runners, cyclists, triathletes, swimmers, and weight lifters, generally show clinically normal resting lymphocyte concentrations, it is probable that long-term exercise training has little effect on lymphocyte number.

Acute Changes in Lymphocyte Number

The extent of lymphocytosis during exercise depends on an interaction of exercise intensity and fitness level. During moderate or very brief exercise, lymphocyte number may remain unchanged (Rose and Bloomberg 1989; Smith et al. 1989) or increase up to 50% above resting levels (Foster et al. 1986; Oshida et al. 1988). During prolonged submaximal exercise, lymphocyte number may increase to values two or three times resting levels. As with leukocytes, lymphocyte number rises progressively with increasing work rate, and the magnitude of lymphocytosis is related to exercise intensity. However, in contrast to findings for leukocytosis, exercise duration may not be an important determinant (discussed below).

Brief Moderate and Intense Exercise

Brief maximal exercise induces large increases in circulating lymphocyte number. Compared with resting levels, lymphocyte concentration has been observed to increase two- to threefold after 6 minutes of all-out rowing in trained oarsmen (Nielsen et al. 1996a), 20-km cycle ergometer time trial in cyclists (Ndon et al. 1992), 5 km race in distance runners (Espersen et al. 1990), and graded cycle test to volitional fatigue in cyclists (Lewicki et al. 1988). Lymphocyte number also increases to the same extent after brief exhaustive interval exercise such as repeated treadmill sprints to exhaustion (Gray et al. 1993b) and repetitive weight lifting to exhaustion (Nieman et al. 1995e).

There is a biphasic response of lymphocyte number to brief maximal exercise: after an initial increase during and immediately after exercise, cell number declines below baseline levels before returning to preexercise values by 6 hours postexercise. A 30-50% decline in circulating lymphocyte number has been noted between 1 and 3 hours after exhaustive weight lifting (Nieman et al. 1995e), 6 minutes of all-out rowing (Nielsen et al. 1996a), repeated run sprints to exhaustion (Gray et al. 1993b), and 5-km running (Espersen et al. 1990). As noted above, this biphasic response does not occur after moderate exercise (e.g., 45 minutes of walking at 60% $\dot{V}O_{2max}$) (Nehlsen-Cannarella et al. 1991b).

Endurance Exercise

Lymphocyte count increases early during prolonged exercise and may then remain at that level or increase gradually to the end of exercise (Gabriel et al. 1992a; Nieman et al. 1989a; Shek et al. 1995). For example, Gabriel et al. (1992a) reported a 30% increase in cell number during the first 10 minutes of a cycling test to volitional fatigue at individual anaerobic threshold (total time about 60 minutes), followed by a further gradual increase of similar magnitude from 10 minutes to the end of the test. Shek et al. (1995) also observed a doubling of cell concentration during the first 30 minutes of 2-hour treadmill running at 65% $\dot{V}O_{2max}$, followed by a slight decline between 30 and 90 minutes and subsequent increase at the end of exercise (figure 3.4). A significant 30% increase in cell number was noted during the first hour of a 3-hour treadmill run, with only minor increases to the end of exercise (Nieman et al. 1989a).

The magnitude of increase in lymphocyte number during prolonged exercise is related to exercise intensity (Kendall et al. 1990; Nieman et al. 1993b; Tvede et al. 1993) and, to a smaller extent, fitness level (Kendall et al. 1990; Oshida et al. 1988). For example, when subjects performed 60 minutes of cycle ergometry at 25%, 50%, and 75% $\dot{V}O_{2max}$, lymphocyte number did not change after the lowest-intensity exercise, increased by 40% after the middle-intensity exercise, and increased by 125% after the highest-intensity exercise (Tvede et al. 1993). In male distance runners, lymphocyte number increased by 50% during 45 minutes of treadmill running at 80% $\dot{V}O_{2max}$, but did not change after moderate exercise at 50% $\dot{V}O_{2max}$ (Nieman et al. 1993b). Kendall et al. (1990) tested individuals of different fitness levels using

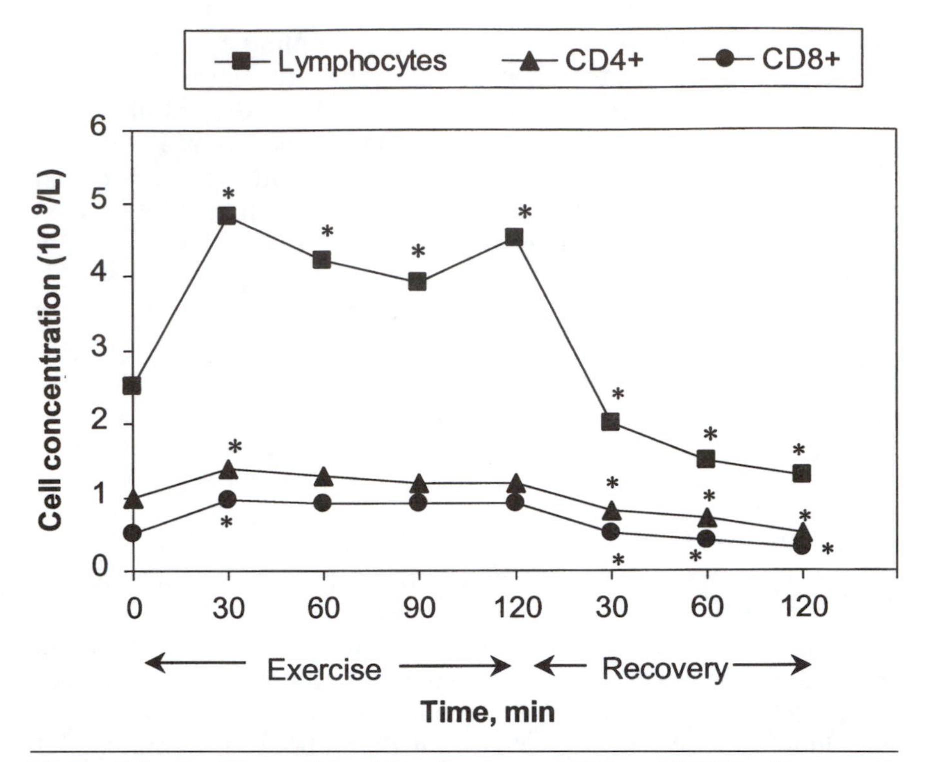

Figure 3.4 Responses of total lymphocyte and CD4 and CD8 T cell counts to endurance exercise. Peripheral blood was sampled from trained men at various times during exercise and recovery; exercise consisted of 2-hour treadmill running at 65% $\dot{V}O_{2max}$. * = Significantly different compared with preexercise values.

Adapted from Shek, P.N., B.H. Sabiston, A. Buguet, and M.W. Radomski. 1995. Strenuous exercise and immunological changes: a multiple-time-point analysis of leukocyte subsets, CD4/CD8 ratio, immunoglobulin production and NK cell response. *International Journal of Sports Medicine* 16: 466-474.

cycle ergometry protocols of various time and intensity from 30 to 120 minutes and from 30% to 75% $\dot{V}O_{2max}$. Regardless of fitness level, larger increases in lymphocyte number were observed after more intense exercise; 60-minute cycling at 30% $\dot{V}O_{2max}$ induced little change in cell number.

Exercise-induced lymphocytosis may be influenced by fitness level, since cell counts increased slightly but significantly more after exercise (2 hours of cycling at 60% $\dot{V}O_{2max}$) in untrained compared with trained subjects (Oshida et al. 1988). However, it appears that the effects of fitness level are much smaller than the acute effects of exercise; that is, acute exercise may double or triple cell count, but at any given time, cell counts differ between trained and untrained individuals by only about

10-20% (Kendall et al. 1990). Exercise mode may also affect the magnitude of exercise-induced lymphocytosis. In distance runners, lymphocytosis was greater after 60-minute treadmill running at 70% $\dot{V}O_{2max}$ using a downhill (–10% grade) compared with level protocol (Pizza et al. 1995).

Biphasic Response of Lymphocytes

As with brief exercise, there is a biphasic response of lymphocyte number to prolonged exercise; cell counts increase during, and decline below baseline levels for several hours after, exercise (see Figures 3.1 and 3.3) (Gabriel et al. 1994a; Nieman et al. 1989a, 1995f; Oshida et al. 1988; Pizza et al. 1995; Shek et al. 1995; Tvede et al. 1993). Lymphocyte concentration is usually 30-50% below preexercise values between 1 and 3 hours after prolonged exercise (Gabriel et al. 1994a; Nieman et al. 1989a, 1995f; Shek et al. 1995; Tvede et al. 1993), gradually returning to baseline levels by 6 hours (Nieman et al. 1989a, 1995f). Just as the magnitude of increase during exercise is related to exercise intensity, so too is the degree of suppression after exercise. For example, although lymphocyte number was lower than resting values for up to 3.5 hours after 45 minutes of running at both 80% and 50% $\dot{V}O_{2max}$, cell counts were suppressed to a smaller extent (30% vs. 45%) after the moderate compared with intense runs (Nieman et al. 1993b).

Exercise Mode

Although exercise mode influences the magnitude of lymphocytosis during exercise (i.e., larger lymphocytosis during downhill running), postexercise lymphocyte number was similar after level and downhill running (60 minutes at 70% $\dot{V}O_{2max}$) (Pizza et al. 1995). Thus, exercise with an eccentric bias appears to induce greater mobilization of cells into the circulation during, and efflux of cells after, prolonged exercise.

Training Responses

In athletes, prolonged periods of training appear to influence neither resting lymphocyte count (Ferry et al. 1990; Fry et al. 1992a; Gleeson et al. 1995; Hack et al. 1992, 1994; Kajiura et al. 1995; Mujika et al. 1996; Ndon et al. 1992) nor the magnitude of lymphocyte response to exercise (Ferry et al. 1990; Hack et al. 1994; Ndon et al. 1992). Resting lymphocyte counts were unchanged by both short-term intense interval training (e.g., 10 days of twice-daily repeated maximal treadmill sprints; Fry et al. 1992a) and longer-term training in elite athletes (e.g., seven-month training season in swimmers; Gleeson et al. 1995). Moreover, the pattern of lymphocytosis during and decreased lymphocyte number after maximal exercise was similar between nonathletes and distance runners, as well as when responses in runners were compared between high- and low-intensity/volume training (Hack et al. 1994). In contrast, there is one report of a significant 12% decline in resting lymphocyte count after a six-month training season (average 500 km · week^{-1}) in cyclists; a nonathlete control group showed no such changes over the same time (Baj et al. 1994). It is possible that very high volume training over several months is required to alter resting lymphocyte number. Resting lymphocyte count does not appear to be influenced by short-term moderate training in previously untrained individuals (Mitchell et al. 1996; Nehlsen-Cannarella et al. 1991a; Nieman et al. 1990c, 1993b).

Exercise and Lymphocyte Subsets

Various lymphocyte subsets may respond differently to acute exercise. In general, all subsets increase in number during exercise, with NK cell number increasing more than B and T cells (Gray et al. 1993b; Nieman et al. 1989a, 1994, 1995a; Pedersen et al. 1988, 1990; Tvede et al. 1993, 1994). Moreover, changes in the ratio of T cell subsets (CD4 and CD8) have been consistently noted after exercise, due to a larger increase in circulating CD8 and smaller increase in CD4 T cell numbers. The end result is that both absolute numbers and relative percentages of lymphocyte subsets change during and after exercise.

In addition, there may be individual variation in the lymphocyte subset response to exercise, which may partially explain inconsistencies in the literature. For example, changes in CD4 and CD8 cell numbers were reported after moderate exercise (60 minutes at 60% $\dot{V}O_{2max}$) on the first of two testing sessions spaced three weeks apart (Ricken et al. 1990). These results may have been due to unfamiliarity with or anxiety about exercise testing on the first occasion and subsequently a more accurate measure of lymphocyte subsets would have been obtained in the second session.

Resting Lymphocyte Subsets in Athletes

The relative proportions of lymphocyte subsets (T, B, and NK cells) obtained at rest generally do not differ between nonathletes and athletes from several sports including distance running, swimming, cycling, and weight lifting (Fry et al. 1992a; Gleeson et al. 1995; Gray et al. 1993b; Mujika et al. 1996; Ndon et al. 1992; Tvede et al. 1989b), or after moderate training in previously sedentary individuals (Nehlsen-Cannarella et al. 1991a; Nieman et al. 1993b), or intense training in athletes (Fry et al. 1992a; Gleeson et al. 1995; Mackinnon et al. 1997;

Mujika et al. 1996). These data suggest that exercise training has no long-term effect on lymphocyte subset distribution.

Acute Responses of T Cells to Brief Exercise

T cells account for up to 70% of total lymphocyte number in the circulation, so it is not surprising that lymphocytosis during exercise is reflected in a large increase in T cell number. Brief maximal exercise recruits T cells into the circulation so that absolute T cell number may double following exercise (table 3.4) (Bieger et al. 1980; Christensen and Hill 1987; Espersen et al. 1990; Ferry et al. 1990; Gray et al. 1993b;

Table 3.4 Acute Effects of Exercise on T Cell (CD3) Number

Exercise	During or immediately after*	Recovery*	Representative references
Brief (< 45 min)			
Moderate	No change	ND	Camus et al. 1992
Intense	↑ 100%	↓ 30% 1-2 h post	Deuster et al. 1988; Lewicki et al. 1988; Nielsen et al. 1996a; Nieman et al. 1992
Intense interval	↑ 60-100%	↓ 30-40% 1-2 h post	Fry et al. 1992b; Gray et al. 1993b; Nieman et al. 1995e
Prolonged (1-3 h)			
Moderate	↑ 20-30%	↓ 20% 2 h post	Oshida et al. 1988; Shinkai et al. 1992
Intense	↑ 30-60%	↓ 30-40% 1-6 h post	Gabriel et al. 1992a; Nieman et al. 1995f; Shek et al. 1995; Tvede et al. 1993
Very long (> 3 h)	↓ 30-40%	↓ 30-40% 3 h post	Gabriel et al. 1994a

* = Percentage change compared with resting or preexercise values; ↑ = increase; ↓ = decrease; ND = no available data.

Hansen et al. 1991; Lewicki et al. 1988; Nielsen et al. 1996a; Nieman et al. 1995e). Normal T cell number may be restored soon after the end of exercise, although the number may decline below baseline before returning to preexercise values (Espersen et al. 1990; Gray et al. 1993b). For example, Gray et al. (1993b) reported a doubling of CD3 and CD8 cell number and a 50% increase in CD4 cell count immediately after repeated 1-minute treadmill sprints to exhaustion in endurance-trained males, followed by declines of 25-50% in T cell and subset counts 1.5 hours postexercise. Normal cell numbers were restored by 6 hours after exercise. Nielsen et al. (1996a) observed similar increases in T cell subsets at the end of, followed by 20-50% decreases in subset counts 2 hours after, 6-minute maximal rowing ergometry in trained rowers. Espersen et al. (1990) noted 50-65% increases in CD4 and CD8 cell counts immediately after a 5-km (15-18 minutes) race in distance runners, followed by a 10-65% decline in cell numbers 2 hours postexercise. Similar changes in T cell subsets have also been reported after weight lifting to exhaustion (Nieman et al. 1995e).

CD4:CD8 Ratio Changes After Brief Exercise

Although both CD4 and CD8 cell numbers increase during and immediately after brief maximal exercise, relative changes in CD8 cell number are greater than for CD4 counts (table 3.5). As a result, the CD4:CD8 ratio may decline during and immediately after brief maximal exercise (Espersen et al. 1990; Frisina et al. 1994; Gray et al. 1993b; Lewicki et al. 1988; Hinton et al. 1997; Nielsen et al. 1996a). For example, Gray et al. (1993b) noted larger increases in CD8 compared with CD4 cell number (100% and 50%, respectively) immediately after repetitive treadmill sprinting to exhaustion. Nielsen et al. (1996a) observed a threefold increase in CD8 compared with a 50% rise in CD4 cell counts after 6 minutes of maximal rowing ergometry. As a result, the CD4:CD8 ratio declines during and immediately after exercise (Espersen et al. 1990; Frisina et al. 1994; Hinton et al. 1997; Lewicki et al. 1988). For example, Lewicki et al. (1988) reported a significant decline in CD4:CD8 ratio, from 1.5 to 1.0, immediately after a graded cycle ergometry test to volitional fatigue in male cyclists. Frisina et al. (1994) also observed a decline in cell ratio, from 1.9 to 1.0, immediately after 25 treadmill sprints to exhaustion in active males.

In contrast to these changes occurring during and immediately after exercise, the delayed decline in CD8 cell number is larger than for CD4 cells. For example, Nielsen et al. (1996a) observed a 50% decline in CD8 cell number compared with a 20% decrease in CD4 cell count 2 hours after 6 minutes of all-out rowing. Gray et al. (1993b) also reported a larger drop in CD8 compared with CD4 cell counts 1.5 hours after

Table 3.5 Acute Effects of Exercise on CD4 and CD8 T Cell Numbers

Exercise	During or immediately after*	Recovery*	Representative references
Brief (< 45 min)			
Moderate	CD4 no change CD8 no change	ND	Camus et al. 1992
Intense	CD4 ↑ 50% CD8 ↑ 50-300%	CD4 ↓ 10-25% 1-2 h post CD8 ↓ 50-65% 1-2 h post	Espersen et al. 1990; Lewicki et al. 1988; Nielsen et al. 1996a; Nieman et al. 1992
Intense interval	CD4 ↑ 50% CD8 ↑ 200%	CD4 ↓ 25% 2 h post CD8 ↓ 50% 2 h post; normal by 6 h post	Fry et al. 1992b; Gray et al. 1993b
Prolonged (1-3 h)			
Moderate	CD4 ↑ 0-25% CD8 ↑ 0-15%	CD4 and CD8 ↓ 10-30% 1-2 h post	Shinkai et al. 1992; Tvede et al. 1993
Intense	CD4 ↑ 20-40% CD8 ↑ 60%	CD4 ↓ 30% 2 h post CD8 ↓ 30% 2 h post	Gabriel et al. 1992a; Kendell et al. 1990; Shek et al. 1995; Tvede et al. 1994
Very long (> 3 h)	CD4 ↓ 40% CD8 ↓ 40%	CD4 ↓ 20% 3 h post CD8 ↓ 20% 2 h post	Gabriel et al. 1994a

* = Percentage change compared with resting or preexercise values; ↑ = increase; ↓ = decrease; ND = no available data.

repetitive 1-minute treadmill sprinting to exhaustion (50% and 25% declines, respectively). Taken together, these data suggest that CD8 cell count is more labile in response to brief maximal exercise. That is, compared with CD4 cells, CD8 cell number increases more during and immediately after, but declines more 1-2 hours after, brief maximal exercise. These data suggest that CD8 cells are preferentially recruited into the circulation during, and removed from the circulation after, brief exercise.

Effects of Endurance Exercise on T Cell Subsets

During prolonged exercise, T cell number increases early (within the first 30 minutes) and then declines slightly throughout exercise (Shek et al. 1995) (figure 3.4, p. 76). For example, in untrained males who performed 2-hour treadmill running at 65% $\dot{V}O_{2max}$, T cell count increased from about 1.7 to $2.7 \times 10^9 \cdot L^{-1}$ during the first 30 minutes, declined to $2.3 \times 10^9 \cdot L^{-1}$ between 30 and 90 minutes, and then increased again to $2.5 \times 10^9 \cdot L^{-1}$ at the end of exercise (Shek et al. 1995). Gabriel et al. (1992a) reported a 30% rise in CD3 cell number during the first 10 minutes of cycle ergometry to exhaustion (about 60 minutes) at individual anaerobic threshold, followed by a more gradual increase to 67% higher than resting levels at the end of exercise. Nieman et al. (1989a) observed a 30% increase in T cell number during the first hour of a 3-hour simulated marathon on a treadmill, after which cell count remained relatively constant until the end of exercise.

In contrast, other studies have found no significant changes in T cell number after prolonged exercise such as marathon running (Gmunder et al. 1988; Moorthy and Zimmerman 1978), and cell number may decline at the end of very prolonged exercise (Gabriel et al. 1994a). For example, in nine male and female distance runners, CD3, CD4, and CD8 cell concentrations were all decreased by 20% to 40% for up to 3 hours after a 100-km race (8 hours) (Gabriel et al. 1994a). Taken together, these data suggest that T cell number increases early in endurance exercise, followed by a delayed decline below resting values toward the end of and up to several hours after exercise. Thus, the time of blood sampling is critical to determining the kinetics of the T cell response to prolonged exercise. A single measurement at the end may miss the increase early during exercise and the trend toward declining numbers toward the end or after exercise.

CD4:CD8 Ratio Changes After Endurance Exercise

As with brief exercise, there are disproportionate changes in CD4 and CD8 numbers during and after prolonged exercise. For example, Gabriel et al. (1992a) noted a 20% rise in CD4 and a 60% increase in CD8 cell number during cycle ergometry to exhaustion (about 60 minutes). Pizza et al. (1995) observed a significant decline in the CD4:CD8 ratio 1.5 hours after 60 minutes of treadmill running. These changes in cell subset ratio appear to be related to increases in plasma osmolality occurring during prolonged exercise (Greenleaf et al. 1995). For example, in four males, CD4:CD8 ratio declined to less than 1.2 (i.e., below the clinically normal range) after experimentally induced dehydration (Greenleaf et al. 1995). Subjects subsequently performed 70-minute supine cycle ergometry at 70% $\dot{V}O_{2max}$ with and without fluid replacement. During exercise the CD4:CD8 ratio increased; this ratio was significantly negatively correlated with changes in plasma osmolality both before and after exercise ($R = -0.76$ for preexercise and -0.92 for postexercise values). Since plasma volume was not significantly correlated with changes in this ratio, it was concluded that recruitment of cells into the circulation is sensitive to changes in plasma osmolality but not volume.

Exercise Training Effects on T Cell Subsets

Exercise training appears to have little influence on resting T cell number and the ratio of CD4 to CD8 cells. In elite swimmers, CD3, CD4, and CD8 cell numbers and the CD4:CD8 ratio were constant, and within normal limits at all times, throughout a seven-month training season leading to major competition (Gleeson et al. 1995). In distance runners, no changes in the CD4:CD8 ratio were observed after four different cycles of intensified training involving manipulation of volume and intensity (Kajiura et al. 1995). Fry et al. (1992a) observed no changes in CD3, CD4, CD8 cell counts or CD4:CD8 ratio after 10 days intense interval training in endurance athletes. In contrast, declines in lymphocyte (12%), CD3 (20%), and CD4 (30%) cell counts were reported after a six-month high-volume training season ($500\ km \cdot week^{-1}$) in 15 male cyclists (Baj et al. 1994). CD8 cell number remained unchanged, resulting in a significant decline in the CD4:CD8 ratio from 2.3 to 1.8, although this latter value was still within the clinically normal range. These data suggest that T cell number and subset ratio are relatively impervious to extended periods of normal exercise training, although some changes may occur with very high volume work.

B Cell Response to Exercise

Resting B cell number appears to be clinically normal in athletes (Fry et al. 1992a; Gleeson et al. 1995; Gray et al. 1993b; Hinton et al. 1997; Nielsen et al. 1996a; Nieman et al. 1995c) and is not altered by short- (e.g., 10 days; Fry et al. 1992a) or long- (e.g., seven months; Gleeson et al. 1995) term intense training. Although total B cell number may not change with training, there is one report of higher proportion of activated B cells ($CD23^{+}$) in male track athletes compared with active matched control subjects (Baum et al. 1994). This increased expression of CD23 was attributed to frequent stimulation by catecholamines during daily exercise training, and may suggest enhanced B cell function without changes in cell number.

Circulating B cell number may increase or remain unchanged during exercise depending on exercise duration, intensity, and time of blood sample (table 3.6). Changes that do occur are considerably smaller and less persistent than for T and NK cells (Deuster et al. 1988; Gray et al. 1993b; Nielsen et al. 1996a; Nieman et al. 1992, 1995e). B cell number either remains unchanged or increases modestly during and immediately after brief exercise. For example, B cell number has been reported to be unchanged immediately after a 5-km race (15-18 minutes) in

Table 3.6 Acute Effects of Exercise on B (CD19, CD20) Cell Number

Exercise	During or immediately after*	Recovery*	Representative references
Brief (< 45 min)			
Moderate	No change	ND	Camus et al. 1992; Iversen et al. 1994
Intense	No change	↓ 0-25% 1-2 h post	Espersen et al. 1990; Nielsen et al. 1996a; Nieman et al. 1992
Intense interval	↑ 0-70%	No change 1-6 h post	Gray et al. 1993b; Hinton et al. 1997; Nieman et al. 1995e
Prolonged (1-3 h)			
Moderate	No change	No change	Nieman et al. 1993a; Shinkai et al. 1992
Intense	No change	No change	Gabriel et al. 1992a; Pizza et al. 1995; Shek et al. 1995
Very long (> 3 h)	↓ 10%	↑ 30% 3 h post	Gabriel et al. 1994a

* = Percentage change compared with resting or preexercise values; ↑ = increase; ↓ = decrease; ND = no available data.

trained runners (Espersen et al. 1990), 15 1-minute treadmill sprints at 95% $\dot{V}O_{2max}$ (Hinton et al. 1997) in runners, and progressive treadmill running to volitional fatigue (Deuster et al. 1988). Increases in cell number on the order of 30-70% over resting values have been reported after repetitive weight lifting (squats) to exhaustion in weight lifters (Nieman et al. 1995e), 6 minutes of all-out rowing in trained rowers (Nielsen et al. 1996a), and 16 1-minute treadmill sprints to exhaustion in distance runners (Gray et al. 1993b). Compared with untrained individuals, athletes may exhibit a smaller increase in B cell number during brief maximal exercise (Ferry et al. 1990). Normal B cell concentration is usually restored by 1 to 2 hours after brief exercise (Gray et al. 1993b; Nielsen et al. 1996a; Nieman et al. 1995e).

B cell number does not appear to change appreciably during and after prolonged exercise. Circulating B cell number was not significantly altered during and for several days after 2-hour treadmill running at 65% $\dot{V}O_{2max}$ despite large changes in T and NK cell numbers (Shek et al. 1995), nor was B cell number different from resting values after cycle ergometry to exhaustion at 100% individual anaerobic threshold (approximately 60 minutes; Gabriel et al. 1992a). Moreover, B cell number did not change in response to 60-minute level or downhill (–10% grade) running at 70% $\dot{V}O_{2max}$ despite different responses of T cell subsets to the two exercise protocols (Pizza et al. 1995).

Lymphocyte Activation by Exercise

Lymphocytes may be activated by acute exercise and exercise training. There are several reports showing increased expression of cell surface activation markers such as CD25 (low-affinity IL-2 receptor subunit), CD122 (high-affinity IL-2 receptor subunit), and HLA-DR antigen. It is not currently known whether this occurs because activated cells are selectively recruited into the circulation or because cells are activated during exercise; it is likely that both occur simultaneously during exercise. For example, Gray et al. (1993a) noted a twofold increase in the number of T cells expressing $CD3^+HLA\text{-}DR^+$ immediately after intense interval running (16 1-minute runs at 100% $\dot{V}O_{2max}$) that paralleled the increase in total T cell number. The data suggested that both activated and nonactivated cells were recruited into the circulation. Gabriel et al. (1993) reported a significant increase in the number of $CD3^+$ T cells expressing $CD45RA^+$ and $CD45RO^+$ antigens after brief (about 20 minutes) maximal exercise; these antigens identify subsets of activated T cells involved in memory. The number of $CD45RA^+RO^+$ cells nearly doubled after, compared with before exercise. Moreover, expression of $CD45RA^{+}RO^{+}$ was increased both at rest and after a second exercise test performed eight days after a 12-hour (240 km) cycle race. These data suggest that both brief and very prolonged intense exercise induce activation of T cells. In contrast, moderate exercise does not appear to alter T cell activation. In moderately trained women, 45 minutes of walking at 60% $\dot{V}O_{2max}$ did not significantly change the number of $CD3^{+}$ T cells expressing CD25 (IL-2R) (Nehlsen-Cannarella et al. 1991a).

The number and relative proportion of activated T cells may be altered at rest in well-trained athletes depending on the activation marker used and cell population studied (Baum et al. 1994; Gabriel et al. 1992a, 1992b, 1993; Gleeson et al. 1995; Nieman et al. 1995b; Rhind et al. 1994). For example, the number of T cells expressing $CD3^{+}HLA\text{-}DR^{+}$ was not significantly different at rest in male distance runners compared with matched nonathletes (Nieman et al. 1995b). In another study, $HLA\text{-}DR^{+}$ cell number and percentages were similar in elite swimmers compared with matched nonathletes and did not change over a seven-month competitive season in the swimmers (Gleeson et al. 1995). In contrast, greater expression of the high-affinity IL-2 receptor subunit (CD122) was observed at rest in distance runners compared with nonathletes (Rhind et al. 1994). Compared with matched nonathletes, in runners CD122 expression was about 10-20% higher in NK cells ($CD16^{+}$ and $CD56^{+}$ cells) and nearly two times higher in CD8 T cells, but not different for CD4 T cells. Gabriel et al. (1992a) reported an acute 25-50% increase in $CD3^{+}HLA\text{-}DR^{+}$ cell number during and after intense endurance exercise (cycle ergometry at 100% individual anaerobic threshold until exhaustion, > 60 minutes). $CD3^{+}$ and $CD3^{+}HLA\text{-}DR^{+}$ increased proportionately, whereas $CD4^{+}RO^{+}$ (activated $CD4^{+}$) cell number did not change. These data suggest that the increase in the number of activated $CD3^{+}$ cells was due to an increase in activated $CD8^{+}$ cell number.

Fry et al. (1992a, 1994) observed significant increases in the number of lymphocytes expressing CD25 (2.5 to 2.9-fold increase) and $HLA\text{-}DR^{+}$ (1.3 to 1.8-fold increase) despite no changes in the number of lymphocytes or T cells during 10 days of twice-daily intense interval training in well-trained males; these increases were maintained during 5 days of recovery training, suggesting a long-lasting effect of very intense training. An increase in the proportion of $CD4^{+}CD25^{+}$ cells (i.e., T cells expressing the low-affinity IL-2 receptor) and increased appearance of soluble IL-2 receptor in blood were also observed in track athletes at the end compared with the start of a competitive season (Baum et al. 1994). Taken together, these data suggest that lymphocyte activation occurs acutely in response to intense exercise. This effect may be transitory and not reflected in substantial increases in activation markers in cells obtained at rest in athletes during normal training, although very rigorous training may lead to persistent activation of lymphocytes.

Effects of Exercise on Lymphocyte Proliferation

As discussed in chapter 2, lymphocytes are activated upon exposure to antigen, after which they enter the cell cycle and proliferate. Lymphocyte proliferation can be assessed in an in vitro assay (described in chapter 2) in which cells are first incubated with polyclonal mitogens and then pulsed with ^{3}H-thymidine that is incorporated into DNA within the newly formed lymphocytes; radioactivity is then quantified as a measure of cellular proliferation. This assay has been used by several groups to study the effects of exercise on lymphocyte proliferation, in both human and experimental animal models.

As with many immune parameters, lymphocyte proliferation appears to be sensitive to exercise intensity and duration (table 3.7) (Kendall et al. 1990; MacNeil et al. 1991; Nieman et al. 1994; Tvede et al. 1993). In general, brief moderate exercise appears to have little effect on, or may stimulate, lymphocyte proliferation, whereas intense or prolonged exercise may cause suppression of proliferative responses. Fitness level and the total amount of work performed may also be important variables influencing the lymphocyte proliferative response to exercise (MacNeil et al. 1991; Oshida et al. 1988; Tvede et al. 1993; Soppi et al. 1982). For example, Tvede et al. (1993) exercised untrained male subjects for 60 minutes at three work rates, equal to 25%, 50%, and 75% $\dot{V}O_{2max}$. PHA- and IL-2-stimulated lymphocyte proliferation were unchanged for up to 2 hours after exercise at the two lowest intensities. In contrast, proliferation declined by 35% immediately after exercise at 75% $\dot{V}O_{2max}$, but was restored by 2 hours postexercise. This decline in proliferation was related to a similar decrease in CD4 cell number.

Brief Exercise

Brief exercise may have little influence on, or may stimulate, T cell responsiveness to mitogens (Edwards et al. 1984; Eskola et al. 1978; Espersen et al. 1991; Hedfors et al. 1976; Nielsen et al. 1996a; Robertson et al. 1981; Soppi et al. 1982; Verde et al. 1992). For example, in runners, compared with preexercise values, responses to ConA and PHA were unchanged in blood samples obtained 30 minutes after a 7 km run, in contrast to the 50% suppression of proliferative responses observed 30 minutes after a marathon run (Eskola et al. 1978). Nielsen et al.

Table 3.7 Acute Effects of Exercise on Lymphocyte Proliferation

Exercise	During or immediately after*	Recovery*	Representative references
Brief (< 45 min)			
Moderate	↓ 0-10%	20% ↓ 1-2 h post#	MacNeil et al. 1991; Nieman et al. 1994
Intense	No change or ↑ 20-30%#	↑ 0-40% 1-2 h post	Eskola et al. 1978; Espersen et al. 1990; Nielsen et al. 1996a; Nieman et al. 1992
Intense interval	↑ 30-60%#	Normal 2 h post	Frisina et al. 1994; Fry et al. 1992b; Hinton et al. 1997
Prolonged (1-3 h)			
Moderate	↓ 0-15%	No change	MacNeil et al. 1991; Tvede et al. 1993
Intense	↓ 25-35%#	Normal 2-6 h post	Eskola et al. 1978; MacNeil et al. 1991; Nieman et al. 1995f; Tvede et al. 1989, 1993, 1994
Very long (> 3 h)	ND	ND	

* = Percentage change compared with resting or preexercise values; ↑ = increase; ↓ = decrease; ND = no available data; # indicates that the change in proliferation was correlated with the change in T cell number or percentage.

(1996a) reported no effect of 6-minute all-out rowing ergometry on IL-2- and PHA-stimulated proliferation in trained rowers. Verde et al. (1992) noted no effect of 30-minute submaximal exercise (treadmill running at 80% $\dot{V}O_{2max}$) on PHA- and ConA-stimulated proliferation during normal training in distance runners, although there was an acute suppression after exercise during heavy training (discussed below). Espersen et al. (1991) reported a 40% to 100% increase in PHA-, ConA-, and PWM-stimulated responses in samples obtained 2 hours after a 5-km race in distance runners. Soppi et al. (1982) noted a 5-15% increase in PHA- and ConA-stimulated proliferation in naval recruits after a cycle ergometry test to volitional fatigue.

Endurance Exercise

In male endurance athletes, proliferation in response to the T cell mitogens ConA and PHA was suppressed by 40-50% after a marathon (Eskola et al. 1978; Gmunder et al. 1988) and after endurance cycling (Oshida et al. 1988; Tvede et al. 1989a). In male distance runners, ConA-stimulated proliferation was suppressed by about 25% for up to 3 hours after a 2.5-hour treadmill run at 75%$\dot{V}O_{2max}$, with resting values restored by 6 hours postexercise (Nieman et al. 1995f). In the latter study, the time course of suppression of proliferation was similar to that for increases in plasma cortisol and decreases in CD3 T cell number, suggesting that hormonal changes may influence proliferation via altering circulating T cell number during and after exercise.

Maximal Exercise

Maximal exercise, performed either continuously or in intervals, appears to suppress lymphocyte proliferation, and the duration of suppression is related to exercise duration (Frisina et al. 1994; Fry et al. 1992b; Smith et al. 1993). For example, Frisina et al. (1994) and Fry et al. (1992b) used a similar exercise protocol in distance runners and kayakers, 25 1-minute all-out sprints on a treadmill and kayak ergometer, to elicit postexercise decreases in T cell proliferation. Con A- and PHA-stimulated proliferation declined by 30-60% immediately after exercise but was restored by 2 hours after exercise. Responses were similar regardless of fitness level as well as for treadmill running and kayak ergometry, suggesting no effects of training level or mode of exercise (i.e., upper vs. entire body exercise) (Fry et al. 1992b). In this latter study, addition of indomethacin to cultures had no effect on exercise-induced suppression of proliferation, suggesting no role for prostaglandins. Furthermore, addition of IL-2 to cells obtained at various times after exercise stimulated proliferation to a similar extent but did not attenuate the suppression of proliferation after exercise, suggesting that cells retained the ability to respond to IL-2. Postexercise suppression of lymphocyte proliferation was correlated with changes in T cell number, and it was concluded that such changes may result from redistribution of lymphocyte subsets (Frisina et al. 1994; Fry et al. 1992b).

A recent study using weight lifting was undertaken to compare the lymphocyte proliferative response in an activity requiring less demand on the cardiorespiratory system (Nieman et al. 1995e). Experienced male weight lifters performed repetitive leg squats to exhaustion (mean time 37 minutes). ConA-stimulated proliferation increased 50% immediately after exercise, with preexercise values restored by 2 hours postexercise. A similar pattern of increased CD3 T cell number was observed, and when proliferation was adjusted to account for changes in T cell number, there was no effect of exercise on proliferative capacity of each lymphocyte. There were only modest changes in circulating levels of cortisol and epinephrine, in contrast to the large increases observed during endurance exercise. These hormones have been implicated as mediators of suppression after intense endurance exercise, and it was suggested that the absence of an effect on proliferative capacity per cell after weight lifting reflected the modest changes in these hormones.

Influence of Fitness Level

As noted above, two studies found no relationship between fitness level and degree of suppression of proliferation after maximal exercise in trained runners and kayakers (Frisina et al. 1994; Fry et al. 1992b). In contrast, some studies have reported a relationship between fitness level, exercise intensity, and T cell responses to exercise (MacNeil et al. 1991; Oshida et al. 1988). For example, in five untrained and six trained subjects, PHA-stimulated proliferation was suppressed by about 40% after 2-hour cycling at 60% $\dot{V}O_{2max}$, with significantly greater and prolonged suppression in athletes compared with nonathletes (Oshida et al. 1988). MacNeil et al. (1991) compared lymphocyte proliferative responses to exercise at two intensities in 24 individuals of varying fitness level from low to highly fit athletes. Subjects completed four tests of cycle ergometry for 30-120 minutes at 30-65% $\dot{V}O_{2max}$, and lymphocytes obtained before and after exercise were incubated with ConA. For data analysis, subjects were grouped into four fitness levels according to $\dot{V}O_{2max}$ and current exercise habits. Compared with resting values, lymphocyte proliferation was depressed immediately after exercise under all conditions. However, within each fitness level, proliferation was lower after the higher-intensity exercise. Moreover, comparison between groups showed that postexercise suppression of proliferation was directly related to fitness level, regardless of exercise intensity, with the highly fit group exhibiting the greatest postexercise suppression. These results were attributed to differences in the total amount of work performed in the exercise tests. That is, despite exercising at the same relative work rate (i.e., a given percentage of individual $\dot{V}O_{2max}$), because of their higher aerobic power the highly fit groups completed more work than less fit groups. Thus, the response of lymphocyte proliferation to exercise may depend on both exercise intensity relative to an individual's maximum capacity and total amount of work completed during exercise.

Mechanisms Related to Exercise-Induced Effects on Lymphocyte Proliferation

The suppression of lymphocyte proliferation after prolonged exercise does not appear to be related to increased core temperature. For example, hyperthermia induced by 2-hour immersion in a water bath until core temperature reached 39.5 °C had no effect on IL-2- or PHA-stimulated lymphocyte proliferation (Kappel et al. 1991a). Several studies have suggested that exercise-induced changes in lymphocyte proliferation may be secondary to redistribution of lymphocytes in the circulation (Hinton et al. 1997; Nieman et al. 1992, 1994; Oshida et al. 1988; Tvede et al. 1993). As detailed above, total lymphocyte and T cell counts and the CD4:CD8 ratio often decline after exercise. The in vitro proliferation assay uses either a given volume of whole blood or a given number of peripheral mononuclear cells, both of which include a mixture of cells. Any reduction in lymphocyte number, or proportion of T to other cell types, may thus reduce the number of cells capable of proliferating; since CD4 cells are central to stimulation of lymphocyte responses, decreasing CD4 cell number may significantly impair the ability of lymphocytes to proliferate.

Nieman et al. (1994) estimated the role of cell redistribution on exercise-induced changes in lymphocyte proliferation following 45 minutes of moderate- and high-intensity exercise (treadmill walking at 50% and running at 80% $\dot{V}O_{2max}$, respectively). Con A-stimulated lymphocyte proliferation declined significantly 1 and 2 hours after both exercises, although the decrease was larger after intense exercise. However, when proliferative responses were adjusted for changes in $CD3^+$ cell number, proliferation remained unchanged up to 3.5 hours after moderate exercise, but was lower than baseline levels only at 1 hour after high-intensity exercise. These data suggest that exercise-induced changes in lymphocyte proliferation after moderate exercise are mediated entirely via changes in cellular redistribution, in particular the decline in T cell number. In contrast, intense exercise appears to transiently suppress T cell proliferation independently of changes in cell number.

In a more recent study, Hinton et al. (1997) addressed this question directly by culturing purified T lymphocytes in blood samples obtained before and after exercise (figure 3.5). Five male distance runners completed an intense interval training session (15 1-minute treadmill sprints at 95% $\dot{V}O_{2max}$). CD4 and CD8 T cells were isolated by

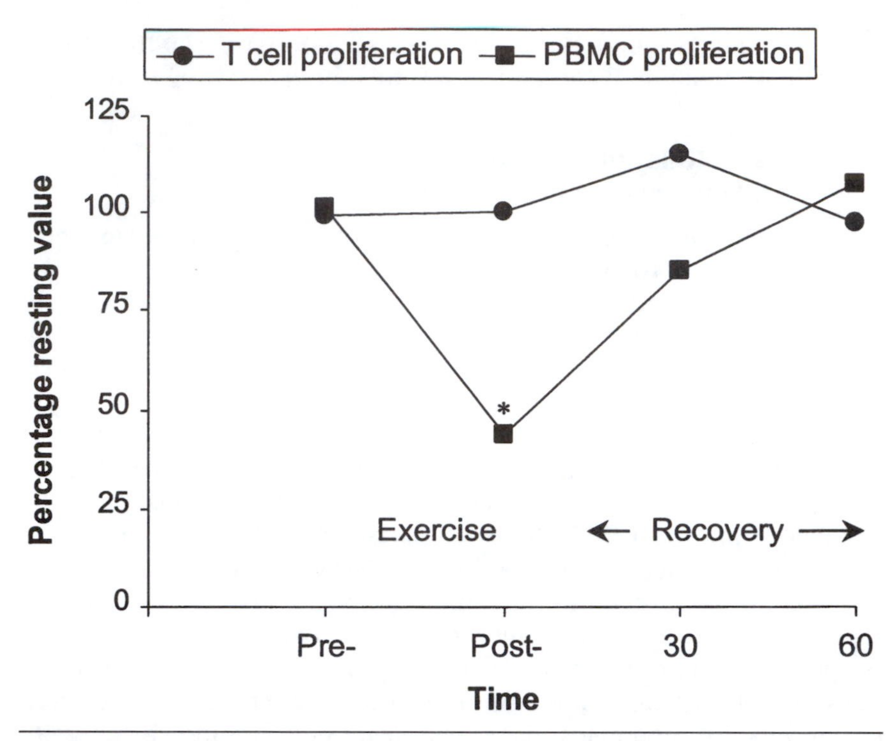

Figure 3.5 Lymphocyte proliferative response to intense interval exercise. Peripheral blood was sampled pre- and postexercise and then 30 and 60 minutes postexercise in endurance-trained men. Exercise consisted of 15 × 1-minute treadmill running intervals at 95% $\dot{V}O_{2max}$. PHA-stimulated proliferative responses were measured and compared in mononuclear cells (PBMC) and T cells separated by magnetic separation using antibodies to CD4 and CD8. * = Significantly different from preexercise value in PBMC cultures but not purified T cell cultures.

Adapted from Hinton, J.R., D.G. Rowbottom, D. Keast, and A.R. Morton. 1997. Acute intensive and interval training and in vitro T-lymphocyte function. *International Journal of Sports Medicine* 18: 130-135.

a magnetic separation technique using monoclonal antibodies to these surface antigens conjugated to magnetic beads (antibody to CD3 was not used because binding to the CD3 receptor can directly stimulate lymphocyte proliferation). In PHA-stimulated mixed (nonseparated) cell cultures, an expected 60% suppression of proliferation was noted immediately after exercise. In contrast, no such suppression was observed in CD4 and CD8 isolated cultures. Moreover, T cell number declined immediately after exercise, and the proliferative responses in mixed cell cultures were significantly correlated with cell number, in particular NK and CD4 cell number and the CD4:CD8 ratio.

Catecholamines appear to mediate at least part of the exercise-induced suppression of lymphocyte proliferation by influencing the proportion of $CD3^+$ and $CD4^+$ cells in the circulation. For example, Tvede et al. (1994) infused epinephrine into healthy subjects to match plasma concentrations observed during exercise (1 hour of cycle ergometry at 75% $\dot{V}O_{2max}$). Changes in $CD3^+$ and $CD4^+$ cell percentages and PHA-stimulated proliferation were similar during and after exercise and epinephrine infusion. Suppression of proliferation was correlated with changes in $CD3^+$ and $CD4^+$ percentages. Together with evidence discussed above, these data strongly suggest that changes in cellular distribution, in particular the increase in NK and corresponding decrease in T cell numbers mediated by catecholamines, underlie exercise-induced changes in lymphocyte proliferation.

Training Responses

Lymphocyte proliferation in response to mitogenic challenge appears normal in athletes during moderate training (Baj et al. 1994; Nieman et al. 1995c) and may be enhanced during periods of intense training (Baj et al. 1994; Verde et al. 1992). For example, a cross-sectional comparison between male distance runners and matched nonathletes showed no differences in resting PHA- or ConA-stimulated lymphocyte proliferation (Nieman et al. 1995c). Proliferation was, however, negatively correlated with age (range 22-57 years; $R = -.41$) and was lower in older (> 40 years) compared with younger runners.

PHA-stimulated proliferation was similar in 15 male endurance cyclists assessed at the start of the season compared with matched nonathletes (Baj et al. 1994). However, proliferation increased by 50% after six months of training (500 km per week) in cyclists while remaining unchanged in controls. The enhanced proliferation could not be explained by changes in T cell number, since CD3 and CD4 T cell numbers declined during the season in cyclists. Verde et al. (1992) also reported a significant increase in lymphocyte proliferation after three-week increased training in male distance runners. Training volume was increased by 35%, from 100 to 130 km per week^{-1}, resulting in persistent fatigue in the runners. Resting PHA- and ConA-stimulated proliferation increased by 25% and 32%, respectively, after the three weeks and was maintained after an additional three-week recovery (86 km $\cdot$ week^{-1}) training. Although proliferation declined 12-18% after a single exercise session (30 minutes at 80% $\dot{V}O_{2max}$) during high-intensity training, these postexercise values were still higher than levels observed after exercise during normal training, because of the new higher resting values. Moreover, the effect was transitory, since proliferation returned to resting values by 30 minutes postexercise. Thus, it appears that resting lymphocyte proliferation is unimpaired, and possibly enhanced, during prolonged periods of intense training in athletes. Enhancement of proliferative capacity may be related to the apparent activation of lymphocytes, as described above.

Moderate exercise training does not appear to alter resting lymphocyte proliferation. For example, in previously untrained young adult males, PWM- and PHA-stimulated proliferation was unchanged after 12-week moderate exercise training consisting of 30-minute cycle ergometry at 75% $\dot{V}O_{2max}$, three days per week (Mitchell et al. 1996).

Exercise and NK Cell Number

In general, resting NK cell number in the circulation appears to be similar in athletes compared with nonathletes (Baj et al. 1994; Nieman et al. 1995b, 1995c; Tvede et al. 1991). However, there are reports of higher or lower NK cell number in some athletes compared with matched nonathlete control subjects. These differences may be due to small sample size or to effects of varying training intensity at different times in the competitive season. For example, NK cell number was reported to be higher in a small group (N = 7) of male distance runners compared with matched nonathletes (Rhind et al. 1994). In athletes, resting NK cell number may vary with intensity and volume of training during the season (Gleeson et al. 1995; Tvede et al. 1991), which may partially explain this observed difference between trained and untrained subjects (Rhind et al. 1994). For example, Gleeson et al. (1995) reported 50-60% higher NK cell number and percentage in elite swimmers during a moderate training phase compared with matched nonathletes. NK cell number declined significantly as the seven-month season progressed through high-intensity phases leading to major competition, so that NK cell number was lower in swimmers compared with nonathletes at the end of the season. In another study of shorter-term training, NK cell number declined progressively over 10 days of intense training (twice-daily all-out sprinting) and remained low throughout 5 days of recovery in male endurance athletes (Fry et al. 1992a, 1994). Although not conclusive at this point, these data suggest that, in athletes, NK cell number is normal or perhaps slightly elevated during moderate training phases but that NK cell number may decline during periods of very intense exercise training.

Exercise causes profound changes in both the percentage and number of circulating NK cells (table 3.8). NK cell percentage relative to total lymphocytes increases 50% to 300% following brief (< 45 minutes) submaximal and maximal as well as prolonged (> 45 minutes) submaximal exercise (Berk et al. 1990; Gabriel et al. 1992a; Gray

Table 3.8 Acute Effects of Exercise on Natural Killer Cell Number

Exercise	During or immediately after*	Recovery*	Representative references
Brief (< 45 min)			
Moderate	↑ 0-50%	Normal by 1 h post	Camus et al. 1992
Intense	↑ 100-200%	Normal by 1-2 h post;	Nielsen et al. 1996; Nieman et al. 1992
		↓ 40% 2-4 h post	Gabriel et al. 1992c
Intense interval	↑ 100-200%	Normal by 1-2 h post	Fry et al. 1992b; Gray et al. 1993b; Nieman et al. 1995e
Prolonged (1-3 h)			
Moderate	↑ 70-100%	↓ 0-50% 1-2 h post	Nieman et al. 1993a; Shinkai et al. 1992; Tvede et al. 1993
Intense	↑ 100-200%	↓ 30-60% 1-2 h post	Mackinnon et al. 1988; Nieman et al. 1989a; Pedersen et al. 1988, 1990; Shek et al. 1995
Very long (> 3 h)	↓ 50%	↓ 33% 3 h post	Gabriel et al. 1994a

* = Percentage change compared with resting or preexercise values; ↑ = increase; ↓ = decrease; ND = no available data.

et al. 1993b; Mackinnon et al. 1988; Nielsen et al. 1996b; Nieman et al. 1993a; Shek et al. 1995; Tvede et al. 1989).

NK cell number rises very early in exercise and continues to increase or possibly decrease during prolonged exercise (Berk et al. 1990; Shek et al. 1995). In untrained males, NK cell number rose by more than 50% during the first 30 minutes of a 2-hour treadmill run at 60% VO_2max, then declined slightly between 30 and 60 minutes, and increased a further 20% by the end of exercise (Shek et al. 1995). Berk et al. (1990) noted a 60% increase in NK cell number during a 3-hour treadmill run in distance runners, with a decline to resting values by the end of exercise.

The magnitude of increase in NK cell number is related to exercise intensity. Tvede et al. (1993) noted increases in NK cell number after 60-minute cycle ergometry at 25%, 50%, and 75% $\dot{V}O_{2max}$, but the magnitude of changes in NK cell number was directly proportional to exercise intensity. In male runners, 45 minutes of treadmill running at 80% $\dot{V}O_{2max}$ produced a much higher increase in circulating NK cell number than did walking at 50% $\dot{V}O_{2max}$ (Nieman et al. 1993b).

Discontinuous exercise also appears to mobilize NK cells into the circulation. In male endurance athletes, repeated 1-minute treadmill sprints to exhaustion produced large increases (up to 400%) in NK cell number (Gray et al. 1993b). A threefold increase in NK cell number was also observed immediately after repetitive weight lifting to exhaustion in experienced weight lifters (Nieman et al. 1995e).

After brief or moderate exercise, NK cell number generally returns to preexercise levels soon after the end of exercise (Fry et al. 1992b; Gabriel et al. 1991b; Lewicki et al. 1988; Nielsen et al. 1996b; Nieman et al. 1992). For example, Nielsen et al. (1996b) noted a significant 400% increase in NK cell number immediately after 6 minutes of all-out rowing in oarsmen, but preexercise values were restored within 2 hours. In contrast, NK cell number may decline below baseline values for up to several hours, and perhaps days, after very intense or prolonged exercise (Berk et al. 1990; Espersen et al. 1990; Gray et al. 1993b; Mackinnon et al. 1988; Nieman et al. 1989a, 1993b, 1995e; Pedersen et al. 1988, 1990; Shek et al. 1995; Shinkai et al. 1992; Tvede et al. 1993). For example, Berk et al. (1990) noted a 30-60% decrease in NK cell number between 1 and 21 hours after a 3-hour treadmill run with the lowest levels observed 6 hours postexercise. NK cell number may be restored by 24 hours postexercise (Berk et al. 1990; Mackinnon et al. 1988), although in one recent study a long-lasting 40% suppression of circulating NK cell number was observed for up to seven days after a 2-hour treadmill run at 60% $\dot{V}O_{2max}$ (Shek et al. 1995).

The responses to exercise of NK cell number and function are discussed in more detail in chapter 7.

Exercise and Monocyte Number

Monocyte number may increase acutely by up to 100% during and after intense exercise of both short and long duration (table 3.9). As with other cell subsets, the magnitude of change may be related to

Table 3.9 Acute Effects of Exercise on Monocyte Number

Exercise	During or immediately after	Recovery	Representative references
Brief (< 45 min)			
Moderate	No change	ND	Iversen et al. 1994; Foster et al. 1986
Intense	↑ 0-20%	↑ 0-50% 2 h post	Espersen et al. 1990; Nielsen et al. 1996a
Intense interval	↑ 40-50%	↑ 15-60% 2-6 h post	Fry et al. 1992b; Gray et al. 1993b; Nieman et al. 1995e
Prolonged (1-3 h)			
Moderate	No change	No change	Shinkai et al. 1992; Tvede et al. 1993
Intense	↑ 50-100%	↑ 50-100% 2-3 h post	Gabriel et al. 1992a; Nieman et al. 1989a; Shek et al. 1995; Ullum et al. 1994a
Very long (> 3 h)	↑ 250%	↑ 250% 3 h post	Gabriel et al. 1994a

↑ = increase; ND = no available data.

fitness level and exercise duration. Because monocytes secrete a number of regulatory factors (e.g., cytokines), recruitment of monocytes into the circulation during exercise may account for increases in the concentrations of these factors (see chapter 6) and may also significantly influence the function of various immune cell subsets, such as NK cells (see chapter 7). Moreover, monocytes (macrophages) have been shown to localize to and infiltrate skeletal muscle damaged during exercise, and it has been suggested that recruitment of these cells into the circulation during exercise, followed by their rapid removal after exercise, may reflect the homing of these cells to damaged tissue.

Resting Monocyte Number

Resting monocyte number appears to be clinically normal in a variety of athletes including distance runners (Espersen et al. 1990; Fry et al. 1992a; Gray et al. 1993b; Nieman et al. 1989a, 1995b, 1995e), track runners (Baum et al. 1994), cyclists (Keen et al. 1995; Ndon et al. 1992; Nieman et al. 1995b), swimmers (Mackinnon et al. 1997; Mujika et al. 1996), rowers (Nielsen et al. 1996a), and weight lifters (Nieman et al. 1995e).

Effects of Brief Exercise on Monocyte Number

Monocyte number may not change or may increase during and after brief exercise, depending on exercise duration and intensity. Monocyte concentration was unchanged immediately after 5 minutes of cycle ergometry at 200 W in untrained individuals (Iversen et al. 1994), 15-20 minutes of cycle ergometry to 60-85% of maximum workload in active subjects (Foster et al. 1986), and 5-km running (15-18 minutes) in distance runners (Espersen et al. 1990). In contrast, more intense brief exercise may increase monocyte number during and after exercise. Monocyte concentration was reported to increase by about 50% immediately after repeated treadmill sprints to exhaustion in endurance athletes (Gray et al. 1993b), 6 minutes of all-out rowing in trained rowers (Nielsen et al. 1996a), and repetitive weight lifting (squats) to exhaustion in experienced weight lifters (Nieman et al. 1995e). Elevation of monocyte number may persist for 2 to 6 hours after exhaustive exercise (Gray et al. 1993b; Nieman et al. 1995e) but may return to normal levels by 2 hours after shorter maximal exercise (e.g., 6 minutes of rowing; Nielsen et al. 1996a).

Effects of Endurance Exercise on Monocyte Number

Prolonged exercise appears to increase monocyte concentration only if exercise is above a certain threshold. For example, monocyte number did not change after 60 minutes of cycle ergometry at 25% and 50% $\dot{V}O_{2max}$, but did increase after exercise at 75% $\dot{V}O_{2max}$ (Tvede et al. 1993). Monocyte number increases early and remains elevated during and after prolonged exercise. In active males of various fitness levels, monocyte number increased by 50% during the first 10 minutes and remained elevated throughout cycle ergometry to volitional fatigue (approximately 60 minutes at 100% individual anaerobic threshold) (Gabriel et al. 1992a). Shek et al. (1995) reported a 50% increase in monocyte concentration during the first 30 minutes of 2 hours of treadmill running at 65% $\dot{V}O_{2max}$, after which cell number increased to twice resting levels at the end of exercise. Very prolonged exercise (100 km running race, about 8 hours) induces large increases (2.5-fold) in monocyte count (Gabriel et al. 1994a). Elevation of monocyte number is transitory, with higher values maintained for 2-3 hours after exercise (Nieman et al. 1989a, 1995f; Shek et al. 1995; Tvede et al. 1989, 1993; Ullum et al. 1994a) but usually returning to preexercise values by 6 hours after endurance exercise such as 2.5-3.0 hours of treadmill running (Nieman et al. 1989a, 1995f).

Effects of Training on Monocyte Number

There are inconsistent data on the monocyte response to periods of training in athletes; these differences between studies may relate to the types of athletes and their training programs. No changes in monocyte number were reported during 12 weeks normal training followed by a 4-week taper (reduced training) in competitive athletes (Mujika et al. 1996), after 12 days cycle racing in road cyclists (Keen et al. 1995), or after 10 days intense interval training in endurance runners (Fry et al. 1994). On the other hand, Baum et al. (1994) observed a significant increase in circulating monocyte number between an early-season endurance training phase and later competition phase in male track athletes. Benoni et al. (1995a) reported a significant increase in monocyte concentration during a four-month training season in professional basketball players.

In contrast, others have found a reduction in monocyte number during periods of intense training in endurance athletes (Mackinnon et al. 1997; Ndon et al. 1992). Ndon et al. (1992) observed a 50% decrease in concentration from 0.3 to $0.18 \times 10^9 \cdot L^{-1}$ after 28 days increased training (50% increase in volume) in eight male cyclists; the latter value would be considered clinically low. Mackinnon et al. (1997) reported a significant 45% decline in monocyte number within the first two of four weeks intensified training in elite swimmers (progressive 10% increase in volume in each of the four weeks); monocyte number remained low at the end of the four weeks. Taken together, these data suggest that monocyte number is not significantly altered during normal training conditions or during relatively short periods of intense training, but that monocyte number may decrease during longer periods (e.g., four weeks) of intense training. The significance of such changes is unknown, but may represent a positive adaptation to high training loads since monocytes produce a number of inflammatory mediators (e.g., IL-1, prostaglandins); inhibition of monocyte number may help to limit secretion of these factors. Alternatively, lower monocyte numbers in the blood may reflect a redistribution of these cells to tissue sites of injury during intense training.

The effects of exercise on monocyte function are discussed in more detail in chapters 4 and 7.

Mechanisms Underlying Leukocyte Redistribution

At any given time under resting conditions, less than half of the body's mature leukocytes are circulating in the vascular system. The remainder are sequestered in underperfused microvasculature in the lungs, liver, and spleen. The exact mechanisms by which leukocytes are released into the circulation during exercise, and then removed from the circulation after, are unknown at present, but most likely involve mechanical factors such as increased cardiac output and perfusion of the microvasculature as well as changes in the interactions between leukocytes and vascular endothelium. Immature leukocytes may also be released from the spleen and bone marrow during very intense exercise. Many of these changes are mediated by stress hormones and inflammatory mediators such as cytokines and acute phase proteins (discussed below).

Role of the Lungs

The lung is a major source of marginated leukocytes that are recruited into the circulation during exercise. In one study, in which six subjects were assessed under various stresses, cycle ergometry and infusion of epinephrine or the β-agonist isoproterenol induced similar increases in circulating leukocyte number (Muir et al. 1984). When resting subjects were infused with indium-labeled neutrophils, up to 25% of cells were retained by the lungs. Exercise and catecholamine administration caused a similar pattern of progressive release of labeled cells from the lung into the circulation. It was suggested that leukocytosis during exercise is due, in part, to release of cells from marginated pools in the lungs. In another study, trained subjects exercised for up to 20 minutes at 63-80% maximum work rate, with and without administration of the β-adrenergic receptor antagonist propranolol (Foster et al. 1986). Propranolol had no effect on cardiac output or plasma catecholamine concentration, nor did it influence the magnitude of exercise-induced increases in total leukocyte and neutrophil number. It was concluded that exercise-induced mobilization of leukocytes results at least in part from demargination of cells from the lungs in response to increases in cardiac output and pulmonary blood flow. Neither of these studies, however, discounted the possible role of changes in cell binding to the endothelium (discussed further below).

Since minute ventilation may increase more than 20 times during intense exercise, it is possible that circulating leukocyte number may change in response to alterations in the ventilatory pattern. To test whether the exercise-induced increase in leukocyte number occurs in response to increased ventilatory movement, Fairbarn et al. (1993) compared the leukocyte response during exercise and during a nonexercise but ventilation-matched condition in six untrained males. Subjects performed progressive cycle ergometry exercise in 5-minute increments from 50 to 200 W for 20 minutes total. Final heart rate was about 90% of age-predicted maximum, and minute ventilation increased about sevenfold from the start to end of exercise. On a separate occasion, subjects also simulated ventilatory patterns (both tidal volume and frequency) without exercising while seated on the cycle ergometer; heart rate was not significantly elevated above resting in this condition. During exercise, the expected pattern of change in cell number was observed, with leukocyte, lymphocyte, and neutrophil counts increasing about 30-70%. In contrast, cell counts remained unchanged during the matched ventilation nonexercise condition. It was concluded that the exercise-induced increase in circulating cell number due to recruitment of cells from the marginated pool in the lung could not be attributed to ventilatory movement, but it was most likely related at least partially to an increase in cardiac output.

Role of Stress Hormones

There is evidence to implicate stress hormones as mediators of exercise-induced changes in leukocyte number and subset redistribution, although such relationship appears to be complex. It has long been known that hormones such as epinephrine and cortisol influence leukocyte distribution between the circulation and various body compartments such as the spleen, liver, and bone marrow.

Increases in both epinephrine and cortisol are a function of exercise intensity relative to individual exercise capacity, with an apparent threshold of about 60% $\dot{V}O_{2max}$ for release of epinephrine (Brooks and Fahey 1995; Wilmore and Costill 1994). There is more individual variation in the glucocorticoid response to exercise, and cortisol is usually released only during rigorous exercise, especially in well-trained subjects. Exercise training results in lower responses of circulating epinephrine and cortisol to the same absolute amount of exercise. Epinephrine level increases rapidly during exercise and returns to baseline levels quickly (within 30 minutes), whereas cortisol often exhibits a lag before increasing during exercise and may continue to increase or remain elevated longer after cessation of exercise (McCarthy and Dale 1988; Wilmore and Costill 1994). Moreover, cortisol levels show diurnal variation, peaking early and then declining throughout the morning; in the exercise literature, appropriate nonexercise controls have not always been included to account for possible diurnal variations. It is generally accepted that exercise-induced leukocytosis observed during and for some time after exercise reflects a combined effect of both epinephrine and cortisol, with the relative contribution of each hormone depending on exercise mode, intensity and duration, and time point of blood sample.

Glucocorticoids

Several studies have reported significant correlations between blood cortisol level and changes in cell counts during and after exercise (Eskola et al. 1978; Gabriel et al. 1992b, 1992c; Hansen et al. 1991; Moorthy and Zimmermann 1978; Nieman et al. 1989a, 1995f), although other studies have failed to find a relationship (Gimenez et al. 1987; Gray et al. 1993b; Nieman et al. 1995e; Pizza et al. 1995; Smith et al. 1989). Increases in leukocyte and granulocyte numbers were significantly correlated with an increase in serum cortisol concentration (R = 0.78 and 0.87, respectively) after marathon running (Moorthy and Zimmerman 1978). A negative correlation was also observed between training distance or degree of leukocytosis with the increase in cortisol (R = −0.60 for each). Nieman et al. (1989a) also reported significant correlations between the increase in cortisol concentration and leukocyte or neutrophil counts from before to 1.5 hours after a 3-hour treadmill run (R = 0.78 and 0.81, respectively). In a later study from the same laboratory (Nieman et al. 1995f), a significant correlation (R = 0.68) was observed between plasma cortisol concentration and the neutrophil:lymphocyte ratio 3 hours after 2.5 hours of treadmill running at 75% $\dot{V}O_{2max}$. Gabriel et al. (1992b) noted significant correlations (R = 0.65-0.79) between plasma cortisol levels measured immediately after exercise (87 minutes at 100% individual anaerobic threshold) and leukocyte or neutrophil counts 1 and 2 hours postexercise. Lymphocyte number at 1 hour postexercise was significantly inversely correlated (R = −0.625) with immediate postexercise plasma cortisol level.

In contrast, other studies have failed to find a relationship between blood cortisol and leukocytosis occurring after exercise (Gimenez et al. 1986; Gray et al. 1993b; Smith et al. 1989; Nieman et al. 1995e; Pizza et al. 1995) or during prolonged periods of intense training (Fry et al. 1992a; Mujika et al. 1996). For example, Gray et al. (1993b) reported significant elevation of total and free plasma cortisol concentrations immediately and 1.5 hours after intense interval exercise (16 1-minute maximal treadmill sprints to exhaustion) in trained triathletes, but no significant correlations between hormone concentration and leukocyte, granulocyte, or lymphocyte numbers. Pizza et al. (1995) also found no significant correlation between serum cortisol level and leukocyte or neutrophil numbers for up to 24 hours after 60 minutes of downhill and level running at 70% $\dot{V}O_{2max}$, despite differences in the pattern of leukocytosis between exercise protocols (discussed further below). Similarly, Nieman et al. (1995e) reported no significant change in plasma cortisol concentration immediately after, and a decline 2 hours after, 37 minutes of repeated leg squats to exhaustion in weight lifters, despite significant a doubling of leukocyte and neutrophil counts for up to 2 hours after exercise.

Inconsistencies between studies may be explained by several factors, such as exercise intensity and duration, subject fitness level, whether total or free cortisol concentration is measured, and whether an appropriate nonexercise control condition is included to account for diurnal variation. Studies showing a relationship between cortisol and leukocytosis have often used prolonged intense exercise, such as distance running (Eskola et al. 1978; Moorthy and Zimmerman 1978; Nieman et al. 1989a, 1995f), whereas studies reporting no correlation have used shorter continuous exercise in untrained subjects (Gimenez et al. 1986; Smith et al. 1989), more moderate exercise in trained subjects (Pizza et al. 1995), or maximal interval exercise in athletes (Gray et al. 1993b; Nieman et al. 1995e). Because blood cortisol level may not always increase during brief (i.e., < 30 minutes) or moderate (i.e., < 60-70% $\dot{V}O_{2max}$) exercise, it appears that cortisol may contribute to leukocytosis (primarily through granulocytosis) only during and after prolonged intense exercise (> 60 minutes at $> 70\%$ $\dot{V}O_{2max}$).

Compared with total cortisol, free cortisol levels increase more during and decline more after intense exercise (Gray et al. 1993b). Few studies have measured free cortisol (Gray et al. 1993b), with most measuring only total hormone levels. Free cortisol is the biologically active form, but represents about 4-10% of the total amount in the circulation (Gray et al. 1993b). Subtle changes in free cortisol concentration may not always be obvious from measurement of total cortisol. Few studies have included nonexercise conditions to control for possible diurnal variations (Nieman et al. 1995e, 1995f), which may also obscure a relationship between cortisol and cell counts.

Changes in circulating cortisol concentration do not appear to be related to suppression of lymphocyte proliferation after prolonged exercise (Nieman et al. 1995f). Exogenous administration of glucocorticoids does not affect the magnitude of leukocytosis or distribution of lymphocyte subsets following intense interval exercise. For example, in active subjects, oral administration of dexamethasone (4 mg) or hydrocortisone (100 mg) 4 hours before exercise (10 × 30-second treadmill runs at 90% $\dot{V}O_{2max}$) did not alter the rise in leukocyte, lymphocyte, and NK cell numbers or the relative percentages of NK, CD3, CD4, and CD8 cells induced by exercise (Singh et al. 1996). Compared with levels after placebo administration, plasma cortisol levels were lower before and after exercise in dexamethasone treatment but higher at both times in hydrocortisone treatment. These data suggest that, if glucocorticoids are involved in exercise-induced mobilization of leukocytes, there is no additive effect of exogenous glucocorticoids to that due to exercise. It was noted, however, that the CD4:CD8 ratio declined to very low values (< 0.6) after exercise under both treatments, and it was suggested that intense exercise be avoided in individuals on corticosteroid therapy.

Changes in plasma cortisol levels have been observed in response to periods of intense training in well-trained athletes (Kirwan et al. 1988; Fry et al. 1992a), but such changes do not appear to influence resting circulating leukocyte or subset numbers. For example, in five well-trained male athletes, resting plasma cortisol concentration declined significantly during the 5-day recovery period following 10 days intense interval training (repeated 1-minute maximal treadmill sprints) (Fry et al. 1992a). However, there were no significant changes in total leukocyte, neutrophil, lymphocyte, or monocyte counts during this time. In a study on elite swimmers followed over a six-month competitive season, resting plasma cortisol concentration was not significantly correlated with leukocyte or any subset count (Hooper et al. 1995). In another study, competitive swimmers were followed over a 12-week period of high-intensity training and subsequent 4-week taper (reduced training) (Mujika et al. 1996). Although a reduction in training volume was significantly correlated with an increase in lymphocyte count during the taper, lymphocyte and other cell counts were not correlated with plasma cortisol concentration, which did not change significantly during training and the taper. Taken together, these data suggest that blood cortisol level is not related to circulating immune cell counts over prolonged periods of intense exercise training.

Catecholamines

Exercise increases β-adrenergic activity, causing local release of norepinephrine, especially in blood vessels and the spleen, which may influence leukocyte migration as well as splenic output of cells. In addition, exercise may cause changes in leukocyte adrenergic receptor density (Frey et al. 1989; Landmann et al. 1988; Maisel et al. 1990), which may alter cell activity despite the absence of change in circulating hormone level. Catecholamines may indirectly affect leukocyte trafficking via inducing release of cytokines such as IL-1 that in turn modulate expression of adhesion molecules (Janeway and Travers 1996). Epinephrine and norepinephrine concentrations increase markedly during intense exercise such as 1-3 hours of treadmill running at 75% $\dot{V}O_{2max}$ (Kappel et al. 1991c; Nieman et al. 1989a, 1994, 1995f; Tvede et al. 1994) and shorter intense interval exercise such as weight lifting to exhaustion (mean time 37 minutes) (Nieman et al. 1995e). More variable responses are observed with shorter exercise such as treadmill sprinting to exhaustion (Gray et al. 1993b). Since catecholamines have a short half-life in the blood (approximately 1-3 minutes) (Berne and Levy 1988), resting values are quickly restored within 1-2 hours postexercise (Gray et al. 1993b; Nieman et al. 1989a, 1994, 1995e, 1995f). Thus, any effect of catecholamines would be expected to occur during or immediately after exercise.

Compared with the role of cortisol, fewer studies have specifically addressed the role of catecholamines in leukocyte trafficking during exercise. Leukocytosis comparable to that observed during exercise has been simulated by exogenous administration of epinephrine at appropriate concentrations seen with exercise (Foster et al. 1986; Kappel et al. 1991c; Muir et al. 1984). For example, administration of physiological amounts of epinephrine to the level observed during exercise (60 minutes at 75% $\dot{V}O_{2max}$) caused a similar increase in leukocyte number as was observed during exercise (Kappel et al. 1991c). Leukocytosis during exercise may be attenuated by β-adrenergic blockade during exercise (Ahlborg and Ahlborg 1970), although other studies have reported inconsistencies (reviewed in McCarthy and Dale 1988). Gabriel et al. (1992b) reported significant correlations ($r = 0.57$-0.67) between plasma epinephrine concentration immediately after exercise (87 minutes at 100% individual anaerobic threshold) and leukocyte and neutrophil counts 1-2 hours postexercise. The immediate postexercise epinephrine concentration was significantly negatively ($r = -0.62$-0.67) associated with lymphocyte and CD8 cell counts 2 hours postexercise. These data are consistent with the view that epinephrine mediates at least part of the delayed granulocytosis and lymphocytopenia in the postexercise recovery period. However, release of epinephrine appears to be most related to recruitment of lymphocytes and NK cells, and to a lesser extent neutrophils, into the circulation during exercise (McCarthy and Dale 1988; Pedersen 1997). This may relate to the relatively high number of β-adrenergic receptors on these cells (Landmann et al. 1988; Maisel et al. 1990).

Hyperthermia

Hyperthermia may also contribute to changes in leukocyte number and subset distribution during and after exercise. For example, an increase in peripheral blood leukocyte and neutrophil counts was demonstrated in healthy subjects during and after hyperthermia induced by 2-hour immersion in a water bath until core temperature reached 39.5 °C (Kappel et al. 1991a). During immersion, leukocyte and neutrophil numbers increased in proportion to the increase in core temperature from 37 °C to 39 °C; core temperature often rises to 39-40 °C during intense exercise (Wilmore and Costill 1994). Plasma catecholamine concentrations also increased during immersion. However, the increases in cell numbers and catecholamine concentrations during immersion were less than those noted during prolonged exercise, suggesting that hyperthermia alone cannot account for all of the increase in cell number during exercise. In contrast, leukocyte and neutrophil numbers continued to rise, and lymphocyte counts fell, by 2 hours after subjects were removed from the water bath. These data suggest that, during exercise, hyperthermia may contribute to some of the increase in total leukocyte, neutrophil, and lymphocyte counts, possibly mediated via catecholamine release; hyperthermia induced by exercise may exert a more profound effect on redistribution of leukocyte subsets in the first few hours of recovery after exercise.

Models of Exercise-Induced Leukocytosis

At least two quite similar models have been put forward to explain the role of catecholamines and cortisol in leukocyte trafficking during and after exercise (McCarthy and Dale 1988; Pedersen 1997). McCarthy and Dale first proposed that exercise-induced changes in leukocyte number and subset proportions can be explained by a combined effect of epinephrine and cortisol. Leukocytosis during brief exercise (< 1 hour) is due to epinephrine release. Because there is a lag in the appearance of cortisol in response to exercise, cortisol-induced increases in cell number occur 1 hour after the onset of brief exercise, hence the biphasic or delayed response of leukocyte number to brief exercise. During exercise lasting longer than 1 hour, the two hormones act simultaneously with maximum leukocytosis occurring 3 hours from the onset of prolonged exercise. At the end of exercise, the initial rapid decline in leukocyte number is due to rapid removal of epinephrine (within 30 minutes), whereas the later, more gradual return to baseline levels is due to slower return of cortisol to resting values. During very prolonged exercise, depletion of catecholamines or cortisol may cause a decline in cell number during exercise.

Based on this previous model along with more recent data, a comprehensive and elegant model has recently been put forward to account for the complex pattern of changes among leukocyte subsets during and after exercise (Pedersen 1997). It is proposed that the initial rise in leukocyte, neutrophil, lymphocyte, and NK cell numbers is due to rapid increases in epinephrine (neutrophil, lymphocytes, NK and T cell numbers) and growth hormone (neutrophil number). It is also proposed that delayed release of cortisol during- and after-exercise is responsible for the continued postexercise increase in neutrophil count and decline in NK and lymphocyte numbers. This model is well supported by the current literature.

Role of Inflammatory Mediators and Adhesion Molecules

During and after exercise, leukocyte redistribution may also be mediated indirectly by other factors such as altered expression of adhesion molecules on leukocytes and vascular endothelium; release of cellular components from damaged skeletal muscle, which may be chemotactic, stimulating leukocyte homing to injured tissue; and release of inflammatory mediators such as IL-1, IL-6, TNFα, or acute phase proteins.

Exercise with an eccentric bias, such as downhill running, is associated with enhanced recruitment of leukocytes and lymphocyte subsets into the circulation compared with normal exercise without a large eccentric component. For example, a larger and more persistent elevation of leukocyte number was reported after 60-minute downhill (–10% grade) compared with level running (figure 3.6) (Pizza et al. 1995). Whereas leukocyte, neutrophil, and lymphocyte numbers were elevated by about 50% immediately after both types of exercise, counts were significantly higher after downhill compared with level running. Moreover, leukocyte and neutrophil numbers remained significantly higher up to 12 hours after downhill running, suggesting a sustained recruitment of these cells into the circulation. In addition, peak creatine kinase activity in plasma (an indication of skeletal

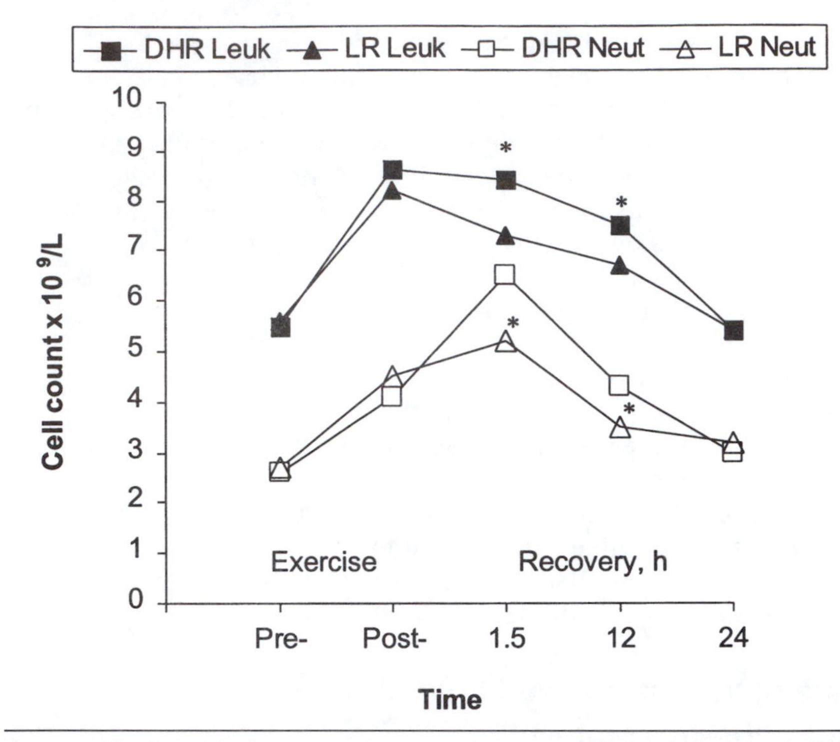

Figure 3.6 Total leukocyte and neutrophil responses to downhill and level treadmill running. Peripheral blood was sampled pre- and postexercise and then 1.5, 12, and 24 hours after exercise. Endurance runners completed two treadmill protocols in a crossover design—60 minutes level running and downhill running at –10% grade, both at 70% $\dot{V}O_2$max. DHR = downhill running; LR = level running; Leuk = leukocyte count; Neut = neutrophil count. * = Significantly higher after downhill compared with level running.

Adapted from Pizza, F.X., J.B. Mitchell, B.H. Davis, R.D. Starling, R.W. Holtz, and N. Bigelow. 1995. Exercise-induced muscle damage: effect on circulating leukocyte and lymphocyte subsets. *Medicine and Science in Sport and Exercise* 27:363-370.

muscle damage) was significantly correlated with circulating neutrophil number (R = 0.71-0.83), suggesting a relationship between the extent of cellular damage and recruitment of neutrophils into the circulation.

Exercise with an eccentric bias is also associated with mobilization of neutrophils into the circulation and infiltration of macrophages and neutrophils into damaged skeletal muscle (Cannon et al. 1994; Fielding et al. 1993; reviewed in Tidball 1995). Although macrophages appear in damaged muscle sooner after injury than neutrophils, neutrophils respond in far greater number; T cells are sometimes found, but B and NK cells are rarely found in damaged skeletal muscle (Tidball 1995). In untrained subjects, 45 minutes of downhill running or eccentric cycling produced a significant, nearly twofold elevation in circulating neutrophil number (Cannon et al. 1994). The magnitude of neutrophil mobilization was significantly correlated with blood concentrations of complement component C3a (R = 0.66) and plasma creatine kinase (R = 0.52), indicators of inflammation and damage to skeletal muscle, respectively. In another study from the same laboratory (Fielding et al. 1993), histological quantification revealed significant correlations between the appearance of neutrophils and structural damage in skeletal muscle (R = 0.66), and the presence of IL-1β and neutrophil accumulation in muscle (R = 0.38). Taken together, these data suggest that eccentric exercise enhances mobilization of leukocytes, primarily macrophages and neutrophils, during and for several hours after exercise. Phagocytes accumulating within damaged skeletal muscle are presumably involved in degradation and repair of damaged proteins and cellular components. The exact signals from skeletal muscle that initiate leukocyte migration to damaged muscle are not currently known, but may involve local release of mediators such as cytokines or fragments of contractile proteins from skeletal muscle cells damaged by excessive loading during eccentric exercise. As discussed in chapters 4 and 6, inflammatory mediators such as IL-1, IL-6, TNFα, and acute phase proteins are released during and for several hours after prolonged load-bearing exercise such as running; these factors may be involved in sustained increases in circulating leukocyte number and their localization to damaged tissue during and after prolonged exercise.

As discussed in chapter 2, complementary adhesion molecules expressed on leukocytes and the vascular endothelium are involved in leukocyte movement throughout the body, primarily by facilitating anchoring of leukocytes to the endothelium, allowing subsequent diapedesis through blood vessel walls and into tissues or lymphoid organs. It has been suggested that changes in adhesion molecule expression and/or avidity for ligands may be one factor mediating the movement of leukocytes into and out of the circulation during and after exercise (Gabriel et al. 1992a, 1992b, 1994a). To test this hypothesis, Gabriel et al. (1994a) reasoned that adhesion molecule expression should decline in those cells that rapidly leave the circulation after exercise; expression of CD11a, the α chain of the integrin LFA-1, was measured on various leukocyte subsets before and after exercise. Nine ultramarathon runners completed a 100-km run (mean time 8 hours), with blood sampled before and 5 and 30 minutes postexercise. Indirect evidence supporting the role of adhesion molecules in leukocyte trafficking included the observations of rapid postexercise declines in cells expressing a high density of CD11a, in particular monocytes, $CD16/56^+$ NK cells, and $CD3^+CD8^+CD45RO^+$ T cells. These cells normally have the highest expression of CD11a, and were those that declined the most and the fastest after exercise. It was proposed that exercise-induced increases in IL-1, TNFα, and body temperature enhance the avidity of CD11a (LFA-1) for its ligand (the ICAMs on endothelial cells). IL-1 and TNFα are known to increase during prolonged exercise (see chapter 6). This implies that cells remaining in the circulation are those expressing only low levels of LFA-1. Gabriel et al. (1994a) acknowledged that this cannot completely explain leukocyte trafficking during and after exercise and that other factors are most likely involved.

Inflammation increases expression of adhesion molecules such as ICAM-1, which is also released into the blood. In 20 male track runners tested at three times in a season (endurance phase, eight-week sprint/strength phase, competition), the proportion of monocytes expressing ICAM-1 (CD54) increased significantly from the endurance to competition phases; monocyte ICAM-1 expression was significantly higher than in active control subjects at this latter time (Baum et al. 1994). Increased ICAM-1 expression was also accompanied by a significant rise in number of circulating monocytes. It was suggested that the increase in ICAM-1 expression reflected an activation of monocytes. Moreover, since ICAM-1 also appears to be a receptor for some rhinoviruses that cause the common cold (Marlin et al. 1990), it was further suggested that enhanced resistance to viral infection may occur in response to training in athletes. That is, exercise may increase monocyte uptake and degradation of virus via enhancing expression of ICAM-1. Whether such enhanced expression of ICAM-1 is related to resistance to infection is yet to be tested experimentally.

Summary and Conclusions

Circulating leukocyte number increases markedly during exercise. The magnitude of increase is related to exercise duration and intensity, with the largest and more persistent changes occurring with intense prolonged exercise. The increase in leukocyte number during and immediately after exercise is due primarily to an increase in granulocyte (neutrophil) number, although lymphocyte and monocyte numbers may also increase; the persistent elevation of leukocyte number for several hours after exercise can be attributed almost entirely to increased neutrophil number. These changes in total leukocyte and granulocyte counts are related to changes in circulating stress hormones, in particular epinephrine and cortisol. The exercise-induced increase in lymphocyte number is associated with disproportionate changes in subsets; NK and CD8 T cell numbers increase far more than B and CD4 T cell counts. Although leukocyte and granulocyte counts may remain elevated for several hours after exercise, lymphocyte number may decline below baseline values for up to 6 hours postexercise before returning to preexercise levels.

The marked changes in leukocyte and lymphocyte numbers are for the most part transitory, and normal levels are generally restored within 24 hours after a single bout of exercise. These changes reflect only redistribution of existing cells and do not represent synthesis of new cells (which requires days). Athletes generally exhibit clinically normal cell numbers, although prolonged periods of intense, high-volume training, at least in distance runners, may lead to clinically low leukocyte counts. Intense exercise has been associated with an acute suppression of lymphocyte proliferation in response to mitogenic stimulation. This suppression appears to be related to epinephrine-mediated redistribution of lymphocyte subsets, in particular to changes in the relative proportion of T cells (due to an increase in NK cell number) and the CD4:CD8 T cell ratio. Whether such transitory effects of acute exercise on leukocyte distribution and lymphocyte function are related to susceptibility to illness in athletes is still open to speculation, and will be discussed further throughout the rest of this book. Certainly, medical practitioners who treat athletes should be aware of possible perturbations in leukocyte counts and subset proportions that may persist for some time after exercise.

Research Findings

- Leukocyte and subset counts are generally normal in athletes, although values at the low end of the clinically normal range have been noted during prolonged periods of high-volume intense training.
- Circulating leukocyte and subset counts increase dramatically during exercise, even very brief exercise; cell counts may remain elevated for several hours after the cessation of exercise.
- Changes in total leukocyte counts often follow a biphasic pattern, characterized by an initial increase during exercise, return to normal value 30-60 minutes postexercise, and a delayed increase 2-4 hours postexercise. Changes in neutrophil count follow this pattern, whereas lymphocyte counts may decline below resting values 1-4 hours postexercise.

- NK cells exhibit the largest proportionate increases in cell number during exercise, changing the relative proportion of lymphocyte subsets (i.e., decreasing T cell while increasing NK cell relative percentages).
- Changes in the relative proportion of lymphocyte subsets contribute at least partially to the observed suppression of lymphocyte proliferation observed after intense exercise.
- Both acute intense exercise and intense training appear to increase the number of circulating lymphocytes expressing activation markers.
- During exercise, recruitment of cells into the circulation reflects a complex interaction of effects of stress hormones (e.g., corticosteroids, catecholamines), hemodynamic changes (e.g., increased cardiac output), and hyperthermia.

Practical Applications

- Clinicians treating athletes must be aware of acute and chronic effects of exercise on circulating leukocyte and subset counts, since these variables are often used in diagnosis.
- Modifying perturbations in circulating leukocyte number and subset relative proportion may attenuate some of the potentially adverse effects of intense exercise on immune cell function (e.g., suppression of lymphocyte function).

Yet to Be Explored

- Does intense exercise training in athletes increase turnover of leukocytes (and various subsets), possibly causing decreases in cell number in the circulation and throughout the body?
- Where do leukocytes recruited into the circulation during exercise go after the end of exercise? Do they simply return to their preexercise sites or do they migrate to specific locations?
- To what extent are acute exercise changes in the relative distribution of lymphocyte subsets responsible for observed changes in functional activity of these cells? That is, is transitory redistribution of cells of clinical relevance?
- Is exercise-induced suppression of lymphocyte proliferation related to alterations in immune function in athletes?
- What is the mechanism underlying the apparent activation of lymphocytes by intense exercise training, and does this partially compensate for alterations in other aspects of immune function in athletes?

Chapter 4

Exercise and Innate Immunity: Phagocytes, Complement, and Acute Phase Proteins

Innate immunity is the first aspect of immune function encountered by an infectious agent. Since the more complex adaptive immune system requires a few days to become fully activated, innate immunity is vital to the early response to any infectious agent, especially bacteria. In addition, phagocytic cells of the innate immune system play a central role in presenting foreign antigen to lymphocytes and in initiating the adaptive immune response.

Innate immunity is mediated by a variety of structural and chemical barriers that limit entry into the host, as well as phagocytic cells that kill foreign microorganisms, process and present antigen to other immune cells, and release soluble factors that stimulate other immune cells. As described in chapter 2, the major cells mediating innate immunity are the phagocytes—monocytes in blood, macrophages in tissues, and neutrophils. In addition, soluble mediators such as complement and the acute phase proteins are involved in myriad immune functions by stimulating phagocytosis, chemotaxis, and release of cytokines in response to infection or inflammation.

Phagocytic Cells

In the past, the phagocytes—monocytes/macrophages and neutrophils—have received relatively less attention than lymphocytes in the exercise literature, although recent work has focused on the role of these cells in mediating both long- and short-term responses to exercise (Pyne 1994; Smith 1994; Smith et al. 1996; Woods and Davis 1994). Studies on exercise and phagocytes have focused on several parameters of phagocytic function, usually measured in vitro, including cellular adherence, phagocytosis of particles, bactericidal activity, oxidative respiration, activation, intracellular enzyme content and/or release into the circulation, and cytotoxic and cytostatic activities. Cells from a variety of sources have been used to study phagocytic activity, such as peritoneal murine or rat macrophages (Fehr et al. 1988; Lotzerich et al. 1990; Woods et al. 1993, 1994b), human connective tissue macrophages (Fehr et al. 1989), human peripheral blood monocytes and neutrophils (Gabriel et al. 1994b; Gray et al. 1993a, 1993b; Hack et al. 1992, 1994; Smith et al. 1990, 1996; Suzuki et al. 1996), human nasal fluid neutrophils (Muns et al. 1996), and equine alveolar macrophages (Wong et al. 1990).

Phagocytic cells from different lymphoid tissues and from different species may not necessarily respond similarly to exercise. Phagocytic functions may increase, decrease, or remain unchanged in response to a single bout of exercise, depending on the type of exercise, state of training of the subject, and source of cells. Because phagocytic cells are also involved in the process of inflammation, via their localization to damaged tissue and release of soluble factors (e.g., proteolytic enzymes, cytokines, and chemotactic factors), exercise-related effects on these cells have important implications for immunity and long-term adaptations to high-level exercise training in athletes. As discussed below, there is some evidence suggesting downregulation of at least neutrophil function; this has been attributed to suppression of chronic inflammation secondary to intense daily exercise training.

Neutrophils

As detailed in chapter 3, circulating neutrophil numbers are generally within the clinically normal range in athletes (Pyne et al. 1995), although low resting neutrophil counts have been noted in some athletes compared with nonathlete controls or clinical norms (Blannin et al. 1996; Keen et al. 1995) (see table 3.2). The acute response to exercise includes a dramatic increase in circulating neutrophil number, which may persist for several hours after exercise and appears to be mediated by hormonal factors such as cortisol. This rise in cell number acutely with exercise is accompanied by neutrophil activation during and for several hours after exercise. However, there also appears to be an overriding downregulation of neutrophil function in athletes that is apparent both at rest and during and after exercise.

Assessing Neutrophil Function

A variety of parameters have been used to assess the effects of exercise on neutrophil function (see Pyne 1994 for review). Circulating cell number gives an estimate of recruitment of neutrophils into the circulation but does not give information about the functional capacity of the cells, since neutrophils may exist in different states of activation. To present a comprehensive view of the current research literature on the neutrophil response to exercise, it is helpful to first give a cursory overview of some of the diverse techniques used to assess neutrophil function, most of which are in vitro assays. As detailed in chapter 2, effective neutrophil action against a foreign agent or microorganism involves a complex series of steps consisting of migration to the affected site in response to chemotactic factors, adherence to the endothelium and diapedesis through the vessel wall, phagocytosis of the foreign agent or particle, release of proteolytic enzymes from intracellular granules (degranulation), and priming or activation in which toxic reactive oxygen species (ROS) and nitrogen species (RNS) are released with an associated increase in respiration (respiratory or oxidative burst).

Migration of neutrophils can be assayed in vitro by incubating cells in the presence of chemotactic factors and then measuring the distance cells move in a given time or the number of cells moving a given distance in a certain time. Phagocytic activity is estimated as the ability of neutrophils to ingest particles, often latex beads, in an in vitro assay in which fluorescently labeled beads are incubated with neutrophils and internalized fluorescence is quantified. Phagocytic activity can be expressed as a phagocytic index (PI) per cell or per given number of cells, and thus is independent of changes in circulating neutrophil number. In addition, the number of phagocytically active (i.e., positive) cells can also be quantified. Once activated to kill microorganisms or foreign agents (called priming), neutrophils exhibit an oxidative burst, characterized by generation of ROS in the form of an oxygen free radical (O_2^-). To measure killing or priming, neutrophils are first incubated with a substance that stimulates phagocytosis, such as bacterial components, opsonized zymosan (OZ), or phorbol myristate acetate (PMA). Production of ROS (such as H_2O_2 or O_2^-) is then quantified using fluorescent probes or spectrophotometric assays. Activation of neutrophils involves release of intracellular granules (degranulation) containing proteolytic enzymes such as elastase or myeloperoxidase (MPO), which can be measured in plasma as an indicator of activation. Degranulation can also be estimated from 90° light scatter in flow cytometry, which gives a measure of cellular granularity. Once activated, neutrophils express certain cell surface antigens or receptors, such as CD16 (FcγR) and CD11b (complement receptor CR3), which can be quantified using flow cytometry and fluorochrome-labeled monoclonal antibodies to the cell surface antigens.

Resting Neutrophil Function

Although neutrophil number appears to be within the clinically normal range in athletes, several recent studies have observed lower neutrophil functional capacity, measured in a variety of assays, in athletes compared with untrained control subjects (table 4.1) (Blannin et al. 1996; Hack et al. 1994; Smith et al. 1990). Smith et al. (1990) noted lower resting and postexercise OZ-stimulated neutrophil ROS production in male cyclists compared with untrained controls. Hack et al. (1994) reported significantly lower resting and postexercise PMA-stimulated ROS production in neutrophils from distance runners during intense training compared with their values during moderate training and those in untrained control subjects. A 20-30% lower capacity for migration in response to chemotactic stimulation with autologous serum and fLMP has been observed in neutrophils sampled at rest in distance runners compared with nonathletes (Espersen et al. 1991). A recent report also showed lower number of phagocytically active neutrophils at rest in trained male cyclists compared with age-matched untrained subjects (Blannin et al. 1996).

Neutrophil oxidative capacity on a per-cell basis, measured as intracellular H_2O_2 production in PMA- or OZ-stimulated isolated cells, was significantly lower at rest in 12 elite male and female swimmers compared with matched nonathlete controls at the start of a 12-week intense training cycle (Pyne et al. 1995) (discussed below). Oxidative capacity was, on average, 50% lower in swimmers compared with controls. Moreover, the percentage of cells responding to stimulation by either substance was significantly lower in swimmers. In swimmers, both oxidative capacity and percentage of positively respond-

Table 4.1 Comparison of Neutrophil Function in Athletes and Nonathletes

Subjects	Exercise	Neutrophil function	Main results	Reference
20 cyclists, 19 UTr	Graded cycling to VF	Adherence	Lower at rest and post in cyclists	Lewicki et al. 1987
		PA	↑ post in UTr only	
11 cyclists 9 UTr	60 min at 60% $\dot{V}O_{2max}$	Oxidative burst	Smaller ↑ post and 6 h post in cyclists	Smith et al. 1990
10 runners 10 triathletes	Graded run to VF	PA	No difference at rest or post	Hack et al. 1992
		Oxidative burst		
10 UTr		Migration		
7 runners during IT/MT	Graded run to VF	PA Oxidative burst	Smaller ↑ 24 hr post IT vs. MT and UTr	Hack et al. 1994
7 active 6 inactive	15 min at HR = 150	Adherence Bacterial activity	Larger ↑ post in active vs. inactive	Benoni et al. 1995b
		Oxidative burst	Larger ↓ post in active vs. inactive	
8 cyclists 8 UTr	45 min submax ex at 150 W	PA	Similar ↑ in no. of active cells post	Blannin et al. 1996
		CPC	↓ post in cyclists	

Abbreviations: CPC = circulating phagocytic capacity; HR = heart rate; IT = intense training; MT = moderate training; PA = phagocytic activity; UTr = untrained; VF = volitional fatigue.

ing cells declined during intense training later in the season (discussed in more detail below).

It has been proposed that a long-lasting activation of neutrophils by a single bout of exercise (discussed below in more detail) may cause depletion of granule contents (e.g., enzymes), impairing the capacity of neutrophils to respond to microbial challenge (Smith et al. 1990; Pyne et al. 1995; Blannin et al. 1996). Repeated daily or more frequent exercise, as performed by athletes, may thus result in a population of circulating neutrophils with suboptimal antimicrobial activity. It is a matter of some debate whether chronic downregulation of neutrophil function is related to susceptibility to infection among athletes.

Blannin et al. (1996) reported significantly lower circulating neutrophil phagocytic capacity at rest and after a standard submaximal exercise in trained male cyclists (35-45 years) compared with age-matched nonathletes. Although the relative percentage of neutrophils ingesting particles after bacterial stimulation was not different between groups, neutrophil count, the number of circulating active phagocytes, and circulating phagocytic capacity (CPC) were significantly lower at rest in athletes compared with nonathletes. CPC is a quantitative measure of phagocytic activity in response to latex bead stimulation, expressed per unit volume of blood. It was suggested that these lower values in athletes reflected a depletion of neutrophil phagocytic activity resulting from chronic endurance training.

Significantly lower resting neutrophil migration in response to chemotactic stimulation was noted in distance runners compared with matched nonathletes (Espersen et al. 1991). In these runners, migration increased to control levels 24 hours after a 5-km race (discussed below), suggesting a capacity to increase migration in response to short-term stimulation and the possibility of chronic suppression in resting samples.

Acute Responses to Exercise

Exercise has been associated with appearance of an activated cell in the circulation in both exercise-trained and untrained subjects, although the magnitude of the acute response may vary with the subject's fitness level, as well as exercise duration and intensity (discussed below). It has been debated whether this acute response reflects primarily an activation of existing cells in the circulation or recruitment into the circulation of an activated cell, although there is evidence in support of both views. As discussed on the following page, it is possible that the mechanism(s) underlying the appearance of activated cells in the circulation differ between exercise-trained and untrained subjects.

Exercise and Neutrophil Priming

Although there are conflicting reports, brief high-intensity and prolonged submaximal and intense exercise have all been reported to induce priming of neutrophils, as indicated by an increase in oxidative burst activity (production of ROS) (tables 4.2 and 4.3) (Hack et al. 1992, 1994; Huupponen et al. 1995; Smith et al. 1990; Suzuki et al. 1996). Enhanced oxidative activity may persist for 6 and possibly 24 hours after exercise. For example, in both cyclists and untrained individuals, oxidative burst activity in OZ-stimulated cells increased immediately and 6 hours after 60-minute cycling at 60% $\dot{V}O_{2max}$ (figure 4.1) (Smith et al. 1990). Hack et al. (1992, 1994) observed significant increases in PMA- and OZ-stimulated oxidative burst activity 24 hours after brief exhaustive exercise (14-21 minutes of graded exercise to volitional fatigue) in male distance runners, triathletes, and nonathletes (figure 4.2). Suzuki et al. (1996) reported significant increases in OZ-stimulated oxidative burst activity immediately and 1 hour after 1.5-hour cycling at 70% $\dot{V}O_{2max}$ in untrained subjects.

In contrast, decreases in oxidative burst activity have been noted immediately after exercise, including a 40% decrease in PMA-stimulated oxidative activity after 60-minute cycling at 50% maximum workload in untrained subjects (Macha et al. 1990); decrease in PMA- and OZ-stimulated activity up to 30 minutes after a brief progressive cycling test to volitional fatigue in male runners and nonathletes (Hack et al. 1994); and a 40% decrease in fLMP-stimulated neutrophil bactericidal activity after an 80 km walking race in healthy males (Fukatsu et al. 1996). Moreover, other studies have shown no effect of exercise on oxidative burst activity. For example, oxidative burst activity was unchanged after 15-minute cycling at a heart rate of 150 beats per minute in active and inactive subjects (Benoni et al. 1995b), and immediately after 20 minutes of cycling at 110% individual anaerobic threshold in cyclists one week before but not after major competition (Gabriel et al. 1994b). These conflicting results may be due to different exercise protocols, time of blood sampling, subject fitness level, and intensity of training in the days prior to testing. Considering all the published data to date, it would appear that moderate or brief exercise may not alter or may increase oxidative burst activity immediately after exercise, whereas intense or very prolonged exercise induces an immediate suppression of oxidative burst activity. The few studies that have measured oxidative burst activity during recovery after exercise suggest that both moderate (Smith et al. 1990) and intense exercise (Hack et al. 1992, 1994) induce a delayed stimulation that may persist for up to 24 hours after exercise (Hack et al. 1992, 1994).

Table 4.2 Effects of Brief Exercise on Neutrophil Function

Subjects	Exercise	Neutrophil function	Main results	Reference
20 cyclists 19 UTr	Graded cycling test to VF	Adherence PA	No change ↑ in UTr	Lewicki et al. 1987
10 runners	2 and 10 km runs	Plasma elastase	↑ after both runs; larger ↑ after longer run	Schaefer et al. 1987
11 runners	5 km race	Migration	Progressive ↑ up to 24 h post	Espersen et al. 1991
10 runners, 10 triathletes, 10 UTr	Graded run to VF	PA Oxidative burst migration	↑ up to 24 h post ↑ 30 min post and ↓ 24 h post No change	Hack et al. 1992
10 UTr	20 min uphill walking and downhill running	MPO, elastase release	↑ only after downhill running	Camus et al. 1992
8 Tr	16 × 1 min runs at 100% $\dot{V}O_{2max}$	CR3, CD16 Plasma elastase 90° light scatter	↑ 6 and 24 h post ↑ 1 h post ↓ 1-6 h post	Gray et al. 1993a

11 UTr	20 min cycling at 80% $\dot{V}O_{2max}$	MPO, C5a	↑ post	Camus et al. 1994
7 runners, 10 UTr	Graded run to VF	PA Oxidative burst	↑ post, highest 24 h post ↓ 30 min post; ↑ 24 h post	Hack et al. 1994
10 cyclists	20 min at 110% IAT before and 8 d after competition	PA Oxidative burst	No change ↓ posttest only after competition	Gabriel et al. 1994b
7 active 6 inactive	15 min at HR = 150	Adhesion Oxidative burst Bactericidal activity	No change No change ↑ post	Benoni et al. 1995b
9 UTr	Graded ex to VF	Oxidative burst	↑ post	Huupponen et al. 1995

Abbreviations: CR3 = complement receptor 3; C5a = complement component 5a; IAT = individual anaerobic threshold; MPO = myeloperoxidase; see table 4.1 for other abbreviations.

Table 4.3 Effects of Prolonged Exercise on Neutrophil Function

Subjects	Exercise	Neutrophil function	Main results	Reference
11 cyclists, 9 UTr	60 min cycling at 60% $\dot{V}O_2max$	Oxidative burst	↑ post and 6 h post	Smith et al. 1990
7 UTr	60 min cycling at 50% max work load	Oxidative burst	↓ post	Macha et al. 1990
7 active	4 h at 45%, 3 h at 60%, 2 h at 75% $\dot{V}O_2max$	Plasma MPO	↑ to similar extent regardless of duration	Bury and Pirnay 1995
21 UTr	45 min eccentric cycling exercise or downhill running, on placebo or fish oil supplements	Plasma elastase	↑ post in placebo, prevented by fish oil supplement	Cannon et al. 1995

8 cyclists, 8 UTr	45 min submax cycling to 150 W	PA CPC	↑ no. of active cells and CPC	Blannin et al. 1996
10 healthy	80 km walking race (14 h)	Bactericidal activity	↓ activity postrace; correlated with cortisol levels	Fukatsu et al. 1996
36 runners	42 km marathon	Chemotactic activity in nasal fluids	↑ activity post-ex	Muns et al. 1996
10 UTr	1.5 h at 70% $\dot{V}O_2max$	Mobility Oxidative burst	No change ↑ post and 1 h post	Suzuki et al. 1996

Abbreviations: CR3 = complement receptor 3; C5a = complement component 5a; IAT = individual anaerobic threshold; MPO = myeloperoxidase; see table 4.1 for other abbreviations.

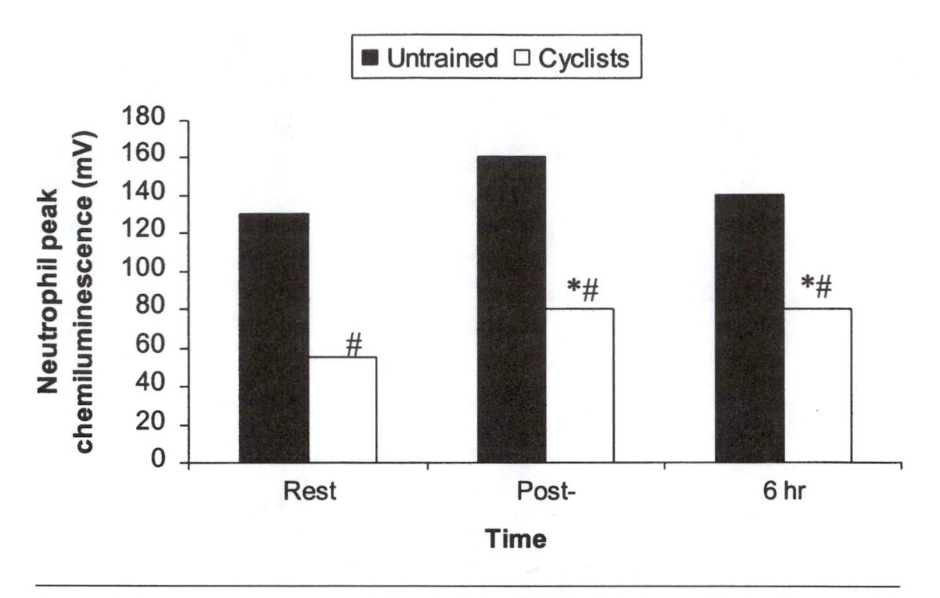

Figure 4.1 Effects of moderate exercise on neutrophil oxidative burst. Neutrophil oxidative burst activity, measured by peak chemiluminescence, before (rest), immediately after exercise, and 6 hours postexercise in untrained subjects and trained cyclists. Exercise consisted of 60 minutes cycle ergometry at 60% $\dot{V}O_{2max}$. * = Significantly elevated compared with resting values; # = significantly lower at each time point in cyclists compared with untrained subjects.

Adapted from Smith, J.A., Telford, R.D., Mason, I.B., and Weidemann, M.J. 1990. Exercise, training and neutrophil microbicidal activity. *International Journal* of Sports Medicine, 11, 179-187.

Exercise and Other Neutrophil Functions

Exercise also appears to alter other aspects of neutrophil function such as increases in plasma elastase and MPO activity, and complement receptor expression (table 4.2). Gray et al. (1993a) reported additional evidence for neutrophil activation, besides an increase in oxidative burst activity, for up to 6 hours after intense interval exercise. Eight endurance-trained males ran repeated 1-minute treadmill sprints to volitional fatigue (mean 16.5 sprints); blood was obtained before and 1, 6, and 24 hours after exercise. Expression of the complement receptor component CR3 increased 6 and 24 hours postexercise, and expression of CD16 antigen increased 24 hours postexercise. Plasma concentration of elastase increased 1 hour after exercise, and 90° light scatter measured by flow cytometry decreased 1 and 6 hours postexercise. Expression of CR3 is considered to reflect degranulation of secondary granules, and degranulation of primary granules is evidenced by increases in plasma elastase

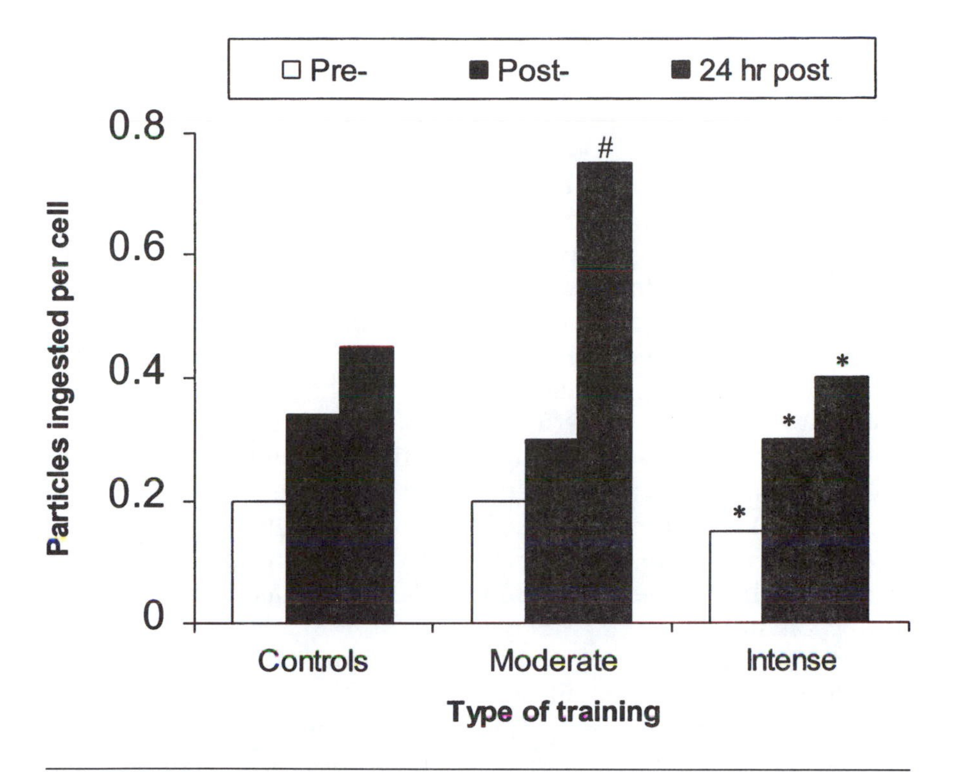

Figure 4.2 Effects of exercise on neutrophil phagocytic activity, measured by the number of ingested latex particles, in untrained subjects (controls) and in male distance runners assessed during periods of moderate and intense training. Neutrophil phagocytic activity was measured in cells obtained at rest (pre-) and immediately (post-) and 24 hours postexercise. Exercise consisted of a cycle ergometry test to volitional fatigue. * = Significantly different from same time point during moderate training; # = significantly different from same time point for controls.

Adapted from Hack, B., G. Strobel, M. Weiss, and H. Weicker. 1994. PMN cell counts and phagocytic activity of highly trained athletes depend on training period. *Journal of Applied Physiology* 77: 1731-1735.

and decreases in light scatter. These data suggest that degranulation occurs within neutrophils and that such degranulation may last at least 6 and possibly up to 24 hours after intense interval exercise. By 24 hours postexercise, normal cell number and light scatter were restored, which was attributed to removal from the circulation of degranulated cells and a recruitment into the circulation of a new population of neutrophils, possibly from bone marrow. Smith et al. (1996) reported a twofold increase in plasma elastase concentration and 20% increase in complement receptor expression after 60 minutes of moderate cycle ergometry (60% APMHR) in untrained males.

Prolonged submaximal exercise also appears to induce release of MPO, suggestive of neutrophil degranulation (Bury and Pirnay 1995). Seven active but not specifically trained males exercised on three occasions: 4 hours at 45% $\dot{V}O_{2max}$, 3 hours at 60% $\dot{V}O_{2max}$, and 2 hours at 75% $\dot{V}O_{2max}$ (exercise mode unspecified) (Bury and Pirnay 1995). MPO concentration in plasma increased during and immediately after exercise to a similar extent, about 85%, in all three exercise sessions (figure 4.3). MPO concentration and neutrophil number were not significantly correlated, suggesting that neutrophil degranulation occurs independently of neutrophilia (increase in cell number) during exercise. Since these subjects did not regularly perform prolonged exercise, it is possible that structural damage to muscle fibers occurred during the three exercise sessions; it has been suggested that neutrophil degranulation occurs in response to muscle cell damage (discussed below).

Release of polymorphonuclear elastase, another indicator of neutrophil degranulation, may be related to the amount of exercise. For example, in 10 male distance runners, plasma elastase increased after both 2-km and 10-km runs (Schaefer et al. 1987). However, a much larger release was evident after the longer run, 300% compared with 6%. These data are consistent with the notion that neutrophil degranulation may occur in response to structural damage elicited by mechanical loading of skeletal muscle. In contrast, a recent study suggests that neutrophil degranulation during exercise may reflect a generalized phenomenon in response to metabolic activity within tissues rather than mechanical loading within working skeletal muscle (Belcastro et al. 1996) (discussed further below).

Neutrophil migration from blood to tissue occurs in response to various chemotactic factors released during inflammation, infection, tissue damage, and other physical stresses such as exercise. Granulocyte migration can be measured in an in vitro assay and expressed as the distance cells moved under agarose when exposed to two known chemoattractants, FMLP and autologous serum; measured this way, migration is independent of cell number. Migration was compared in cells obtained from elite distance runners before and after a 5 km race on a track (Espersen et al. 1991). Although migration was 20-30% lower at rest in runners compared with nonathletes, the ability of cells to migrate was stimulated 20% beginning 2 hours after the race, reaching highest values, similar to those observed in nonathletes, at 24 hours postexercise. It was suggested that the progressive increase in neutrophil migration may be related to cytokines released during exercise.

Muns et al. (1996) reported an increase in neutrophil chemotactic activity (NCA) recovered from nasal secretions immediately after a 42-km

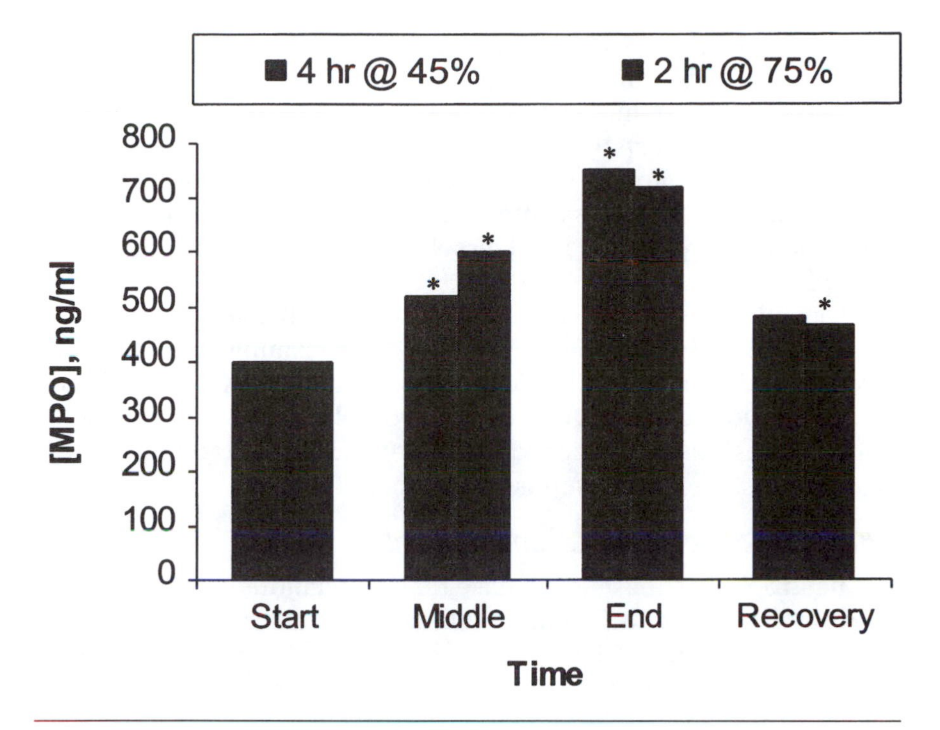

Figure 4.3 Effects of prolonged exercise on plasma myeloperoxidase activity (MPO). An indicator of neutrophil degranulation, MPO was measured at the start, middle, and end of, and during recovery after, prolonged exercise. Two exercise tests were used: 4 hours at 45% $\dot{V}O_{2max}$ and 2 hours at 75% $\dot{V}O_{2max}$ (exercise mode unspecified). * = Significantly elevated compared with preexercise values.

From Bury, T.B., and F. Pirnay. 1995. Effect of prolonged exercise on neutrophil myeloperoxidase secretion. *International Journal of Sports Medicine* 6: 410-412.

marathon race in 36 recreational runners. Nasal fluids obtained via nasal lavage (rinsing the nasal surfaces with saline) were incubated with blood neutrophils obtained at rest from a nonathlete. NCA was estimated from the ability of nasal fluids from the runners to stimulate movement of these neutrophils. NCA increased significantly 2.5 times immediately after the race, returned to pre-race values within 24 hours, and was not different from values observed in nonathletes except for those immediately after the race. The number of neutrophils recovered in nasal lavage increased by 2.7 times immediately after the race and remained elevated for three days, returning to pre-race levels by seven days post-race. It was suggested that increased NCA in nasal fluids after the race stimulated neutrophil migration to the nasal mucosa, where these cells remained for several days after the race. Neutrophil migration to nasal and other airway mucosa occurs after challenge with irritants such as pollutants and allergens, and is thought to contribute to local inflammation. Although the presence of neutrophils in the nasal mucosa may help counteract decrements in aspects of mucosal immunity after intense exercise (discussed in chapter 5), it was suggested that migration of neutrophils to the nasal mucosa after exercise may also induce a local inflammatory process that adversely affects the normal mucosal barrier to infection.

Neutrophil adherence does not appear to change acutely in response to exercise (Lewicki et al. 1987). For example, Lewicki et al. (1987) found no significant effect of brief maximal exercise (graded cycling to volitional fatigue) on neutrophil adherence in 20 well-trained cyclists and 19 matched nonathletes, although adherence was higher at rest and after exercise in the nonathletes.

Differences Between Athletes and Nonathletes

The neutrophil response to exercise may be quantitatively different in exercise-trained versus untrained subjects (see tables 4.1, 4.2, and 4.3). For example, Smith et al. (1990) measured OZ-stimulated respiratory burst activity before, after, and 6 hours after exercise (60 minutes of cycling at 60% $\dot{V}O_{2max}$) in 11 well-trained cyclists and 9 nonathletes. Although activity increased significantly after exercise and remained elevated for 6 hours in both groups, the response was of much smaller magnitude in athletes (see figure 4.1). Moreover, analysis of the kinetics of response indicated that, although athletes and nonathletes exhibited similar rates of oxidative burst activity in maximally stimulated neutrophils, activity in response to lower particle concentration was significantly attenuated in athletes compared with controls, suggesting a lower sensitivity of neutrophils to stimulation. The authors suggested that, in athletes, attenuation of neutrophil priming at low-stimulus concentration may be a beneficial adaptation for limiting the inflammatory response to chronic tissue damage caused by daily exercise training.

Hack et al. (1994) noted quantitatively different neutrophil oxidative burst activity in response to exercise in distance runners during moderate compared with intense training phases (figure 4.2) (discussed further below). Lewicki et al. (1987) also noted differences between neutrophil phagocytic and bactericidal activity in trained and untrained subjects. Twenty experienced cyclists and 19 age-matched nonathletes performed a graded cycling test to volitional fatigue. Neutrophil phagocytic activity against *staphylococcus aureus* increased after exercise in nonathletes, but remained unchanged in cyclists. Bactericidal activity against the same microorganism and neutrophil adherence remained unchanged after exercise in both groups, but were lower at rest and after exercise in cyclists compared with nonathletes. The authors suggested that evidence for suppression of neutrophil adherence, and phagocytic and bactericidal activity in the cyclists, may reflect chronic downregulation of neutrophil function in response to the rigorous training performed by these athletes over many years (more than five years averaging 26,000 km per year).

Blannin et al. (1996) studied the neutrophil response to exercise in eight male cyclists (ages 35-45 years) and eight untrained matched control subjects who performed the same absolute amount of exercise (consecutive 15-minute bouts of cycling at 50, 100, and 150 W); blood was obtained immediately before and after the 45 minutes of exercise. To avoid any confounding effect of prior exercise, athletes rested for 36-48 hours before the start of the study. As noted above, compared with nonathletes, cyclists exhibited significantly lower resting neutrophil counts, number of circulating active phagocytic cells, and circulating phagocytic capacity (CPC) per unit volume of blood. The percentage of phagocytic cells activated by exposure to bacterial extracts increased more after exercise in the cyclists compared with controls (threefold vs. twofold increases) and was about 75% higher after exercise in the cyclists. Although the number of active cells was lower at rest in cyclists, postexercise values were not different from those observed in controls. In contrast, CPC was lower in cyclists both at rest and after exercise (75% and 64% lower, respectively). These data are consistent with other studies suggesting decreased neutrophil phagocytic activity both at rest and after exercise in trained athletes compared with nonathletes. On the other hand, differences in postexercise data may be confounded by the fact that all subjects completed the same absolute, rather than relative, amount of work. That is, although exercise was submaximal for all subjects, untrained subjects would have been exercising at a higher relative percentage of their maximum functional capacities, which may be reflected in differences in neutrophil responses to exercise. A work rate of 150 W would generally be quite low in well-trained cyclists (e.g., $< 60\% \dot{V}O_{2max}$), but much higher (e.g., $> 60\% \dot{V}O_{2max}$) in untrained subjects.

Chronic Responses to Exercise Training

Neutrophil number appears to remain within the clinically normal range during training seasons in most athletes (Hack et al. 1992, 1994; Pyne et al. 1995), although there are some studies suggesting lower neutrophil number in athletes compared with nonathlete control subjects (Blannin et al. 1996; Keen et al. 1995). Despite clinically normal cell counts, however, recent evidence suggests chronic downregulation of several aspects of neutrophil function in athletes during prolonged periods of intense exercise training (table 4.4). As discussed before, cross-sectional comparisons between athletes and nonathletes have shown lower indices of neutrophil function in athletes measured at rest and after exercise. Data from several recent studies following elite athletes during different training phases are also consistent with the concept of chronic suppression of neutrophil function in response to prolonged periods of intense training.

In 12 elite swimmers, neutrophil oxidative burst activity, measured by ROS production in cells stimulated with OZ or PMA, decreased significantly during the 12-week intense training period, with lowest values observed during the peak training phase consisting of 4 weeks of highest-intensity work (figure 4.4) (Pyne et al. 1995). Oxidative burst activity recovered partially during a 2-week rest phase (taper) at the end of the intense training period before major competition. Activity was measured on a per-cell basis, thus is independent of changes in neutrophil number (which tended to increase throughout the season). The percentage of positive cells responding to PMA stimulation also declined significantly during the peak training phase. The authors suggested two possible mechanisms to explain why fewer neutrophils reacted positively to stimulation in the in vitro assay system: release into the circulation of "newly matured" neutrophils with intrinsically lower activity or chronic reduction in the ability of neutrophils to respond to pathogens. However, despite the marked decrease in neutrophil oxidative activity, there was no correlation with the appearance of URTI during the 12-week intense training, suggesting that such changes may not always compromise immune function in athletes.

Another study on male distance runners also provides evidence that both resting and postexercise neutrophil function are compromised during periods of intense training (Hack et al. 1994). Seven male distance runners were studied during periods of moderate and intense training (89 and 102 km $\cdot$ week^{-1}, respectively; the latter also included high-intensity interval work). Ten nonathletes were included for comparison. In neutrophils obtained at rest, phagocytic index, measured by latex bead ingestion, and ROS production in PMA-stimulated cells were significantly lower in runners during intense training compared with their own values during moderate training and values in controls (figure 4.2). Resting plasma epinephrine con-

Table 4.4 Effects of Exercise Training on Neutrophil Function

Subjects	Exercise	Neutrophil function	Main response	Reference
10 M cyclists	20 min cycling at 110% IAT 8 d before/after competition	PA Oxidative burst	No effect ↓ post-ex after competition	Gabriel et al. 1994b
7 runners during MT and IT	Graded cycling to VF	PA Oxidative burst	↓ 50% at rest and ↓ 40% 24 h post during IT	Hack et al. 1994
12 elite swimmers 11 controls	12 wk training	Oxidative burst	↓ during IT; partial recovery in taper	Pyne et al. 1995
7 basketball	Training season	Oxidative burst Adhesion Bactericidal activity	↑ mid vs. start and end ↓ mid vs. start and end	Benoni et al. 1995a
8 cyclists, 8 UTr	45 min submax cycling to 150 W	PA CPC	Lower no. active cells and CPC at rest in cyclists	Blannin et al. 1996
10 UTr	7 d 1.5 h at 70% $\dot{V}O_{2max}$	Mobility Oxidative burst	No effect Less ↑ post-ex on d 4 and 7 vs. d 1	Suzuki et al. 1996

See previous tables for abbreviations.

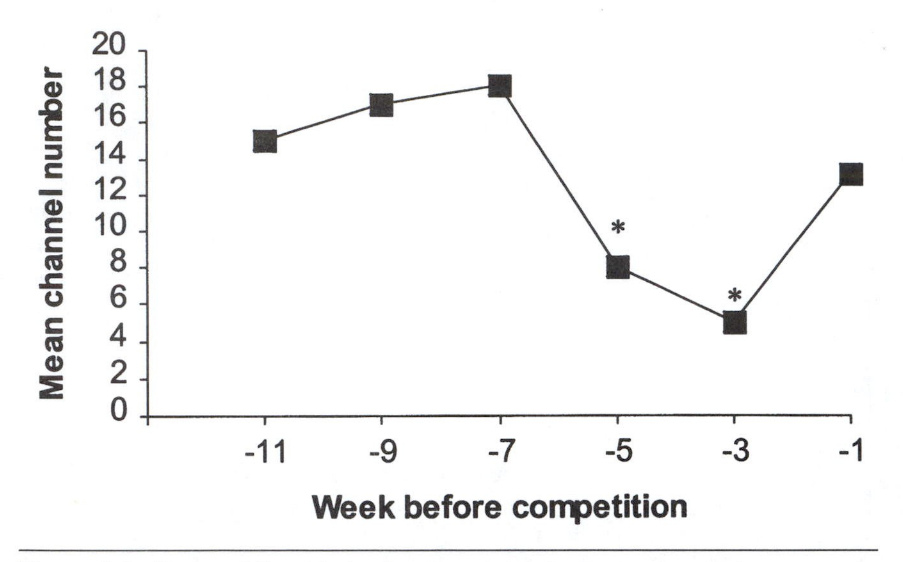

Figure 4.4 Neutrophil oxidative burst activity during a competitive season in swimmers. Long-term changes in neutrophil oxidative burst activity were measured by flow cytometry in PMA-stimulated neutrophils; mean channel number corresponds to oxidative activity. Cells were obtained from elite swimmers at rest every 2 weeks for 12 weeks leading to major competition. * = Significantly lower compared with early-season (week –11) values.

Adapted from Pyne, D.B., M.S. Baker, P.A. Fricker, W.A. McDonald, R.D. Telford, and M.J. Weidemann. 1995. Effects of an intensive 12-wk training program by elite swimmers on neutrophil oxidative activity. *Medicine and Science in Sport and Exercise* 27: 536-542.

centration was 120% higher in athletes during intense compared with moderate training. Since catecholamines are capable of inhibiting some aspects of neutrophil function (e.g., chemotaxis, ROS production, and release of proteolytic enzymes) and epinephrine concentration was significantly negatively correlated with ROS production, it was suggested that high catecholamine concentrations may be one mechanism related to suppression of neutrophil function during periods of intense exercise training.

Moreover, in this study, intense training was associated with large decrements in phagocytic activity and ROS production 24 hours postexercise compared with moderate training and in controls (see figure 4.2). These untrained subjects and distance runners completed a standard graded test of aerobic power lasting 14-21 minutes. Blood was obtained immediately before and after, then 30 minutes and 24 hours after exercise; runners completed the test twice—once during the moderate training period and again during the intense training period. Phagocytic activity, measured by latex ingestion, increased significantly and progressively after exercise in all subjects, with highest values obtained 24 hours postexercise. However, compared to that in nonathletes, activity was higher 24 hours postexercise during moderate training and lower during intense training, suggesting greater activation of phagocytic activity by exercise during moderate training, yet suppression of activation during intense training. ROS production declined significantly immediately after exercise in runners at both times and at 30 minutes after exercise in controls. As for phagocytic activity, ROS production increased significantly 24 hours postexercise, but again, the response was blunted in runners during intense compared with moderate training and in controls. Taken together, these data suggest that moderate training does not alter, or may even enhance, some aspects of neutrophil function, but prolonged periods of intense training appear to cause suppression of both resting and postexercise neutrophil function. These data suggest that daily intense exercise training exerts a prolonged suppressive effect on neutrophil function lasting at least 24 hours; since athletes are usually rested for 12-36 hours prior to laboratory testing, this prolonged effect may account for the lower resting neutrophil activity observed in athletes (Blannin et al. 1996; Keen et al. 1996). It has been suggested that decreased neutrophil function in athletes may reflect partial suppression of the inflammatory response to chronic low-level tissue damage resulting from intense daily exercise (Smith 1994; Smith et al. 1990, 1996).

Short-term intense training may also attenuate the neutrophil response to exercise. For example, in a recent study (Suzuki et al. 1996), 10 previously untrained men completed daily exercise consisting of 1.5 hours cycling at 70% $\dot{V}O_{2max}$ for seven consecutive days; this is fairly intense exercise training for untrained subjects. Neutrophil mobility and OZ-stimulated oxidative burst activity were measured immediately pre- and post- and at 1 hour postexercise on days 1, 4, and 7. Exercise-induced neutrophilia was of lower magnitude on days 4 and 7 compared with day 1. Neutrophil mobility, either spontaneous or in response to stimulation with fMLP, was not significantly altered by exercise on any day. Oxidative burst activity increased significantly after exercise on each day, but the response was attenuated on days 4 and 7 compared with day 1. Enhancement of oxidative burst activity after exercise was significantly correlated with the appearance of band neutrophils, reflecting release of "juvenile" cells into the circulation; this increase in band neutrophil count was also attenuated on days 4 and 7 compared with day 1. It was suggested that these newly matured cells may exhibit a higher capacity for priming in response to exercise, and that the attenuated priming response to exercise after seven days intense training resulted from recruitment of fewer of these cells into the circulation.

Gabriel et al. (1994b) questioned whether the acute neutrophil response to exercise is influenced by endurance competition. Ten well-trained cyclists performed an intense exercise test, cycling at 110% of individual anaerobic threshold to volitional fatigue (mean time 20 minutes) on two occasions—one week before (Ex 1) and one week after (Ex 2) a strenuous 240 km cycling race. Phagocytotic activity against opsonized *E coli* bacteria and the number of phagocytically active cells did not change significantly after either exercise, or between exercise sessions. In contrast, PMA-stimulated oxidative burst activity declined by about 50% after Ex 2 but not after Ex 1, suggesting a blunting of neutrophil priming in response to exercise one week after exhaustive endurance exercise.

In contrast to these studies suggesting downregulation of neutrophil priming and antibacterial activity at rest and after exercise in endurance athletes, significant elevation of neutrophil priming during a training season was reported in seven professional basketball players (Benoni et al. 1995a). Neutrophil antibacterial activity (against *staphylococcus aureus*) and oxidative burst activity were measured at the start of the season, in the middle, and at the end after one-week competition. Antibacterial activity increased about 75% during the season compared with before, and returned to preseason values at the end of competition; similarly, oxidative burst activity increased more than twofold during the season and returned to preseason values after competition. Training throughout the season included 3 hours per day of interval, sprinting, resistance, and skill-related activities. It is likely that differences in training demands account for contrasting results in these studies. That is, prolonged periods of high-volume intense training as performed by elite swimmers (Pyne et al. 1995) and runners (Hack et al. 1994), or shorter periods of intense training in previously untrained subjects (Suzuki et al. 1996), have been associated with suppression of neutrophil function, whereas more moderate training (Hack et al. 1994; Pyne et al. 1995) or lower-volume, interval work interspersed with moderate-intensity, skill-related activities (Benoni et al. 1995a) does not appear to cause long-term suppression of neutrophil function.

Mechanisms Related to Exercise-Induced Changes in Neutrophil Function

Smith et al. (1990, 1996) suggested that the delayed and prolonged increase in neutrophil priming following a single exercise session may be due to stimulation by cytokines released during exercise, such as IL-1, IL-2, TNFα, or GM-CSF, or various hormones such as epinephrine, β-endorphin, or growth hormone. For example, in untrained men, neutrophil activation (as measured by H_2O_2 generation) and growth hormone secretion were elevated to a similar extent after 60 minutes moderate exercise (cycle ergometry at 60% APMHR) (Smith et al. 1996). These data suggest that neutrophil priming during exercise is mediated at least partially by growth hormone release. However, high concentration of growth hormone ($> 20 \text{ ng} \cdot \text{ml}^{-1}$) was associated with a decline in neutrophil priming. It was suggested that the effect of exercise on neutrophil priming may reflect a balance between potentiating factors, such as moderately elevated growth hormone concentration, and other negative mediators, such as corticosteroids. It was further noted that several mediators such as prostaglandins and β-endorphin exert dual effects on neutrophil activation (i.e., stimulatory at low and inhibitory at higher concentrations).

In addition to hormonal influences, enhanced neutrophil activity may be due to redistribution of neutrophils, which may bring into the circulation a more active subset of cells. It is also possible that both occur simultaneously. However, Smith et al. (1990) and Gray et al. (1993b) discounted redistribution of cells as a mechanism behind increased activity, based on their data showing no association between neutrophil distribution and oxidative burst activity after other forms of exercise in exercise-trained athletes.

In contrast to these arguments (Smith et al. 1990; Gray et al. 1993b), it has been suggested that at least part of the exercise-induced stimulation of neutrophil oxidative burst activity may be due to an increased recruitment of "juvenile" cells with a higher responsiveness to agents that induce priming (Suzuki et al. 1996). As described above, untrained men cycled for 1.5 hours at 70% $\dot{V}O_{2max}$ for seven consecutive days, which could be considered intense exercise for untrained individuals. Neutrophil number, especially the number and proportion of "band" neutrophils, increased dramatically immediately after and remained elevated for 1 hour after exercise, resulting in a 50% increase in the ratio of band to total neutrophils. Band neutrophils are newly matured cells. This increase in ratio was significantly correlated with the postexercise increase in ROS production ($R = 0.73$ immediately post- and 0.70 at 1 hour postexercise), suggesting some relationship, although not necessarily causal, between release of newly matured neutrophils and increased oxidative burst activity. It is possible that exercise induces different responses depending on state of training. During exercise in untrained subjects, recruitment of band "juvenile" cells with higher activity may predominate. In contrast, trained athletes appear to exhibit less neutrophilia and recruitment of band neutrophils into the circulation.

Regardless of whether cells are activated or a more active cell is recruited into the circulation, the presence of a more active neutrophil in the circulation after exercise is thought to relate, at least in part, to tissue damage sustained during exercise. Hypothesizing that exercise with an eccentric bias would cause greater tension within and damage to muscle fibers, and thus induce a more profound neutrophil response, Camus et al. (1992) exercised 10 males for 20 minutes using uphill and downhill protocols of equal metabolic cost. Subjects walked uphill at 5% grade and, on a separate occasion, ran downhill at –20% grade at speeds eliciting 60% $\dot{V}O_{2max}$ (approximately 6 and 10 km · hour^{-1}, respectively). Neutrophil number increased to a similar extent in the two exercise conditions. Uphill walking did not induce release of elastase or MPO, two indicators of neutrophil activation. In contrast, downhill running caused significant elevation of MPO and elastase immediately after exercise, with continued elevation of elastase 20 minutes postexercise. The authors suggested that neutrophil activation during exercise occurs in response to mechanical loading and possibly structural damage but is not necessarily related to metabolic factors. Since cell number increased to the same extent during both types of exercise, it was further suggested that activation of cells could not be explained simply by recruitment into the circulation of a more active cell.

Chemotaxis, phagocytosis, and enzyme release are stimulated by a variety of factors including complement components, cytokines (e.g., IL-1, IL-6, and TNFα), hormones (e.g., growth hormone, catecholamines), and inflammatory mediators (e.g., prostaglandins, leukotrienes). To determine whether neutrophil recruitment, activation, and degranulation were related to release of complement, MPO and complement component C5a concentrations were measured in plasma obtained before and after cycling exercise (20 minutes at 80% $\dot{V}O_{2max}$) in 11 male subjects (Camus et al. 1994). C5a concentration increased 2.5-fold and MPO concentration by 30% immediately after exercise, but both variables returned to preexercise values after 20 minutes recovery. Neutrophil number increased during and remained elevated for at least 20 minutes after exercise. The increases in MPO and C5a were significantly correlated ($R = 0.65$), suggesting that neutrophil degranulation may occur in response to increasing C5a levels. The changes in neutrophil number, in contrast, were not significantly correlated with changes in MPO release, suggesting that recruitment of neutrophils into the circulation is not a major factor influencing neutrophil degranulation.

As noted above, it is generally believed that neutrophil infiltration into skeletal muscle occurs in response to cellular damage induced by mechanical loading and contractile activity, in particular eccentric contractions. To distinguish whether neutrophil infiltration into tissue occurs in response to contractile or metabolic activity, MPO activity was measured in rat skeletal muscle (gastrocnemius), heart, and liver before and after 58 minutes of treadmill running to exhaustion (Belcastro et al. 1996). MPO activity increased by 40-50% in each of these tissues after exercise; kinetic analysis showed an increase in maximum velocity of the enzyme in each tissue after exercise. These data suggest that neutrophil infiltration into tissues (as evidenced by the increase in MPO activity) occurred to a similar extent in the three tissues. Since mechanical loading and contractile activity occur only in skeletal muscle, these data are inconsistent with the notion that structural damage secondary to contractile activity is the main stimulus for neutrophil infiltration. It was concluded that these data provide indirect evidence that the neutrophil infiltration of tissues during exercise is a generalized response, and that metabolic status rather than contractile activity within tissues may be a more important stimulus to neutrophil infiltration. It was further suggested that cytokines or other chemotactic mediators released during exercise may be involved in the migration of neutrophils to various tissues during strenuous exercise.

Summary

Neutrophil number appears to be clinically normal in athletes even during periods of intense training. Acute exercise induces marked recruitment into the circulation of neutrophils (neutrophilia) and activation of several aspects of neutrophil function. Acute exercise is associated with enhanced oxidative burst activity (neutrophil priming), phagocytic activity, release of elastase and MPO, expression of complement receptors (degranulation), and granulocyte migration in response to chemotactic factors. Neutrophilia during exercise may be attenuated somewhat by exercise training. Despite normal cell number in athletes, however, recent evidence suggests downregulation of both resting and postexercise neutrophil function during prolonged periods of intense exercise training. Lower resting and postexercise responses of neutrophil oxidative burst activity and phagocytic activity have been reported during periods of high-intensity training in athletes such as swimmers and distance runners; in contrast, moderate training is not associated with such decrements. It has been suggested that downregulation of neutrophil function observed in athletes may serve to limit chronic inflammation resulting from skeletal muscle fiber damage during exercise.

Monocytes and Macrophages

As described in chapter 2, monocytes are mononuclear immune cells produced in bone marrow that appear in the circulation for a brief period (half-life of two to three days) (Woods and Davis 1994). Immune actions of monocytes include phagocytosis, antigen presentation, cytotoxic activity, and production of certain cytokines (e.g., IL-1). Monocytes move from blood into the tissues where they differentiate into larger and longer-lived (i.e., months) macrophages. Monocytes and macrophages thus exist within the body in a variety of functional states (inactive, primed, inflammatory) and in different sites (bone marrow, circulation, tissues) (Woods and Davis 1994). Thus, monocytes/macrophages are important to a wide variety of immune responses, in particular to the early innate response to microbial invasion and tumor growth and to inflammation resulting from tissue injury. This section will describe the responses to exercise of monocytes/macrophages in terms of their role in phagocytosis, antimicrobial activity, inflammation, and cytokine production; the effects of exercise on monocyte/macrophage antitumor or cytotoxic activity will be detailed in chapter 7.

As described in chapter 3, monocyte number generally increases acutely with exercise, although there are studies that have failed to find such an increase (see table 3.9). Exercise-induced increases are transitory, with preexercise values restored soon after cessation of exercise. The few studies focusing on monocytes in highly trained athletes generally report clinically normal resting monocyte number and percentage in the blood, suggesting no long-term decrement in monocyte production and turnover. In contrast, there is a growing recent body of literature suggesting that, despite maintenance of monocyte number, monocyte and macrophage functions are altered both acutely after a single bout of exercise and chronically after a period of exercise training. Because of the need to obtain tissue samples for analysis of macrophage function, much of this work has been performed using experimental animal models.

Assessing Monocyte/Macrophage Function

Monocyte/macrophage functions that have been addressed in the exercise literature include measurement of cellular metabolism, receptor expression and sensitivity, phagocytosis and antimicrobial activity, cytokine production, and antitumor cytotoxic and cytostatic activities (Woods and Davis 1994); many of these assays are performed in vitro, and mainly on monocytes because of their ready accessibility. Metabolic assays include measurement of monocyte glucose metabolism and insulin receptor sensitivity. Phagocytotic and antimicrobial activity is a function of several steps (adherence, chemotaxis, phagocytosis, oxidative burst activity) similar to those detailed in the previous section on neutrophils. Such measurements may include ingestion of particles (e.g., antibody-coated particles, latex beads); monocyte adherence to certain substances (e.g., nylon wool, plastic); chemotaxis in response to certain attractants under agarose or through a semipermeable membrane in a special chamber; lysosomal enzyme activity; and production of oxygen free radical species (ROS, see above) in response to particle or chemical stimulation. Since monocytes produce several cytokines such as IL-1, IL-6, and TNFα (see chapter 6), some studies have measured in vitro secretion of these substances by monocytes isolated from blood samples obtained before and after exercise. Monocytes are capable of producing immunomodulating hormones such as prostaglandins, which can also be assessed in an in vitro system. Finally, monocytes/macrophages are known to exert both cytotoxic (killing) and cytostatic (inhibition of growth) activity against certain tumor cell lines, believed to be related to early immunosurveillance against malignancy and metastatic growth. The cytotoxic and cytostatic activities of monocytes have been assessed in vitro against specific tumor cell lines and in vivo against experimentally induced tumors.

Resting Monocyte/Macrophage Function

Few studies have compared resting monocyte/macrophage function in athletes and nonathletes. Lewicki et al. (1987) compared monocyte adherence in trained cyclists and untrained individuals. Monocyte adherence was lower at rest and after exercise (graded test to volitional fatigue) in cyclists compared with nonathletes.

Osterud et al. (1989) measured LPS-stimulated monocyte thromboplastin activity (a factor involved in blood coagulation) in elite and sub-elite cross-country skiers and nonathletes. Monocytes obtained at rest from the skiers were significantly less responsive to LPS stimulation compared with cells from nonathletes. It is not currently known whether monocytes/macrophages from athletes also differ from those of nonathletes in other aspects of macrophage function. The authors suggested that lower monocyte activation observed in athletes may be beneficial in limiting production of vasoactive substances (e.g., leukotrienes, prostaglandins, thromboxane A_2, lysosomal enzymes), which may in turn result in less capillary permeability and outward shifts of fluid and cells in response to exercise.

Acute Responses to Exercise

Adherence of monocytes may be influenced by expression of adhesion molecules during and after very prolonged exercise. For example,

Gabriel et al. (1994a) reported a 60% decrease in the number of monocytes expressing the adhesion molecule LFA-1 (CD11a) immediately and 3 hours after an ultramarathon race (100 km, approximately 8 hours) in distance runners (figure 4.5). It was suggested that prolonged exercise causes enhanced monocyte adhesion to the vascular endothelium, which presumably increases removal from the circulation (to tissue sites of inflammation/damage possibly). Removal of monocytes expressing the adhesion molecule implies selection of cells expressing less LFA-1, thus the lower number of monocytes expressing this surface molecule after exercise. Whether such changes occur during shorter, less demanding exercise remains to be studied.

Despite enhanced monocyte metabolism (see below), brief exhaustive exercise has been associated with suppression of monocyte phagocytic activity. For example, Bieger et al. (1980) reported significant reduction of OZ-stimulated monocyte phagocytosis after a graded

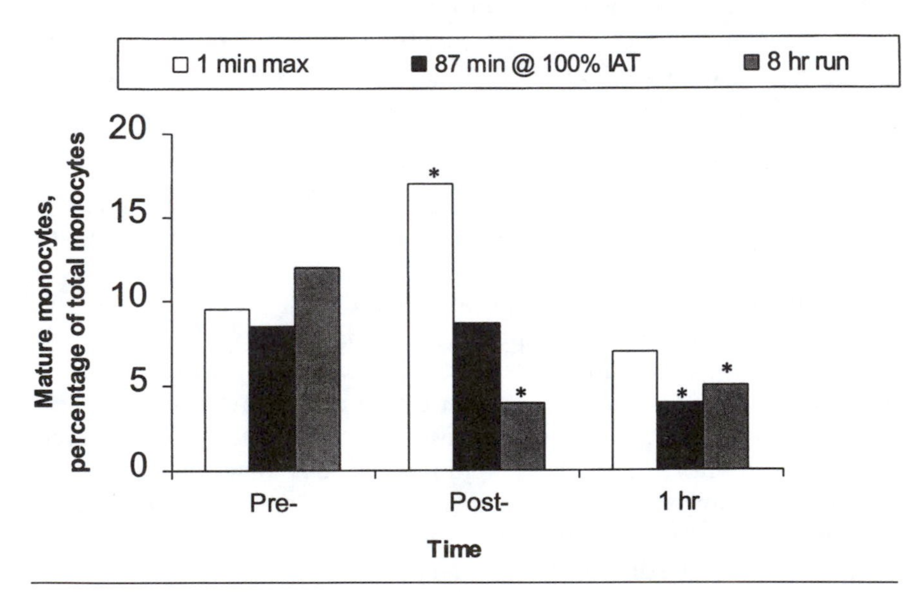

Figure 4.5 Percentage of mature monocytes expressing CD14[low+] in the circulation after different types of exercise was measured pre-, immediately post-, and 1 hour postexercise. Trained endurance athletes completed three exercise conditions: 1-minute maximal cycle ergometer test (1 min max), endurance cycle test to fatigue at 100% individual anaerobic threshold (mean time 87 min) (87 min at 100% IAT), and 100 km (8 hr). * = Significantly different compared with preexercise values.

Adapted from Gabriel, H., L. Brechtel, A. Urhausen, and W. Kindermann. 1994a. Recruitment and recirculation of leukocytes after an ultramarathon run: preferential homing of cells expressing high levels of the adhesion molecule LFA-1. *International Journal of Sports Medicine* 15: S148-S153.

treadmill test to volitional fatigue (mean time 24 minutes) in untrained subjects; suppression of phagocytosis resembled that seen with addition of insulin to monocytes obtained at rest, but was lower than that observed by addition of dexamethasone. It was suggested that changes in the ratio of cortisol and insulin may influence monocyte responses to exercise.

Laegreid et al. (1988) reported significant reductions in alveolar macrophage viability, phagocytic activity, and phagocytic efficiency after a single 4-minute session of intense treadmill exercise in unconditioned thoroughbred horses. Cells were obtained via broncho-alveolar lavage at various times after exercise; to avoid confounding effects of the lavage itself, lavage was performed on each horse only once per week. Phagocyte number recovered in lavages was significantly reduced for three days, phagocytic viability for five days, and phagocytic efficiency (number of particles ingested per cell) for three days after exercise. Using a similar sampling protocol, Wong et al. (1990) also reported significant suppression of OZ-stimulated alveolar macrophage oxidative burst activity after a single maximal treadmill run (about 10-minute duration) in four unconditioned horses. Although cell number remained unchanged, oxidative burst activity declined more than 50% by three days after exercise, recovering to resting values on day 5 after exercise. These data may indicate impaired macrophage function following intense exercise in untrained horses. Because these horses were previously untrained, however, the response to exercise cannot be distinguished from the response to stress of a novel situation (treadmill running); the high cortisol levels measured in both serum and lavage fluids are consistent with a stress response (Laegreid et al. 1988).

In contrast to these seemingly suppressive effects of exercise on monocyte and alveolar phagocytic activity, other tissue macrophages appear to be stimulated both acutely and chronically by exercise. For example, in human connective tissue macrophages, phagocytosis of latex beads (on a per-cell basis) increased 30-60% after an exhaustive 15 km run in trained athletes (soccer and handball players) (Fehr et al. 1989). Moreover, macrophage content and activity of lysosomal and metabolic enzymes (e.g., succinate dehydrogenase, acid phosphatase, glucuronidase assessed histochemically) also increased 20-60% after exercise, suggesting a parallel increase in metabolic and antimicrobial activity. These data on humans were consistent with a previous study from the same laboratory showing enhanced peritoneal macrophage metabolic and lysosomal enzyme activity after exhaustive treadmill running (about 45 minutes) in trained and untrained mice (Fehr et al. 1988).

Monocyte glucose uptake and metabolism appear to be influenced by endurance exercise, although the response may vary with exercise intensity and duration (reviewed in Woods and Davis 1994). Bieger et al. (1980) reported slight enhancement (about 12-13%) of monocyte glucose uptake and oxidation via the hexose monophosphate shunt, but no effect on anaerobic glucose metabolism, after a progressive treadmill exercise test to volitional fatigue in 20 untrained subjects (mean 24 minutes duration). Enhancement of glucose oxidation via the hexose monophosphate shunt may be important to monocyte function since this pathway is used to generate peroxide, a ROS produced during the respiratory burst in phagocytic cells. The exercise-induced increases in glucose uptake and oxidation were qualitatively similar to, but of smaller magnitude than, the increases observed by addition of insulin to resting monocytes. It was suggested that hormonal changes occurring during exercise (e.g., changes in the ratio of cortisol to insulin) may be related to the lower response of monocytes to exercise compared with insulin stimulation, although the possibility of redistribution of monocyte subsets during exercise was not discounted as a possible explanation.

As described in detail in chapter 6, cytokines and other inflammatory mediators such as IL-1, IL-6, and TNFα and prostaglandin E_2 (PGE_2) appear to be released from monocytes during exercise. Although plasma levels of cytokines may not change appreciably during and after exercise, due to rapid clearance from the blood, large increases in the concentration of these cytokines in urine indicate increased secretion during prolonged exercise (discussed further in chapter 6). Since monocytes/macrophages are major producers of these cytokines, especially IL-1, it is likely that at least some of the exercise-induced increase in cytokine appearance is due to enhanced production by these cells (Woods and Davis 1994). There are few studies looking specifically at this issue. Rivier et al. (1994) measured release of IL-6 and TNFα from monocytes isolated from blood samples taken before and after exercise. Ten male young adult and masters athletes performed a graded cycle ergometer test to volitional fatigue; blood was sampled immediately before, immediately after, and 20 minutes after exercise. A constant number of monocytes were cultured with or without epinephrine, and the concentrations of cytokines released into the supernatant were measured. Release of IL-6 and TNFα was slightly but significantly elevated immediately after exercise, and the increases in the output of these cytokines were correlated with each other and with maximum power output and $\dot{V}O_{2max}$. Since a constant number of cells was used in the culture system, these changes could not be attributed to differences in monocyte number; these data suggest slight elevation of the release of some cytokines from monocytes during exercise.

Chronic Responses of Monocyte/Macrophage Function to Exercise Training

There has been little work on the effects of exercise training on human monocyte/macrophage function, and there are only a few studies using animal models. Phagocytic activity of human connective tissue macrophages increased 30-60% after a 15 km exhaustive run in endurance-trained men (Fehr et al. 1989). Forner et al. (1994) reported that peritoneal macrophages from old mice and guinea pigs exhibited higher adherence but lower chemotactic capacity compared with those of younger animals. It was suggested that higher adherence capacity limited the ability of macrophages to migrate (chemotaxis) in response to a stimulus. Four weeks vigorous swim training (25 minutes per day) altered these variables in old but not young mice, with reduced adherence and increased chemotaxis activity in response to an exercise session after as compared with before training in the older animals. It was suggested that macrophage mobility was enhanced following training as a result of the decreased adherence, which may counteract the normal aging response (i.e., higher adherence, lower chemotaxis without exercise) of these variables. It is currently unknown whether such changes occur in aging humans and to what extent they may influence macrophage activity in inflammation and infection.

In mice, exhaustive daily swim training for 20 days increased peritoneal macrophage functions including mobility, chemotaxis, and bactericidal activity (De la Fuente et al. 1993). Moreover, the chronic training effect on these variables exceeded that observed after a single exhaustive exercise bout, suggesting a true training effect. In a training study on seven thoroughbred horses, six weeks of interval training was associated with long-term suppression of alveolar macrophages (Laegreid et al. 1988). Macrophages were obtained via bronchial lavage three days after an interval training session after the six weeks of training. A progressive decline in the number of macrophages recovered via lavage was observed. Moreover, significantly lower phagocytic activity against opsonized sheep red blood cells, and oxidative burst activity, were also noted; these changes were associated with increasing serum and pulmonary cortisol levels after exercise. It was suggested that such changes may alter the integrity of pulmonary mucosal defenses against microbial infection.

Soman et al. (1979) reported a significant enhancement, by 35%, of monocyte insulin receptor number after six weeks of endurance training in previously untrained subjects (1 hour of cycling at 70% $\dot{V}O_{2max}$ four days per week). This increase in receptor number was positively correlated with a 30% increase in insulin-mediated glucose uptake by monocytes and a 20% increase in $\dot{V}O_{2max}$. These data suggest that vigorous exercise training may enhance glucose uptake by monocytes, which may be related to increased monocyte function, although such conclusion awaits further experimentation.

Soluble Mediators of Innate Immunity

As discussed in chapter 2, soluble mediators represent a diverse range of factors carried in solution (hence the name) in blood and other body fluids. These factors act as chemical messengers, capable of initiating or modulating various aspects of immune function. Soluble mediators mediating aspects of natural immunity include cytokines, complement, and acute phase proteins. Each of these classes represents a complex and heterogeneous range of factors; for example, complement represents at least 20 proteins, and there are more than 40 cytokines identified in humans. The effects of exercise on complement and acute phase proteins are discussed in this chapter because they are integral factors in innate immunity; the effects of exercise on cytokines are discussed in a separate chapter (chapter 6) because of their diversity of structures and influence on virtually all aspects of immunity.

Complement

As described in chapter 2, complement represents a complex system of at least 20 proteins found in serum. Complement is active in resistance to bacterial infection via the ability to directly kill bacteria and also to opsonize microbes and thus stimulate phagocytosis. Some complement components act as peptide mediators of inflammation, recruiting inflammatory cells to sites of infection and injury. As listed in table 4.5, not many studies have looked at serum complement levels in response to exercise. Although prolonged periods of intense exercise training may lower complement levels, there does not appear to be a major acute response of complement to exercise.

Resting Levels in Athletes

Lower resting complement components C3 and C4 were observed in 11 male distance runners compared with matched nonathletes (Nieman

Table 4.5 Effects of Exercise on Complement Levels

Subjects	Exercise	Variable	Main results	Reference
UTr	20 min cycling	Serum complement titer	Slight ↑ post	Eberhardt 1971
	1 h run	Serum C3, C4 concentrations	No change	Hanson and Flaherty 1981
8 moderately trained	2.5 h run	Plasma C3a, C4a, C5a concentrations, adjusted for plasma volume changes	↑ C3a and C4a, but no change C5a	Dufaux and Order 1989
	2 h uphill walking at 40% $\dot{V}O_{2max}$	Serum C3 mass and concentration	No change in C3 mass	Sawka et al. 1989
11 runners and 9 non-athletes	Graded run to VF	Serum C3 and C4 concentration	Lower at rest and post-ex in runners; ↑ 11-15% post; normal 45 min post	Nieman et al. 1989c

(continued)

Table 4.5 ***(continued)***

Subjects	Exercise	Variable	Main results	Reference
11 moderately trained	Graded cycling to VF	Plasma C3a and C4a concentrations, adjusted for plasma volume changes	↑ C3a and C4a post; normal by 30 min post	Dufaux et al. 1991
11 runners	5 km track race	Plasma C3d concentration	Athletes in clinically normal range; no change post-ex	Espersen et al. 1991
8 endurance athletes	16 × 1 min TM sprints to exhaustion	CR3 (CD11b) receptor on neutrophils	No change 1 h post; ↑ 6 and 24 h post	Gray et al. 1993a
11 active	20 min at 80% $\dot{V}O_{2max}$	Plasma C5a concentration	↑ post, normal by 20 min post	Camus et al. 1994
12 runners	42 km marathon	Plasma C5a concentration	↑ post, normal 1-16 h post	Castell et al. 1997

Abbreviations: C3, C4, C5 = serum complement components; TM = treadmill; see previous tables for other abbreviations.

et al. 1989c). C3 and C4 levels were about 20% lower in athletes than nonathletes, and near the low end of the clinically normal range. However, no correlation was found between training distance and resting C3 or C4 levels, suggesting that there may be a threshold of training volume related to reduced complement levels, or alternatively, that there is not a straightforward relationship between endurance training volume and complement levels. Since expansion of plasma volume by up to 15% is a common adaptation to endurance training (Wilmore and Costill 1994), it is possible that the difference between athletes and nonathletes or clinical norms may relate in part to the larger plasma volume in athletes.

Acute Response to Exercise

Complement levels may remain unchanged (Hanson and Flaherty 1981; Espersen et al. 1991) or increase after a single session of exercise (Castell et al. 1997; Dufaux et al. 1991; Nieman et al. 1989c). Hanson and Flaherty (1981) observed no significant change in complement C3 and C4 components after 60 minutes of running in male runners. Espersen et al. (1991) also observed no significant change in C3d concentration after a 5 km run in elite distance runners. On the other hand, the concentrations of C3 and C4 components were reported to increase significantly after a graded exercise test to volitional fatigue in untrained and endurance-trained individuals (Dufaux et al. 1991; Nieman et al. 1989c) and after 2.5-hour running in moderately trained males (Dufaux and Order 1989a). Although the absolute C3 and C4 levels were lower at rest and after exercise in runners compared with nonathletes, the pattern of increase in these components was similar in these two groups (Nieman et al. 1989c). C3 and C4 levels increased 11-15% immediately after exercise and returned to preexercise values by 45 minutes following exercise. Dufaux et al. (1991) reported significant 35-45% increases in plasma C3a and C4a concentrations (corrected for plasma volume changes) after a graded cycling test to volitional fatigue in 11 moderately trained males.

Exercise of longer duration has been associated with greater and more prolonged elevation of C3 and C4 levels (Dufaux and Order 1989a). In eight moderately trained (10-40 km · week^{-1} training) males, 2.5 hours of running (25-33 km) induced significant increases in serum C3 concentration during exercise and in C4 concentration during and after exercise; C3 and C4 levels were adjusted for plasma volume changes. C3 levels increased by about 50% during the race and returned to preexercise values within 1 to 3 hours. C4 concentration increased to the same extent, but remained significantly elevated for 3 hours after exercise. Complement C5 component remained below the level of detection during and after exercise. It was suggested that activation of the complement system by both classical and alternative pathways (evidenced by increases in C4 and C3, respectively) may reflect inflammation induced by prolonged load-bearing exercise, and a potential role of complement in clearing proteolytic fragments released from damaged muscle. In contrast to undetectable levels of C5 reported by Dufaux and Order (1989), more recent studies suggest that this component may also increase in response to exercise (Camus et al. 1994; Castell et al. 1997). For example, in 11 active males who completed 20 minutes of cycling at 80% $\dot{V}O_{2max}$, C5a concentration in plasma increased threefold immediately after exercise, returning to preexercise values within 20 minutes. Complement C5a concentration increased nearly fivefold immediately after, and remained higher than resting values for up to 16 hours after, marathon running (figure 4.6) (Castell et al. 1997). It was suggested that the variable response of this complement component may relate to its rapid binding to leukocytes and removal from plasma.

Although changes in complement activation have not been consistently observed, there is evidence for increased expression of the complement receptor on neutrophils. Such increased expression of the receptor may enhance the action of complement without appreciable changes in circulating complement levels. For example, Gray et al. (1993a) reported a significant delayed increase in expression of CR3, the receptor for complement component C3bi (CD11b antigen), on neutrophils isolated 6 and 24 hours after, but not immediately or 1 hour after exercise. Subjects were well-trained endurance athletes who completed repeated 1-minute high-intensity treadmill sprints (at 100% $\dot{V}O_{2max}$ pace; mean 16.5 sprints). Since high-performance athletes train daily, there may be persistent upregulation of complement receptor expression on neutrophils.

In contrast, no changes in serum C3 concentration were noted after more moderate exercise consisting of 2 × 45-minute sessions of uphill walking at 40% $\dot{V}O_{2max}$ in the heat (35°C) (Sawka et al. 1989). C3 concentration was higher in the hypohydrated (equivalent to 5% of body mass) compared with euhydrated states. However, when intravascular C3 mass was calculated to adjust for changes in plasma volume, there were no differences between the two conditions, suggesting that dehydration itself can influence serum complement concentration without appreciable changes in the absolute amount of complement.

Significance of Changes in Complement Levels

The patterns of lower resting and postexercise complement levels in athletes compared with nonathletes, and elevation following pro-

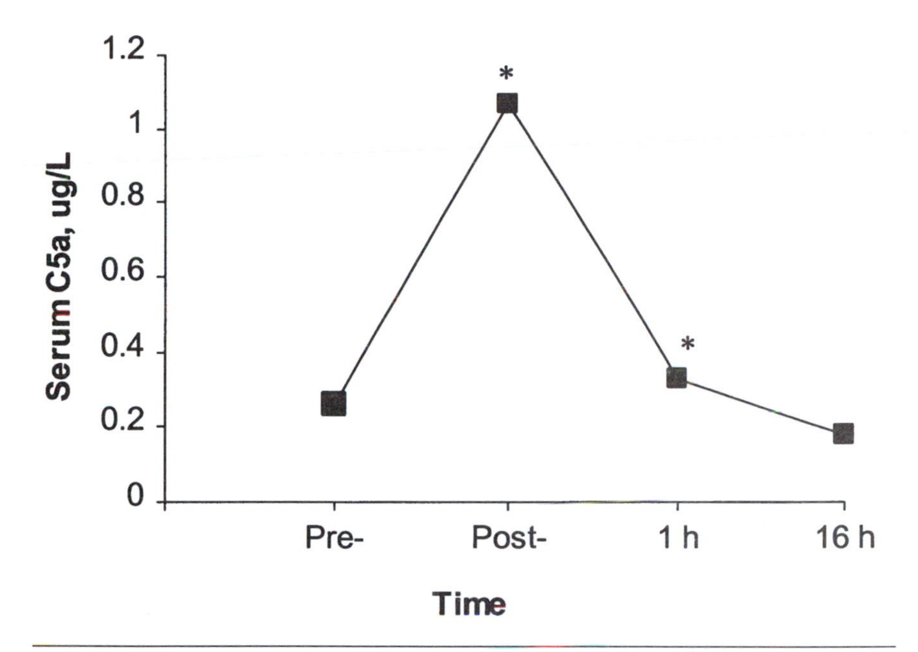

Figure 4.6 Serum complement component C5a concentration before and immediately post-, 1 hour post-, and 16 hours postexercise (42-km marathon run). * = Significant elevation immediately after and 1 hour postexercise compared with preexercise values.

Data from Castell, L.M., J.R. Poortmans, R. Leclercq, M. Brasseur, J. Duchateau, and E.A. Newsholme. 1997. Some aspects of the acute phase response after a marathon race, and the effects of glutamine supplementation. *International Journal of Sports Medicine* 75: 47-53.

longed exercise, are similar to those observed for neutrophil function (discussed above) and acute phase proteins such as C-reactive protein (discussed below). It has been suggested that the lower absolute concentrations in athletes represent long-term adaptation to chronic inflammation resulting from intense daily exercise.

As discussed above, a significant correlation between neutrophil degranulation (evidenced by an increase in plasma MPO level) and appearance of complement C5a in blood observed after exercise has been taken as evidence for involvement of complement activation in the process of neutrophil degranulation, possibly via inducing release of cytokines and other inflammatory mediators such as prostaglandins and leukotrienes (Camus et al. 1994; Castell et al. 1997). As discussed in chapter 6, there is evidence that several pro-inflammatory cytokines, such as IL-1, IL-6, and TNFα, are released during and for some time after exercise; prostaglandins and leukotrienes are also released during exercise.

The delayed exercise-induced expression of the complement receptor on neutrophils has been attributed to inflammation resulting from degranulation of neutrophils during and shortly after exercise (Gray et al. 1993a). Thus, it is possible that during and after exercise, the complex complement system is active in altering several aspects of immune function, possibly through an immediate increase in complement components that induce neutrophil degranulation and cytokine release, followed by a later upregulation of complement receptor expression. There is some agreement that exercise-induced changes in the complement system occur in response to local inflammation and/or release of inflammatory mediators (Camus et al. 1994; Castell et al. 1997; Dufaux and Order 1989a, 1989b; Dufaux et al. 1991; Gray et al. 1993b). It has been suggested that increased complement activity is involved in removal of immune complexes formed from proteolytic cleavage of damaged tissue by phagocytic cells; presumably damaged tissue originates in skeletal muscle disrupted during prolonged or high-intensity exercise (Castell et al. 1997; Dufaux and Order, 1989a, 1989b; Dufaux et al. 1991). Whether exercise-induced changes in the complement system alter immunity to infection has not been thoroughly explored, although it is unlikely such changes play a major role, since they are generally of relatively small magnitude.

Acute Phase Proteins

As described in chapter 2, acute phase proteins (APPs) are a group of unrelated serum glycoproteins released from the liver in response to many factors including infection, inflammation, injury, and other trauma such as myocardial infarction and surgery. The acute phase response is thus a generalized response eliciting mostly beneficial effects (e.g., increased core temperature, sleep, decreased iron availability to pathogens) that may enhance host resistance to infection (Baumann and Gauldie 1994; Steel and Whitehead 1994). There are some similarities between the acute phase response and physiological and metabolic responses to exercise, such as increased core temperature, release of IL-1 and IL-6 (discussed in chapter 6), and hematological changes such as alterations in serum trace metal levels (Cannon et al. 1986; Conn et al. 1995b; Taylor et al. 1987). Although most of the early data on the acute phase response suggested that exercise induces a profound acute phase response, more recent work has questioned whether and to what extent such a response exists (table 4.6); some of these discrepancies between studies may relate to recent advances in methods to measure APPs or to the types of subjects studied.

C-Reactive Protein

As discussed in chapter 2, C-reactive protein (CRP) is the most prevalent APP, released from the liver in response to various traumas including surgery, tissue injury, inflammation, and exercise. Intense prolonged exercise may induce a dramatic increase (more than 20-fold) in circulating CRP concentration. Since CRP has a wide range of functions, including stimulating phagocytosis and complement release in response to tissue injury and inflammation, attention has focused on the response of this APP to various types of exercise.

Resting Levels in Athletes

Compared with those of nonathletes or with clinical norms, resting levels of CRP have been reported to be normal, lower, or higher in some types of athletes (Dufaux et al. 1984; Taylor et al. 1987). For example, in 18 male endurance runners, a pre-race mean CRP concentration of 13.9 $mg \cdot dl^{-1}$ was not significantly different from, although certainly at the high end of, the normal range of $< 12\ mg \cdot dl^{-1}$ (Taylor et al. 1987). In contrast, CRP could not be detected at rest in plasma obtained from triathletes and distance runners (detection limit of assay 0.06 $mg \cdot dl^{-1}$) (Smith et al. 1992). Swimmers and rowers were reported to exhibit significantly lower resting CRP concentrations compared with middle and long distance runners, road cyclists, and soccer players (Dufaux et al. 1984). It was suggested that the lower CRP levels in swimmers and rowers could be related to lower mechanical stress on the body during these activities as compared with other activities such as running or cycling. However, this cannot completely explain the lower CRP levels in swimmers and rowers compared with nonathletes. A dual effect of exercise training was suggested, involving an acute effect of a single exercise session that may elevate CRP for several days and a chronic suppression of CRP release due to continued high-intensity training, similar to what has been proposed for neutrophil function and complement levels (discussed above). Thus, because swimming and rowing induce less mechanical stress, the chronic effect (suppression of APPs) predominates, resulting in low CRP levels. Athletes in more mechanically stressful sports such as running or cycling would exhibit both acute (enhancing) and chronic (suppressing) effects, with the end result of normal CRP levels. This hypothesis has yet to be fully tested. It is possible that differences between studies relate to the duration for which subjects refrained from exercise prior to blood sampling, since CRP levels may remain elevated for several days after a single exercise bout.

Table 4.6 Effects of Exercise on Acute Phase Proteins

Subjects	Exercise	APP measured	Main results	Reference
Active	2-3 h run at 65-85% LaT speed	CRP	↑ up to 3 d post, largest ↑ after longest runs	Liesen et al. 1977
Runners	100 km run	α_1-antitrypsin α_1-acid-glycoprotein α_2-macroglobulin	↑ post and 24 h post	Poortmans and Haralambie 1979
Runners	4 d × 25 km run	CRP	↑ on d 3-5	Dufaux et al. 1983
Runners	15-88 km races	CRP	↑ only after races > 21 km; peaked on d 1 after race	Strachan et al. 1984
Triathletes	160K triathlon (kayak, cycle, run)	CRP	↑ 24 h but not 48 h post	Taylor et al. 1984
8 endurance trained, 8 UTr	60 min cycle ergometry at 60% $\dot{V}O_{2max}$	CRP	No change in trained; ↑ 24 h post in UTr only	Smith et al. 1991
UTr	Max eccentric elbow flexion	CRP	No change up to 5 d post	Nosaka and Clarkson 1996
Runners	42 km marathon	CRP	No change up to 1 h post, but ↑ 16 h post	Castell et al. 1997
UTr	1 h run, 45 min downhill run	CRP	No change post or up to 7 d post	Hubinger et al. 1997

Abbreviations: CRP = C-reactive protein; LaT = blood lactate threshold; see previous tables for other abbreviations.

Acute Response to Exercise

Very long duration load-bearing exercise appears to be necessary to induce an acute phase response, at least in well-trained endurance athletes. For example, Taylor et al. (1987) observed a 266% increase in serum CRP concentration, measured by immunoturbimetry, 24 hours after a 160 km triathlon consisting of 21 km canoeing, 97 km cycling, and 42 km running. CRP levels were nonsignificantly higher than preexercise values 48 hours post-race. The concentration of CRP was reported to increase sixfold after 2 and 3 hours of running at 65-85% of the running speed at the lactate threshold (Liesen et al. 1977); CRP levels remained elevated for up to three days after exercise. CRP was assayed by radial immunodiffusion using monospecific antiserum. CRP levels were higher after 2 hours running at a higher compared with lower intensity (65-85% vs. 82-95% lactate threshold running speed). However, the highest CRP levels were observed after 3 hours running at a lower intensity, suggesting that the postexercise increase in CRP reflects combined effects of both running intensity and duration.

Dufaux et al. (1983) observed a threefold elevation of serum CRP on the third day of a four-day footrace (25 km per day); although lower than on day 3, serum CRP concentration was still elevated to twice resting values on day 5. CRP level was found to increase more than 20 times during the day after a 42 km marathon (Weight et al. 1991). Strahan et al. (1984) observed a distance-related increase in CRP peaking 1 hour post-race, but only in races longer than 21 km; CRP was not elevated after shorter races. This observation may explain why more recent studies have failed to show elevation of CRP after shorter exercise (discussed below). Smith et al. (1992) reported an increase in the number of untrained subjects with detectable levels of CRP in plasma after compared with before, 60 minutes of cycling at 60% $\dot{V}O_{2max}$ (six vs. two of eight untrained subjects, respectively). A recent study showed a nearly fivefold increase in CRP concentration, measured with laser nephelometry, 16 hours after marathon running (Castell et al. 1997). Thus, the time of blood sampling, as well as training level and duration of exercise, appears to influence the CRP response to exercise.

In contrast, some recent reports have failed to find changes in APPs after exercise. For example, no changes in CRP were observed immediately or 1, 3, 5, or 7 days after treadmill running including 1 hours of level running at 75% $\dot{V}O_{2max}$ in untrained males, or 45-minute downhill (–16% grade) running at 70% APMHR in untrained males and females (Hubinger et al. 1997). Nosaka and Clarkson (1996) reported no significant change in plasma CRP concentration in untrained males for up to five days after maximal eccentric elbow flexion that induced clear evidence of muscle cell damage including muscle swelling, large increases in plasma CK concentration, and subjective muscle soreness. The authors linked the lack of CRP response to the failure of this exercise protocol to induce a cytokine response (i.e., no change in plasma IL-1, IL-6, or TNFα levels; see chapter 6), and concluded that exercise-induced local inflammation within skeletal muscle may differ from the inflammatory response to other events such as infection or other forms of tissue injury. Conn et al. (1995a) also found no effect of 20 days moderate exercise training (about 5 km per day voluntary wheel running) on the acute phase response to LPS injection in hamsters; body temperature, serum IL-6 bioactivity, serum amyloid A concentration, and serum iron levels were similar in response to LPS injection in animals allowed access to exercise compared with sedentary animals. Thus, prolonged intense exercise utilizing the whole body or a large proportion of muscle mass appears to be required to induce an acute phase response, at least in trained athletes accustomed to load-bearing exercise. Lower levels of exercise may induce an acute phase response in less well-trained or untrained subjects.

Other APPs

Other APPs, such as protease inhibitors and iron-binding proteins, were also reported to be elevated in athletes at rest, and after endurance running, and to remain elevated for up to six days after exercise (Liesen et al. 1977). For example, Liesen et al. (1977) noted significant elevations of haptoglobulin, α-1-acidglycoprotein, transferrin, and α-1-antitrypsin one to two days after 2 hours exhaustive running in male runners. It is not surprising that CRP levels peak before other APPs, since CRP is generally released within the first 48 hours after injury or inflammation, whereas the other APPs require about 72 hours for release. In contrast, serum haptoglobin concentration decreased immediately and 30 minutes after a 160 km triathlon, returning to preexercise values 24 and 48 hours postrace (Taylor et al. 1987).

In a cross-sectional comparison of athletes (runners, boxers, rowers, and skiers) and matched active but untrained subjects, significantly higher serum concentrations were observed in athletes for α_1-antitrypsin, α_2HS-glycoprotein, α_2 macroglobulin, and ceruloplasmin (Haralambie and Keul 1970; Liesen et al. 1977). Athletes were studied three to five days after intense training, and a long-lasting acute effect from this previous exercise cannot be ruled out as a possible cause (Haralambie and Keul 1970). Serum levels of several of these glycoproteins were also reported to increase after a 100 km run, including increases immediately and 24 hours post-race in α_1-antitrypsin, α_1-acid-glycoprotein, and α_2 macroglobulin (values were adjusted for changes in plasma volume) (Poortmans and Haralambie 1979).

Acute Phase Response and Trace Minerals in Blood

The acute phase response also involves hematological changes such as reduced serum iron and increased serum ferritin concentrations (Conn et al. 1995b; Taylor et al. 1987). Reduced serum iron level is thought to enhance host resistance to microbial infection by limiting available iron required by pathogenic microorganisms. Although serum iron and ferritin concentrations were clinically normal at rest in well-trained endurance runners, a 40% decrease in serum iron and an equivalent increase in serum ferritin concentrations were observed immediately and 30 minutes after a 160 km triathlon (canoeing, cycling, running) (Taylor et al. 1987). It can be speculated that reduction in serum iron after intense exercise may help counteract transitory decreases in functional capacity of some immune cells (e.g., neutrophils, NK cells) or other variables (e.g., IgA) helping to maintain host resistance during the critical time after exercise.

Chronic Training Responses of APPs

As mentioned above, it has been suggested that endurance-type training may induce chronic suppression of the acute phase response (Liesen et al. 1977). For example, when previously untrained subjects completed 2 hours running before and after a nine-week training program, postexercise CRP levels were lower after compared with before training, despite a higher running velocity after training (figure 4.7). Similar trends were observed for other APPs such as haptoglobin and α_1-acid-glycoprotein. In contrast, resting serum levels of some protease inhibitors were higher in well-trained athletes than in nonathletes (Liesen et al. 1977). The higher activity of protease inhibitors may help limit proteolytic activity in skeletal muscle and connective tissue after exercise, and is consistent with downregulation of inflammation after endurance exercise training.

The significance of these changes in serum APPs, especially with regard to immune function, is not clear. APPs are released in response to infection as well as inflammation or injury; APP release is mediated at least in part by cytokines such as IL-1, IL-6, and TNFα, which are also produced during infection and inflammation and in response to intense prolonged exercise. It is not clear whether elevated APPs observed after exercise indicates an enhanced ability to respond to bacterial infection, or whether lower resting APP levels observed in some athletes are linked to immunosuppression. If low APP levels influence immune reactivity, then it follows that athletes with the

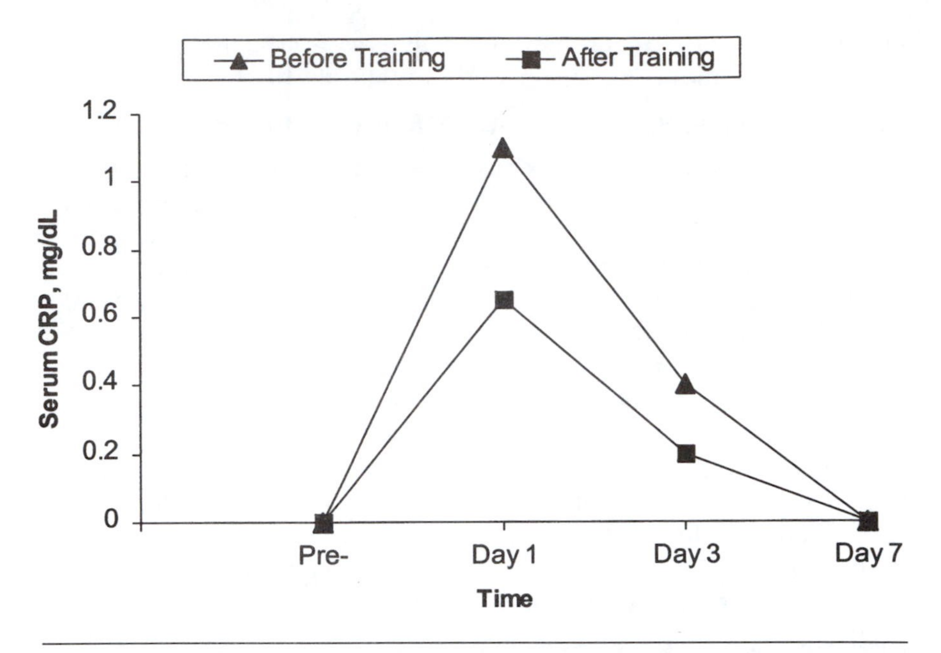

Figure 4.7 Effects of exercise training on serum C-reactive protein (CRP). CRP was measured in serum before (pre-) and one, three, and seven days after a 2-hour treadmill run at 90% individual anaerobic threshold. Subjects repeated the same run after nine weeks exercise training consisting of 10 km running four times each week. The response of serum CRP to acute exercise was attenuated after training.

Adapted from Liesen, H., Dufaux, B., and Hollmann, W. 1977. Modifications of serum glycoproteins the days following a prolonged physical exercise and the influence of physical training. *European Journal of Applied Physiology* 37, 243-254.

lowest APP levels, such as swimmers and rowers, should exhibit the highest rates of infection. This has not, however, been demonstrated.

Summary and Conclusions

Innate immunity represents diverse functions of the early immune response to infectious agents and is also involved in inflammatory processes. Resting neutrophil and monocyte cell counts appear to be clinically normal in well-trained athletes, although there are some reports of clinically low complement and CRP levels in athletes. Acute exercise enhances activity of both neutrophils and monocytes. Despite clinically normal cell counts, recent evidence has suggested downregulation of both resting and postexercise neutrophil functions, including oxidative burst activity and phagocytic activity, during periods of intense exercise training in athletes. In contrast, moderate exercise training does not appear to induce such suppression, and may enhance neutrophil function. Intense prolonged exercise, especially load-bearing exercise such as distance running, may induce an acute phase response, with release of complement components and APPs such as CRP for up to several days after exercise. This acute phase response following exercise may be mediated by release of cytokines during and after prolonged exercise. In athletes, downregulation of neutrophil function and release of inflammatory mediators such as CRP and complement may reflect an attempt to limit chronic inflammation in response to daily intense exercise training.

Research Findings

- Acute exercise stimulates phagocytic activity of neutrophils, monocytes, and macrophages.
- Despite clinically normal cell counts, compared with nonathletes, endurance athletes exhibit suppressed neutrophil function at rest and after intense exercise, suggesting downregulation of this function.
- Intense prolonged exercise induces an acute phase response, causing large increases in some APPs for several days after exercise.

Possible Applications

- In athletes, slightly suppressed natural immunity may reflect a normal downregulation of inflammation in response to chronic tissue injury due to intense daily exercise. Thus, mild suppression of immunity may be part of the normal physiological response to prolonged periods of intense exercise training in endurance athletes.
- Changes in neutrophil function appear related to exercise training intensity and may possibly provide a means of monitoring adaptation to intense training.

Yet to Be Explored

- Is downregulation of neutrophil function in athletes related to suppression of chronic inflammation due to intense daily exercise training?
- Is downregulation of neutrophil function of clinical relevance (i.e., related to susceptibility to illness)?
- What mechanisms are responsible for downregulation of neutrophil function?

- Does migration of neutrophils to the nasal mucosa during exercise play a role in susceptibility to URTI in endurance athletes?
- Is downregulation of phagocytic activity (an early stage in the immune response) compensated by changes in other (perhaps later occurring) immune functions?

Chapter 5

Exercise and Humoral Immunity: Immunoglobulin, Antibody, and Mucosal Immunity[1]

© The Daily Illini

Immunoglobulin (Ig) is an important soluble mediator of humoral immunity to infectious agents such as bacteria, viruses, and parasites. Produced in all mammals, Ig is found in serum, accounting for 20% of all serum proteins. Ig is also found in tissue fluids and mucosal secretions and on some cell surfaces. Igs are glycoproteins that display antibody activity. An antibody is an Ig with ability to bind specifically to a particular antigen. Ig is produced by B lymphocytes and their descendants, the antibody-forming cells and plasma cells, found in blood, lymphoid organs, and other tissues.

Figure 5.1 provides a simplified scheme of the processes leading to antibody (Ab) production in response to a particular antigen (Ag) that

[1] An earlier version of this chapter was previously published in *Exercise Immunology Review* as a review paper (Mackinnon 1996).

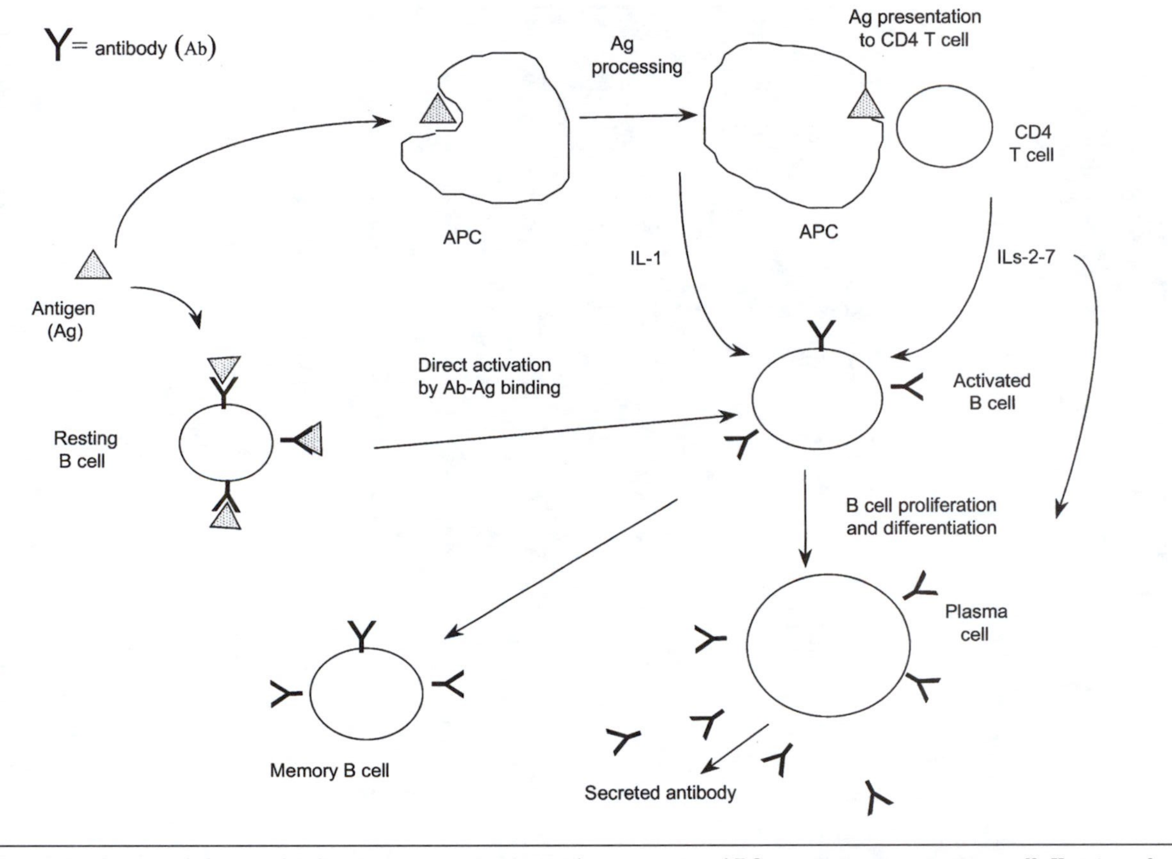

Figure 5.1 General scheme of the antibody response to antigen. Ag = antigen; APC = antigen-presenting cell; IL = interleukin.

the body recognizes as foreign. Antigen can elicit an antibody response via two mechanisms: an indirect mechanism involving stimulation of B cell antibody production by antigen-presenting cells (APCs) and T cells (top scheme), and a more direct stimulation of resting B cells that display cell surface antibody specific to the antigen (middle scheme). In the indirect pathway, Ag on the cell surface of a foreign cell (say a bacterial surface protein) is recognized and engulfed by phagocytic APCs that internally process and then display the Ag on their cell surfaces in conjunction with MHC proteins. This complex is then recognized by CD4 T cells that secrete a variety of stimulating molecules (cytokines), which in turn stimulate B cell proliferation and differentiation into mature Ab-secreting plasma cells. In the direct pathway, a certain proportion of resting B cells will display Ab specific to the Ag on their cell surfaces. Binding of this Ab to the Ag activates these B cells to enter the cell cycle and proliferate and differentiate into more Ab-secreting plasma cells. Both pathways are active during any given infection or response to Ag. Once activated, a subset of B cells persists as "memory" cells that can respond to subsequent exposure to the same Ag by rapidly producing Ab (this is the basis for immunization).

As noted in chapter 2, Igs are bifunctional molecules: on one end is the Fab fragment that is involved in antibody binding, and at the other end is the Fc portion that binds selectively to complement and to Fc receptors on the immune cell surface. As discussed in chapter 2, antibody is an important effector of host resistance to infectious agents, and production of antibody is a major feature of acquired immunity ("memory"). Antibody serves a variety of functions, including neutralizing the deleterious effects of some pathogenic microorganisms, inhibiting the ability of pathogens to gain entry to the body, stimulation of phagocytosis and cytotoxicity, and complement binding (see table 2.4). Also discussed in chapter 2 are the different classes of Ig: IgG, the predominant Ig in serum, and IgA, the most prevalent in mucosal secretions.

Given the central role of antibody in host defense against infectious microorganisms, it is surprising that relatively little has been published on the effects of exercise on Ig and antibody. Although far from conclusive at present, it appears that, as with many immune parameters, moderate physical activity either stimulates or has no effect on Ig and antibody production, whereas heavy exercise and periods of intense exercise training are associated with suppression of Ig and antibody levels. Because serum and mucosal Ig are independently regulated and often respond differently to exercise, these are discussed separately on the following page.

Serum Ig and Antibody

The response of serum Ig to exercise has not been extensively studied, especially in relation to illness among athletes. Although early reports suggested that serum Ig levels are relatively unaffected by acute or chronic exercise, recent work suggests clinically low serum Ig concentrations in some high-performance endurance athletes. Study of serum Ig levels is confounded by changes in plasma volume that occur both acutely and chronically (decreases and increases in plasma volume, respectively), which have not always been taken into account. Since the antibody response to foreign antigens is important to host defense, it might be expected that athletes with low Ig levels might be at risk of infectious illness during intense training. However, although this has not been extensively studied, there are some recent reports in athletes suggesting maintenance of appropriate antibody responses to antigenic challenge despite low Ig levels in serum.

Resting Ig Levels in Athletes

The data are equivocal on resting Ig levels in athletes (table 5.1). Serum concentrations of IgA, IgG, and IgM are typically within clinically normal reference ranges, and values for male and female athletes from diverse sports are not significantly different from those of nonathlete control subjects (Green et al. 1981; Hanson and Flaherty 1981; Haralambie and Keul 1970; Nieman et al. 1989c; Poortmans 1971; Wit 1984). For example, Green et al. (1981) found clinically normal levels of IgA, IgG, and IgM in 20 male distance runners; it was further noted that there was no difference in Ig levels between the fastest and slowest runners (e.g., 2 hours 25 minutes compared with > 3 hours for 42 km marathon). Wit (1974) reported higher IgG, lower IgA, and normal IgM concentrations in a variety of male and female athletes compared with sedentary control subjects.

In contrast, some studies have shown Ig levels in athletes to be lower than those of control subjects or clinically normal ranges. For example, in a prospective study of elite Australian swimmers, Gleeson et al. (1995) reported serum levels of IgA, IgG, and IgM that were each within the lowest 10% of the clinically normal reference ranges. Both Gleeson et al. (1995, 1996) and Gmunder et al. (1990) reported that serum concentrations of the IgG_2 subclass were below clinical norms in the athletes (male and female swimmers and runners, respectively) they studied; Gleeson et al. (1996) also reported significantly lower serum IgG_3 in elite swimmers compared with active, nonathlete control subjects. Rocker et al. (1976) also reported lower serum IgG and

Table 5.1 Observations on Resting Serum Ig Levels in Athletes

Subjects	Variables	Effect	Reference
M runners, boxers, skiers	IgG conc	Clinically normal	Haralambie and Keul 1970
M athletes	IgA,G,M conc	At high end of clini-	Poortmans 1971 cally normal range
M and F athletes (various)	IgA,G,M conc	IgG higher, IgA lower, IgM normal compared with sedentary	Wit 1974
M (?) runners, cyclists	IgA,G,M conc IgA,G.M mass	Lower IgG and M conc, but not mass; higher IgA mass but not conc compared with sedentary	Rocker et al. 1976
M runners	IgA,G,M conc	Clinically normal	Green et al. 1981
M runners	IgA,G,M conc	Not different from sedentary	Nieman et al. 1989c
M and F runners	IgG_2 conc	In low end of clinically normal range	Gmunder et al. 1990
F after 15 wk mod training	IgA,G,M conc	Not different from sedentary	Nehlsen-Cannarella et al. 1991a
M and Felite swimmers	IgA,G,M,IgG_2 conc	Lowest 10% of clinically normal range	Gleeson et al. 1995

Abbreviations in tables 5.1-5.4: M = male; F = female; mod = moderate; conc = concentration; TM = treadmill; $\uparrow\downarrow$ = increase and decrease, respectively.

IgM, but not IgA, concentrations in endurance runners compared with nonathletes.

Since systemic Ig concentration is reported relative to a specific volume of serum (e.g., $g \cdot L^{-1}$), this variable may be influenced by the expansion of plasma volume known to occur in endurance athletes

(Wilmore and Costill 1994). That is, a constant amount (mass) of Ig would appear to have a lower concentration when carried in a larger volume of serum. For example, although Rocker et al. (1976) found significantly lower (by 10%) serum concentrations of IgG and IgM in endurance athletes (runners and cyclists) compared with nonathletes, the differences between groups disappeared after correction of Ig levels for the higher plasma volume in athletes. Moreover, after such correction, IgA intravascular mass was significantly higher in athletes compared with control subjects. Although Gleeson et al. (1995) did not specifically measure plasma volume, there were no differences between groups in serum albumin concentration, suggesting no plasma volume differences. Previous work on other endurance athletes (runners, cyclists) has shown lower concentrations of serum albumin and other proteins in athletes, suggestive of increases in plasma volume (Rocker et al. 1976). It is possible that the type of athlete and training program may influence the degree of plasma volume expansion and thus its contribution to altering serum concentrations of proteins such as Ig.

Serum concentrations of IgA and IgM, but not IgG, were also reported to be significantly lower in male bodybuilders self-administering anabolic steroids compared with nonathlete control subjects and bodybuilders who were not using steroids (Calabrese et al. 1989). IgA and IgM concentrations were approximately 30% to 35% lower in users compared with controls, and IgA levels in four of the users were considered to fall below clinically normal values. Although IgA and IgM levels in nonusers were intermediate between those in the controls and the steroid users, values for nonusers and controls were not significantly different. These data indicate that self-administration of anabolic steroids, as used by some strength and power athletes, possibly in combination with intense weight training, may be associated with compromised immune function. That low Ig levels were observed in steroid users is not surprising, given that some steroids are immunosuppressive (Roitt et al. 1993). In this study, lymphocyte counts and relative subset proportions did not differ between groups, nor did in vitro lymphocyte responsiveness to challenge with the T and B cell mitogens ConA, PHA, and PWM. However, users exhibited a significantly higher in vitro B cell response to a mitogenic challenge with *Staphylococcus aureus*. These data suggest that the capacity for lymphocyte proliferation remains unaffected by steroid use, but that some other aspect of B cell function, possibly differentiation and Ig synthesis, may be adversely affected. The possible contribution of exercise training (e.g., weight lifting) to steroid-induced suppression of IgA and IgM levels is not known.

Ig Responses to Acute Exercise

Serum Ig changes only slightly, if at all, after either brief or prolonged exercise, and several groups have found no change in serum Ig after endurance exercise (table 5.2). For example, in trained cyclists, serum concentrations of IgA, IgG, and IgM were unchanged immediately and at 1 and 24 hours after 2 hours of cycling at 90% of ventilatory threshold (70-80% $\dot{V}O_{2max}$) (Mackinnon et al. 1989). Serum IgA, IgE, IgG, and IgM concentrations were also reported in male distance runners to be unchanged immediately and 24 hours after a 12.8-km run at 70-75% $\dot{V}O_{2max}$ (Hanson and Flaherty 1981). Gmunder et al. (1990) found no significant changes in serum concentrations of IgA, IgE, IgG, IgM, or subclasses IgG_1, IgG_3, and IgG_4 in male and female distance runners immediately after a 21-km run. However, when compared with resting values, total IgG concentration had decreased two days later, and the concentration of subclass IgG_2 had declined immediately after the run. The change in IgG_2 concentration was attributed to an endotoxemia that is known to occur during distance running because of repeated impact microtrauma to the gut endothelium (Bosenberg et al. 1988). IgG_2 interacts specifically with endotoxin, and its levels decrease as levels of endotoxin in the blood increase.

Modest (< 20%), but statistically significant, increases in serum concentrations of some classes of Ig have been reported with endurance exercise. Poortmans (1971) found a significant 14% increase in IgA and a 12% increase in IgG, but no change in IgM, concentrations in male Olympic athletes immediately after a progressive cycle ergometer test to volitional fatigue (mean duration 21 minutes). The magnitude of increases in IgG and IgA levels (14% and 12%, respectively) was larger than that for hematocrit (3%), indicating that only part of the observed changes was due to hemoconcentration. Increases in Ig were transitory, and concentrations of both IgG and IgA had returned to preexercise values by 30 minutes postexercise. Poortmans and Haralambie (1979) also reported a significant 7% increase in IgG, but not IgA concentration, immediately after a 100-km race; this increase exceeded the estimated 2% decrease in plasma volume. IgG concentration returned to preexercise values by the day after the run.

Although slight increases (< 20%) in absolute serum concentrations ($g \cdot L^{-1}$) of IgA, IgG, and IgM were reported in marathon runners following a graded maximal treadmill test, these changes were similar to those observed in blood samples obtained at the same times from sedentary control subjects (Nieman et al. 1989c). The authors attributed these changes to plasma volume shifts during exercise in the athletes and to possible diurnal variation in both athletes and control

Table 5.2 Effect of Acute Exercise and Training on Serum Ig

Subjects	Exercise	Variables	Effect	Reference
Olympic athletes	Graded max cycling test	IgA,G,M conc	IgA ↑ 14%, IgG ↑ 12%, > hemoconcentration; no change IgM	Poortmans 1971
M runners	100 km run	IgA,G conc	IgG ↑ 7%, > hemoconcentration; no change IgA	Poortmans and Haralambie 1979
M runners	12.8 km run	IgA,E,G,M conc	No change	Hanson and Flaherty 1981
M runners	Graded max TM test	IgA,G,M conc	No change when corrected for hemoconcentration	Nieman et al. 1989c
Euhydrated and hypohydrated M	2 × 45 min walk in heat	IgA,G,M intravascular mass	No change when corrected for hemoconcentration	Sawka et al. 1989
M cyclists	2 h cycling 70-80% $\dot{V}O_{2max}$	IgA,G,M conc	No change	Mackinnon et al. 1989

M and F runners	21 km run	IgA,E,G,M conc	No change IgA,G,E,M immediately post-ex; ↓ IgG_2 days post-exercise	Gmunder et al. 1990
		IgG subclasses conc	No change IgG subclasses immediately post-ex except for ↓ IgG_2	
Previously sedentary F	15 wk mod walking training	IgA,G,M conc	↑ 20% after 6 and 15 wk, but not significantly different from controls	Nehlsen-Canarella et al. 1991a
Moderately trained F	45 min walk 60% $\dot{V}O_{2max}$	IgA,G,M conc	↑ 10% post-ex; IgA,M ↓ up to 3 h post-ex	Nehlsen-Canarella et al. 1991b
M subjects	30 s max cycling	IgA,G,M conc	No change when corrected for hemoconcentration	Nieman et al. 1992
Elite M and F swimmers	70 mo swim training	IgA,G,M conc	↑ IgA, no change IgG,M during season	Gleeson et al. 1995
Elite M and F swimmers	12 wk swim training	Pneumococcal antibody response	Clinically normal response	Gleeson et al. 1996

subjects. Serum concentrations of IgA, IgG, and IgM also remain unchanged after 30 seconds of maximal exercise (Wingate cycle ergometry test) when values were corrected for changes in plasma volume (Nieman et al. 1992).

Changes in IgA, IgG, and IgM levels were also observed in overweight premenopausal women after moderate exercise (Nehlsen-Cannarella et al. 1991b). Serum Ig concentrations were measured over a 24-hour period including walking for 45 minutes at 60% $\dot{V}O_{2max}$; to control for possible diurnal variation, serum Ig levels were also measured in all subjects during a nonexercise control period in the laboratory. Compared with control values, serum concentrations of IgA, IgG, and IgM increased significantly immediately after exercise. IgG concentration returned to preexercise values by 1.5 hours after exercise. In contrast, IgA and IgM levels declined significantly below resting concentrations for 3 hours before returning to preexercise values by 5 hours postexercise. The magnitude of these changes was less than 10% in all instances. Plasma volume did not change during exercise, and the authors attributed the modest increases in serum Ig concentrations to influx into the circulation of Ig previously contained in lymph and other extravascular pools.

Sawka et al. (1989) measured the intravascular mass of IgA, IgG, and IgM before and after moderate exercise in the heat to determine the contribution of changes in plasma volume to changes in Ig concentration. Five trained males were first heat acclimated by 120 minutes of daily treadmill exercise at 45°C, 20% relative humidity, for nine days. Subjects then completed two exercise sessions, one in an euhydrated state and the other while hypohydrated by 5% of body mass. Exercise consisted of two 45-minute sessions of walking at 1.34 $m \cdot s^{-1}$ up a 6% grade in a hot humid environment (35°C ambient temperature, 70% relative humidity) with 15 minutes rest between each bout of exercise. Plasma IgA and IgG, but not IgM, concentrations were significantly higher at rest and during and after exercise in the hypohydrated compared with euhydrated state. However, when the intravascular mass of each Ig was calculated taking into account changes in plasma volume, there were no effects of either hydration state or exercise. The authors concluded that previously observed modest exercise-induced increases in serum Ig concentrations (Nehlsen-Cannarella et al. 1991b; Poortmans 1971; Poortmans and Haralambie 1979) were most likely due to a combination of changes in plasma volume and influx of Ig from extravascular pools; they further suggested that body posture and arm position during blood sampling may have influenced serum Ig levels in previous reports.

In Vitro Ig Production

Evidence that exercise influences Ig synthesis is equivocal at present; some studies suggest increased Ig production, whereas others have failed to find any effect. For example, Hedfors et al. (1983) measured in vitro production of IgG in 10 young males in peripheral blood mononuclear cells (PBMC) sampled before and after 15 minutes of moderate exercise (cycle ergometry at a heart rate of 150 beats per minute). Mononuclear cells were isolated from blood and cultured with pokeweed mitogen (PWM), a stimulator of T cell-dependent B cell proliferation and antibody synthesis, for seven days. Measurements were made of soluble IgA, IgG, and IgM concentrations in the supernatant solutions and total Ig appearing on the mononuclear cell surface. Although there was a 50% increase in the number of cells exhibiting surface Ig in cells obtained after exercise compared with before, the in vitro production of all Igs decreased significantly after exercise. The decrease in Ig production in cells obtained after exercise was attributed to a concomitant decrease in the percentage of CD4 lymphocytes and the ratio of CD4 to CD8 cells. Since CD4 cells and the CD4:CD8 ratio are important to B cell proliferation, differentiation, and antibody synthesis, alteration in this ratio would reduce the stimulus for antibody production.

Depression of in vitro Ig production has also been noted after more intense exercise. For example, the number of cells producing IgA, IgG, and IgM in untrained subjects was lower immediately and up to 2 hours after 1 hour of cycling at 80% $\dot{V}O_{2max}$ (Tvede et al. 1989a). PBMC isolated from blood samples before as well as immediately, 1 hour, and 24 hours after exercise were cultured with stimulants of Ig production, including PWM, IL-2, and the Epstein-Barr virus (EBV). The decrease in Ig-producing cell number occurred despite no changes in circulating B cell number, indicating that fewer B cells were producing Ig in culture. Moreover, Ig production in cells isolated immediately and 2 hours after exercise was partially restored by removal of monocytes from the cultures or addition of the prostaglandin inhibitor indomethacin. Since monocyte number generally increases during exercise and prostaglandins are produced by monocytes, these data suggest an indirect effect on Ig production by soluble factors such as prostaglandins released by monocytes during exercise.

In contrast, prolonged, intense exercise did not exert a significant effect on in vitro production of IgA and IgG in PWM-stimulated PBMC obtained from well-trained cyclists (Mackinnon et al. 1989.) Athletes cycled for 2 hours at 90% of ventilatory threshold (equivalent to 70-80% $\dot{V}O_{2max}$), and PBMC were isolated from blood obtained immediately before, immediately after, and then 1 and 24 hours after exercise. Cells were incubated with PWM for seven days, and IgA and IgG secreted by PBMC were measured in the cell culture supernatants. The different responses of in vitro Ig production in these studies (Hedfors et al. 1983; Mackinnon et al. 1989; Tvede et al. 1989a) may relate to differences between the subjects involved—untrained individuals in studies showing suppression of Ig production and well-trained athletes in the study showing no effect on Ig synthesis. That is, a novel exercise task in untrained individuals may represent an immunosuppressive physical stress, whereas the cyclists in our study (Mackinnon et al. 1989) regularly performed that level of exercise.

Serum Ig and Intensive Exercise Training in Athletes

Although most cross-sectional comparisons show that resting serum Ig levels in athletes do not differ from those of nonathletes and clinical norms, decreases in serum Ig concentration and specific antibody titers have been observed during periods of intensive exercise training and major competition in elite athletes (table 5.2). Reports from the former Soviet Union suggest that, while intense exercise training by itself may not necessarily compromise serum Ig and antibody levels, intense training in combination with major competition may lead to reduced concentrations. For example, in athletes from unspecified sports, serum titers of antibodies to tetanus, diphtheria, and staphylococcus decreased during major competition, but were normal during periods of intense training months before and again after recovery from major competition (Pershin et al. 1988).

A recent prospective study of elite Australian swimmers suggests possible long-term suppression of serum Ig in elite athletes (Gleeson et al. 1995). Serum levels of IgA, IgG, and IgM were significantly lower throughout the seven-month season in elite swimmers compared with age-matched nonathlete control subjects. IgG and IgM levels did not change during the season, but the swimmers showed a significant 4.4% increase in serum IgA levels during the taper period (reduced training load) at the end of the season. There was a tendency ($p = .07$) toward lower IgG_2 subclass levels in swimmers compared with control subjects. Although all serum Ig concentrations were within clinically normal ranges, values for the swimmers were in the lowest 10th percentile for their population. Moreover, there were no differences between swimmers and controls in relative or absolute numbers of B, T, or activated lymphocytes ($HLA\text{-}DR^+$), nor were there changes in these cells over the season, indicating that the lower Ig levels were not due to fewer antibody-producing cells or to changes in the relative proportion of T to B cells. These data suggest that elite athletes who train intensely for many years may exhibit compromised serum Ig production.

Another recent report on 60 male and female athletes from various sports (swimming, track and field, cycling, football [soccer], basketball, tennis, and triathlon) showed declining serum Ig levels during three months of intense training (Garagiola et al. 1995). Total Ig, as well as IgM, IgG, and subclass IgG_1 and IgG_2 concentrations, declined progressively from the start to end of the three-month period; training included sessions of approximately 130-140 minutes, five to seven days per week. Concentrations of total Ig, IgG, and IgM decreased approximately 7% to 20%, with much larger decreases observed for the two IgG subclasses (30-45% and 10-30% decreases for IgG_1 and IgG_2, respectively); levels of IgA, IgG_3, and IgG_4 did not change significantly over time. Although nearly one third of the athletes exhibited at least one episode of infectious illness during the three months (mainly upper and lower respiratory tract infection), the authors did not relate changes in Ig levels to the appearance of illness, so it cannot be discerned whether such changes compromised immunity to infection in these athletes.

Serum Ig Response to Moderate Exercise Training

Moderate exercise training has been associated with modest increases in serum Ig levels (table 5.2). In a randomized intervention study, 50 previously sedentary premenopausal, middle-aged, overweight women were randomly assigned to a control or exercise group (Nehlsen-Cannarella et al. 1991a). Exercisers participated in a 15-week program of moderate physical activity consisting of five 45-minute sessions of brisk walking per week at 60% of heart rate reserve; control subjects did not exercise during this time. Compared with measures obtained from control subjects after 6 and 15 weeks, serum concentrations of IgA, IgG, and IgM increased by approximately 20% in the exercised subjects. However, although statistical analysis showed a significant increase over time in the exercisers, Ig concentrations did not differ significantly between groups at any time. Plasma volume did not change with exercise training, and increases in serum Ig concentrations could not be attributed to changes in other immune parameters such as leukocyte, T or B lymphocyte numbers, CD4:CD8 cell ratios, or lymphocyte proliferation. The authors questioned the biological significance of marginal changes in serum Ig levels after moderate exercise training. Mitchell et al. (1996) found no changes in serum IgA, IgG, and IgM concentrations in 11 previously untrained males after 12 weeks moderate exercise training (30 minutes cycling at 75% $\dot{V}O_{2max}$ three times per week).

Specific Antibody Responses to Acute Exercise and Exercise Training

Although circulating levels of Ig may not change after exercise, production of specific antibody appears to be acutely stimulated by exercise and moderate exercise training, whereas extended periods of intense exercise training combined with major competition have been associated with chronic suppression of specific antibody levels in some elite athletes. For example, Eskola et al. (1978) observed higher serum antibody titers to tetanus toxoid in blood obtained from four runners after a 42 km marathon compared with blood sampled at the same time in 59 nonathlete control subjects; both groups had been immunized 15 days before the marathon. The authors attributed the higher antibody titers in runners to an increase in B lymphocyte number in blood obtained after the race. However, since specific antibody levels were not measured immediately before the race, it is not clear whether the higher antibody titers in runners reflected an acute response to the race, a chronic training effect during the 15 days between immunization and the race, or a combination of the two. It is more likely that a chronic training effect was responsible for the increased specific antibody levels. The higher antibody titers observed after the race in runners occurred despite significantly reduced T cell responsiveness to mitogenic challenge, an important step in the initiation of antibody synthesis; these data suggest that acute stimulation of antibody synthesis by T cells was not responsible for the increase in specific antibodies after the race. That the higher antibody titers in runners may relate more to a chronic training rather than acute exercise effect is also consistent with data from animal studies showing enhanced specific antibody levels after moderate exercise training (see later paragraphs). Moreover, in trained cyclists, 2 hours cycling at 90% ventilatory threshold had no acute effect on serum antibody titers to specific antigens from pathogens including adenovirus, influenza A and B, mumps, parainfluenza, respiratory syncytial virus, cytomegalovirus, varicella-zoster, herpes simplex, chlamydia, mycoplasma pneumoniae, and rubeola in well-trained male cyclists (Mackinnon et al. 1989).

A recent study on elite swimmers suggested that although serum Ig concentrations may be lower in elite athletes compared with nonathletes or clinical norms, these athletes are still capable of mounting a clinically appropriate antibody response to antigenic challenge (Gleeson et al. 1996). Twenty elite swimmers and 19 age- and sex-matched nonathlete control subjects were vaccinated with a pneumococcal vaccine, and serum Ig and antibody responses to six common pneumococcal antigens were measured 14 days later. None of the subjects had previously been exposed to the antigens in the vaccine. Vaccination occurred at the end of a 12-week intense training cycle in swimmers. In swimmers, the serum Ig and specific antibody responses were clinically normal, and not significantly different from those observed in control subjects. Although in swimmers compared with controls, prevaccination levels of some IgG subclasses (IgG_2 and IgG_3) were lower, postvaccination levels were similar, indicating a relatively greater response in the swimmers. These data indicate that, despite clinically low prevaccination levels of some IgG subclasses in elite athletes, the ability to respond to a novel antigenic challenge is not compromised after a period of intense training. The authors suggested that vaccination-induced activation of antibody production may provide protection for some elite athletes who experience recurrent infections during periods of intense training.

Experimental animal models suggest that moderate exercise training may stimulate the production of specific antibodies. For example, Douglass (1974) reported a significantly enhanced secondary response to immunization in exercise-trained compared with nonexercised control mice. Mice were trained by swimming for 1 hour twice daily for 11 weeks; control mice sat in 1 cm of water during the exercise training periods. Mice were immunized with diphtheria toxoid 6 weeks after the start of training and boosted by similar injection 21 days later. There was no difference in antibody titers between exercised and control mice during the primary response (after the first immunization) or up to 7 days after the booster injection (secondary response). However, by 14 days after the booster injection when titers were beginning to decline in control mice, antibody levels in exercised mice continued to rise and were 50% higher. These data suggest that prior exercise training may enhance the secondary but not necessarily the primary antibody response to antigen.

Liu and Wang (1986/87) reported significantly higher antibody titers in mice that had been trained to run on a treadmill for 23 days before immunization with the bacteria *Salmonella typhi;* a booster injection of the same antigen was given 10 days after the first injection.

A three- to fivefold elevation in antibody levels, which persisted for at least 16 weeks following the initial immunization, reflects the combined effects of exercise on both primary and secondary responses.

A more recent study by Kaufman et al. (1994) found enhancement of both primary and secondary antibody responses to a novel antigen in exercise-trained rats (figure 5.2). Male rats were moderately trained by swimming, starting with daily activity of 15 minutes and gradually increasing to exercise of 2 hours' duration 5 days per week over a four-week period. At the end of the training period, untrained and trained rats were subjected to a swim to exhaustion and then sacrificed. Untrained and trained animals were injected with a novel antigen, keyhole limpet hemocyanin (KLH), 14 days before the start of the training program (i.e., 42 days before the exhaustive swim and sacrifice of animals) and then boosted with the same antigen on the 1st and 21st days of training. Both IgM and IgG antibodies specific for KLH were significantly higher at all times in trained compared with untrained rats, although the response was higher for antibodies of the IgM class. Moreover, the differences in anti-KLH antibodies between groups increased with time and subsequent exposure to the antigen. The higher IgM levels during the primary response (after the initial injection) were attributed to the kinetics of IgM production in re-

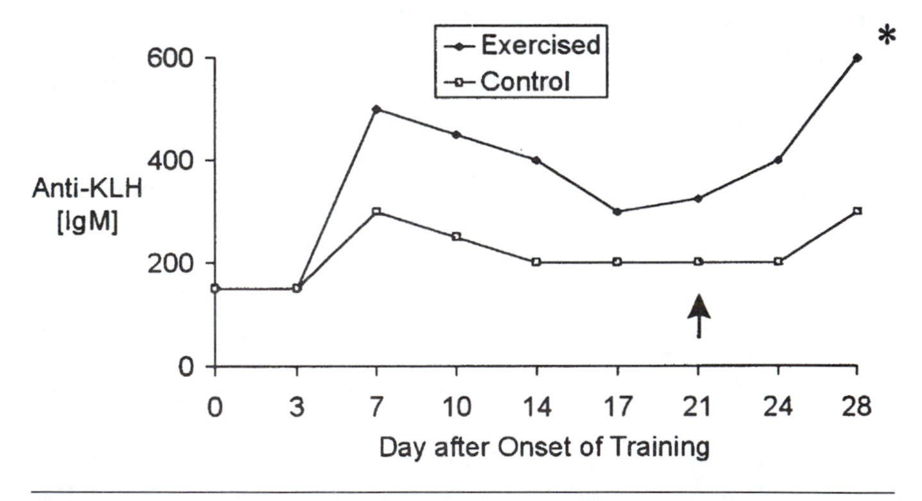

Figure 5.2 Serum-specific antibody response to exercise training. Serum IgM specific for KLH in exercise-trained and control (sedentary) rats. Arrow indicates time of booster injection. See text for details. * = Significantly different from control values.

Data from Kaufman, J.C., T.J. Harris, J. Higgins, and A.S. Maisel. 1994. Exercise-induced enhancement of immune function in the rat. *Circulation* 90, 525-532.

sponse to a novel antigen. As noted above, antibodies of the IgM class are generally seen early in the primary response, with a delayed appearance of IgG. Moreover, IgM has a short half-life in serum, and IgM catabolism is independent of its serum concentration; thus, IgM concentration may continue to rise dramatically over the short term in response to antigenic stimulation. In contrast, IgG appears later and has a longer half-life, and catabolism increases with rising serum concentration; thus, over the short term, IgG levels may not rise as markedly as IgM, due to the lower production of IgG and relatively higher catabolic rate. The prolonged elevation of both IgG and IgM antibodies during the secondary response (after booster injections) appears to be more complex; it was attributed to shifts in the relative proportions of lymphocyte subsets, in particular the percentage of B cells and the ratio of CD4:CD8 lymphocytes in the spleen.

Kaufman et al. (1994) reported significantly higher B cell percentages and ratio of CD4:CD8 cells in the spleens of exercise-trained compared with untrained rats after the exhaustive swim. Since antibody production by B cells and their descendants is dependent on stimulation by CD4 cells, antibody production would be favored by increasing proportions of both B and CD4 cells in the spleen. The authors further suggested that catecholamines released during regular exercise training sessions may have provided continued stimulation of antibody synthesis, since norepinephrine has been associated with enhancement of the early antibody response (Sanders and Munson 1985). In addition, the observed acute enhancement of lymphocyte mitogenic responses and IL-2 receptor expression after exercise observed in this study may have contributed to stimulation of B cell function throughout the study period.

In contrast, Coleman and Rager (1993) found no enhancement of specific antibody production after several weeks of voluntary exercise in rats. Exercised rats were allowed free access to running wheels for 12 hours during the dark portion of the light cycle (i.e., during their normally active period), whereas control rats were prevented from exercising. After five weeks access to exercise, rats were immunized with KLH and permitted a further three weeks voluntary activity before sacrifice. IgG and IgM antibody titers did not differ between control and exercised groups at any of six observation points during the three weeks between immunization and sacrifice, despite significantly higher splenic responses to the T cell mitogen concanavalin A in exercised compared with control rats at the end of the study. Differences in specific antibody responses to the same antigen (KLH) between this study and that by Kaufman et al. (1994) may relate to the nature of exercise and/or time course of study. The study by Kaufman et al. utilized enforced exercise, presumably of a vigorous nature (swimming), whereas Coleman and Rager used voluntary exercise, a less stressful form of exercise that may have induced fewer physiological adaptations in the animals. In addition, the study by Kaufman et al. introduced a secondary response by a booster injection 14 days after the initial immunization. Further work is needed to determine under what conditions exercise training is associated with enhanced specific antibody responses to antigenic challenge.

Taken together, data on the specific antibody response to exercise suggest that despite clinically low serum Ig concentration, the specific antibody response to antigenic challenge (immunization) is not impaired, and may be enhanced, by intense exercise training in athletes. Animal models confirm enhancement of primary and secondary specific antibody responses after nonstressful (voluntary) or moderate exercise. Thus, there is little evidence to suggest that intense exercise training impairs the specific antibody response to antigen.

Clinical Applications

Whether interventions designed to enhance the Ig or specific antibody response to infection are effective in preventing infectious illness associated with intense exercise has not been fully explored. A 15-week moderate exercise training program in women was associated with both reduced incidence of URTI and enhanced serum Ig levels (Nehlsen-Cannarella et al. 1991a). Increases in serum Ig concentration were significantly correlated with enhanced resistance to URTI and improvements in fitness. Whether enhanced resistance to illness was a direct result of increased Ig levels, or resulted indirectly from improved cardiovascular fitness or some other factor(s), remains to be elucidated.

A brief report, in abstract form, on administration of human Ig to athletes suggested that exogenous Ig may modify the body's response to infection (Frohlich et al. 1987). Human Ig or saline placebo was administered intramuscularly at four-week intervals over a six-month period to 20 swimmers in a double-blind manner. Although Ig administration had no effect on the number of infectious episodes, the duration of infection was three times shorter in athletes administered Ig compared with placebo. Although the nature of infectious illness was not specified, it can be assumed that most episodes involved viral URTI since this appears to be the most common infectious illness over the short term among athletes (Hanley 1976; Mackinnon et al. 1993b). If we assume that viral URTI accounted for most of the episodes of illness in the study by Frohlich et al. (1987), it is not surprising that Ig administration reduced the duration but not the number of episodes of illness, for the following reason. As described below, entry of viruses causing URTI occurs via mucosal surfaces of the upper respiratory tract (e.g., mouth, nose, and throat). The mucosal immune system is a major effector of host immunity to these microorganisms, primarily through secretion of IgA into seromucosal fluids bathing these external surfaces. Thus, intramuscular injection of Ig and subsequent appearance of Ig in serum and tissue fluids would do little to prevent colonization by microorganisms at these mucosal surfaces (i.e., no reduction in the number of episodes of infection). On the other hand, increasing serum and tissue levels of Ig may enhance the body's ability to combat the infectious agent once within the body, thus reducing the duration and severity of illness. Obviously, much further work is needed to corroborate these findings and to further understand possible mechanisms by which interventions may assist athletes in resisting frequent infection during periods of intense training and major competition.

In a recent double-blind randomized trial following 60 athletes from various sports over a three-month period of intensive training, oral administration of thymodulin appeared to prevent the marked decrease in serum IgG_2 concentration observed in the placebo group (Garagiola et al. 1995). Thymodulin is a calf thymus derivative associated with enhanced activation and differentiation of T lymphocytes with subsequent enhancement of other immune functions. It was reasoned that administration of this factor may counteract some of the immunosuppressive effects of prolonged periods of intense training in athletes. IgG_2 concentration declined significantly over the three months, by 30%, in the placebo group, but only marginally and not significantly, by 10%, in the thymodulin group; there were no differences between thymodulin-supplemented and placebo groups in other Ig levels. Although the incidence of infectious illness was 25% lower and the number of days of illness nearly 50% lower in the thymodulin compared with placebo group, these differences were not statistically significant; nor were there significant differences between groups in mean duration of illness (5.7 vs. 7.2 days, thymodulin and placebo groups, respectively). It is unclear at present whether supplementation with purported immunoenhancing factors influences immune function and the appearance of infectious illness in athletes undergoing intense training over prolonged periods. Certainly, large well-controlled studies are needed to further explore this possibility.

The Mucosal Immune System

The body's external surfaces provide a large surface area that can be colonized by pathogenic microorganisms. As described in chapter 2, host defense against these microorganisms involves several mechanisms, including biochemical, physical, and immunological barriers protecting the body's surfaces. The secretory immune system is a major effector of host resistance to colonization of external surfaces of the eyes, nose, upper and lower respiratory tracts, gastrointestinal tract, and genitourinary tract by pathogenic microorganisms. Antibodies and other substances in seromucosal secretions interact with a variety of potentially infectious and noxious agents deposited on these external surfaces of the body (Welliver and Ogra 1988).

The humoral immune response of mucosal surfaces is mediated mainly by antibodies of the IgA class, although IgM antibodies are also found in small amounts in seromucosal secretions. Secretory IgA has a number of functions. Secretory IgA has been shown to inhibit attachment and replication of certain viruses and bacteria, thus preventing their entry into the body; to neutralize toxins and some viruses; and to mediate antibody-dependent cytotoxicity (ADCC), another anti-viral defense mechanism (Tomasi and Plaut 1985). Secretory IgA is important to host defense against certain viruses that are not carried in the blood (i.e., do not have a viremic phase), especially those causing URTI. The level of secretory IgA contained in mucosal fluids correlates, more closely than do serum antibodies, with resistance to certain infections caused by viruses, such as URTI (Cate et al. 1966; Liew et al. 1984; Murphy et al. 1982; Perkins et al. 1969; Smith et al. 1966; Tomasi and Plaut 1985).

Lymphoid tissue in the submucosal areas throughout the body, collectively termed mucosa-associated lymphoid tissue (MALT), contains specialized cells that circulate throughout the body but home specifically to MALT. Dimeric IgA is produced by plasma cells residing in the submucosa (figure 5.3). Secretion of IgA into mucosal fluids requires specialized transport of dimeric IgA across the mucosal epithelial barrier, since tight intercellular junctions preclude diffusion of large molecules between epithelial cells (Solari and Kraehenbuhl 1985; Tomasi and Plaut 1985). Secretory component (SC) produced within mucosal epithelial cells acts as both a receptor and transepithelial transporter for IgA. Membrane-bound SC on the basal-lateral surface of the mucosal epithelium binds dimeric IgA. The SC-IgA complex is then internalized and transported across the epithelial cell via endocytosis and transcytosis, respectively, carrying the IgA to

(Levando et al. 1988) (tables 5.3 and 5.4). Moreover, resting secretory IgA levels have been reported to be lower in certain athletes compared with nonathletes (Tomasi et al. 1982).

In the exercise literature, IgA levels have been expressed in three ways: (a) as absolute concentration (μg IgA · ml^{-1} saliva), (b) as concentration relative to total protein or albumin concentration (μg IgA · mg^{-1} protein or albumin), and (c) as secretion rate or the amount of IgA appearing in saliva per minute (μg IgA · min^{-1}) (Mackinnon et al. 1993a; Mackinnon and Jenkins 1993). IgA concentration relative to total protein or albumin is calculated by dividing the absolute concentration of IgA by the concentration of total protein or albumin in saliva. IgA secretion rate is calculated by multiplying absolute IgA concentration times saliva flow rate; the latter value is estimated by collecting saliva over a fixed time (e.g., 4 minutes) and dividing the volume of saliva obtained by sampling time. The concentration of IgA relative to protein or albumin, and IgA secretion rate, give a better indication of the effects of exercise on mucosal Ig levels than does absolute IgA concentration. Exercise, especially of an intense nature, significantly reduces saliva flow. If IgA output and transport remain constant during exercise, absolute concentration of IgA would be artificially increased simply because of the reduced saliva volume. Measuring only absolute IgA concentration can underestimate the magnitude of the effect of exercise on mucosal Ig levels unless it is known that protein or albumin concentrations have not changed over the sampling time period. Thus, correcting IgA concentration for total protein or albumin concentration effectively adjusts for changes in saliva volume. Alternatively, IgA secretion rate (μg · min^{-1}) gives an indication of the total amount of IgA produced each minute or the amount of IgA available on the mucosal surfaces.

Resting Mucosal Ig in Athletes

The first study to report changes in IgA during intense exercise found that elite male and female Nordic skiers (members of the U.S. National Team) exhibited a 50% lower resting salivary IgA concentration compared with age-matched nonathletes (Tomasi et al. 1982) (table 5.3). It was suggested that the low resting IgA levels may reflect chronic suppression due to daily intense training and possibly to psychological stress leading up to major competition. In the skiers, IgA concentration decreased further immediately after major competition (see below). In contrast, a subsequent study from the same laboratory found no difference in resting salivary IgA concentration between trained cyclists and age-matched sedentary control subjects

Table 5.3 Resting Levels and Training-Induced Changes in Secretory IgA Levels

Subjects	Training	IgA measure	Effect	Reference
Elite M and F skiers	Nordic skiing	$\mu g \cdot mg\ prot^{-1}$	Resting level ↓ in skiers vs. nonathletes	Tomasi et al. 1982
Non-elite M cyclists	Road cycling	$\mu g \cdot mg\ prot^{-1}$	Resting level not different from nonathletes	Mackinnon et al. 1989
M swimmers	4-mo season	$\mu g \cdot ml^{-1}$	↓ pre- and post-ex as season progressed	Tharp and Barnes 1990
F hockey players	Major competition	$\mu g \cdot mg\ prot^{-1}$	↓ 20% pre- and post-ex levels over 4 days	Mackinnon et al. 1991
M untrained	10-wk running training	$mmol \cdot L^{-1}$	No change pre- and post-ex levels	McDowell et al. 1992a
M active	10-wk resistance (weight) training	$\mu g \cdot ml^{-1}$; $\mu g \cdot min^{-1}$	No change pre- and post-ex levels	McDowell et al. 1992b

M active	5 × 60 s supramax cycling, 3 × wk	µg · mg $prot^{-1}$ µg · min^{-1}	No change response to supramaximal exercise	Mackinnon and Jenkins 1993
M kayakers	Normal interval training	µg · mg $prot^{-1}$	↓ 20-35% after intense training week	Mackinnon et al. 1993a
M and F hockey and squash athletes	Normal training	µg · ml^{-1}	↓ 25% during exercise associated with URTI	Mackinnon et al. 1993b
Elite M and F swimmers	6-mo season	µg · mg $prot^{-1}$	↓ in stale vs. well trained; no change across season	Mackinnon and Hooper 1994
M runners	90-min at 75% $\dot{V}O_{2max}$ on 3 consecutive days	µg · min^{-1}	No change pre-ex levels ↓ post-ex on days 2 and 3 compared with day 1	
M and F elite swimmers	7 mo season	µg · ml^{-1} (albumin unchanged)	↓ pre- and post-ex as season progressed	Gleeson et al. 1995

Table 5.4 Summary of Acute Changes in Secretory IgA

Subjects	Exercise	IgA measure	Effect	Reference
M and F skiers	2-3 h cross-country skiing	$\mu g \cdot mg\ prot^{-1}$	↓ 40% after races	Tomasi et al. 1982
M and F subjects	Graded max TM test	$\mu g \cdot ml^{-1}$	↑ M; ↓ F	Schouten et al. 1988
M cyclists	2 h at 70-80% $\dot{V}O_{2max}$	$\mu g \cdot mg\ prot^{-1}$	↓ 65% up to 1 h post-ex	Mackinnon et al. 1989
M runners	31 km run	$\mu g \cdot mg\ prot^{-1}$	↓ up to 18 h post-ex	Muns et al. 1989
M basketball players	Normal training and games	$\mu g \cdot ml^{-1}$	No change	Tharp 1991
M active	15-45 min at 60-85% $\dot{V}O_{2max}$	$\mu g \cdot ml^{-1}$	No change	McDowell et al. 1991
M active	TM test to $\dot{V}O_{2max}$	$mmol \cdot L^{-1}$	↓ 23% after max TM run	McDowell et al. 1992a

M active	Resistance (weight) exercise	$\mu g \cdot ml^{-1}$; $\mu g \cdot min^{-1}$	No acute change absolute conc; ↑ 25% flow rate	McDowell et al. 1992b
M active	30 min at 80% $\dot{V}O_{2max}$ at 6, 19, 34 °C	$\mu g \cdot ml^{-1}$	No change	Housh et al. 1992
M active	5 × 60 s supramax cycling	$\mu g \cdot mg\ prot^{-1}$ $\mu g \cdot min^{-1}$	↓ 20% and 50%, respectively, after exercise	Mackinnon and Jenkins 1993
M kayakers	Normal interval training	$\mu g \cdot min^{-1}$	↓ 27-38% after exercise	Mackinnon et al. 1993a
M and F hockey and squash athletes	Normal training	$\mu g \cdot ml^{-1}$	↓ 25% during exercise associated with URTI	Mackinnon et al. 1993b
M and F runners	90 min at 55, 75% $\dot{V}O_{2max}$	$\mu g \cdot mg\ prot^{-1}$; $\mu g \cdot min^{-1}$	No change ↓ 30-40% secretion	Mackinnon and Hooper 1994
M runners	90 min at 75% $\dot{V}O_{2max}$ on 3 consecutive days		rate on days 2 and 3 but not on day 1	

(Mackinnon et al. 1989). The differences between studies were attributed to the status of the athletes; that is, in the study on skiers, elite national team athletes were assessed just before major competition whereas in the study on cyclists, athletes were not elite and were studied midway through the season and not before competition.

A study of elite male and female Australian swimmers reported significantly lower resting salivary IgA concentrations throughout a six-month season in athletes exhibiting symptoms of staleness, or the overtraining syndrome, compared with those considered "well trained" (i.e., not overtrained) (Mackinnon and Hooper 1994). IgA was measured at five times in the season—early, mid-, and late season, during the taper before, and then within three days after, major competition. The lower resting IgA levels attributed to prolonged periods of intense training may be related to the perceived high incidence of URTI among elite athletes during intensive training and overtraining (Tomasi et al. 1982; Fitzgerald 1991).

Acute Exercise Response of Mucosal Ig

Salivary IgA concentration and secretion rate decrease after various types of exercise including sprint, interval, and continuous endurance activity (table 5.4). In the study on Nordic skiers described above (Tomasi et al. 1982), IgA concentration in saliva decreased 40% after 20-50 km races lasting approximately 2 to 3 hours. The authors noted that decreases in IgA levels after the race may have resulted from a combination of intense exercise, the cold temperature (1°C), and competition stress. To delineate the effects of exercise from possible effects of the cold and/or stress, a similar study was undertaken in the controlled noncompetitive laboratory environment (Mackinnon et al. 1989). Competitive cyclists pedaled for 2 hours at 90% of ventilatory threshold, equivalent to 70% to 80% $\dot{V}O_{2max}$. Salivary IgA concentration declined 60% immediately after exercise, remained significantly lower compared with resting levels for 1 hour, and returned to baseline values by 24 hours after exercise (figure 5.4). IgM concentration followed a similar pattern, but there were no changes in salivary IgG concentration, suggesting a specific effect on secretory Ig.

Muns et al. (1989) noted a 40% decrease in salivary IgA concentration in runners immediately and for up to 18 hours after a 31 km race. IgA concentration has also been reported to decrease by 24% immediately after a progressive treadmill running test to volitional fatigue lasting approximately 22 minutes in moderately active males; IgA levels remained significantly lower, by 17%, for 1 hour after exercise (McDowell et al. 1992a). Absolute IgA concentration ($\mu g \cdot ml^{-1}$) was

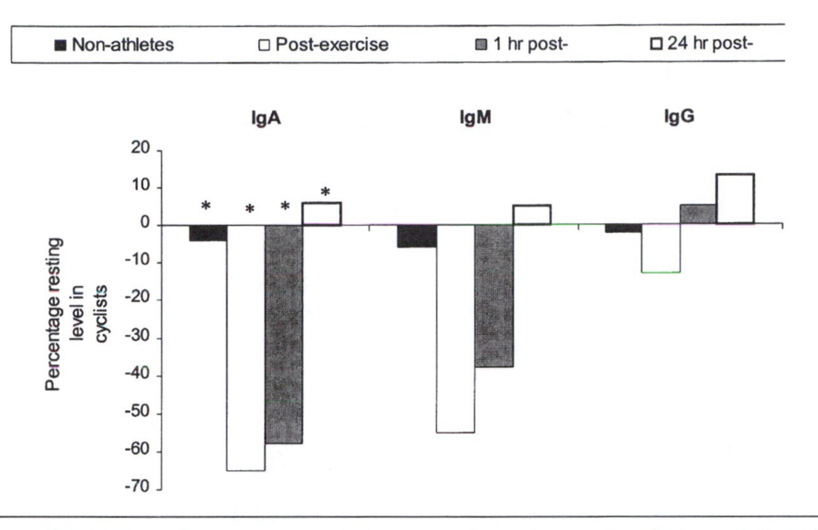

Figure 5.4 Salivary Ig response to intense prolonged exercise. Salivary IgA, IgG, and IgM were measured in cyclists before, immediately after, and 1 and 24 hours after 2 hours of cycling at 90% of ventilatory threshold, and at rest in age-matched nonathletes. Data are expressed as a percentage of cyclists' resting values. See text for details. * = Significantly different from resting values in cyclists.

Data from Mackinnon, L.T., Chick, T.W., van As, A., and Tomasi, T.B. 1989. Decreased secretory immunoglobulins following intense endurance exercise. *Sports Training, Medicine, and Rehabilitation* 1: 209-218.

also shown to decrease after a maximal treadmill test in young healthy adult females (Schouten et al. 1988). In contrast, absolute salivary IgA concentration was reported to increase significantly after a maximal treadmill test in young healthy adult males (Schouten et al. 1988) and after sessions of basketball training and competition in pre- and postpubescent males (Tharp 1991). The latter study did not adjust IgA concentration for changes in salivary protein or albumin. Thus, the increase in absolute IgA concentration may reflect more on the dehydrating effect of exercise, reducing saliva flow and thus increasing the apparent concentration of IgA in saliva, than on any actual stimulation of IgA production and secretion (see above). In the study by Schouten et al. (1988), when data were adjusted for changes in saliva production during exercise, the amount of total IgA did not change after maximal exercise in males and declined only by 10% in females.

Intense interval exercise has also been associated with decreases in salivary IgA levels. For example, in elite male kayakers, IgA secretion rate declined significantly, by 27% to 38%, after normal on-water interval training sessions each lasting up to 30 minutes; exercise was performed at the same time of day on three days over a three-week period

(Mackinnon et al. 1993a). The largest decrease occurred during the most intense exercise session at the end of an especially intense week of training, as rated by the coach. In another study on "recreational" university student athletes (individuals who were active but not training for any particular sport), IgA concentration and secretion rate decreased significantly after five 60-second bouts of supramaximal cycling exercise each separated by 5 minutes passive recovery (i.e., total of 5 minutes supramaximal exercise over a 30-minute period) (Mackinnon and Jenkins 1993). IgA and IgM concentrations relative to total protein declined 21% and 23%, respectively; IgG concentration did not change, indicating a specific effect on mucosal Ig. IgA and IgM secretion rates declined 52% and 65%, respectively, after exercise. Taken together, these data indicate that salivary IgA and IgM production and secretion decline after intense exercise of either short or long duration, whether continuous or interval; this suppressive effect may last for several hours after exercise.

In contrast to the marked decreases in salivary IgA concentration and secretion rate after intense exercise, described above, more moderate exercise does not appear to alter IgA concentration or secretion rate. For example, absolute IgA concentration ($\mu g \cdot ml^{-1}$) did not change significantly in male subjects after 15, 30, or 45 minutes of treadmill running at 60% $\dot{V}O_{2max}$; after 20 minutes running at 50%, 65%, and 80% of $\dot{V}O_{2max}$ (McDowell et al. 1991); or after 30 minutes running at 80% $\dot{V}O_{2max}$ in various ambient temperatures (6, 19, and 34 °C) (Housh et al. 1991). In addition, IgA concentration relative to total protein and IgA secretion rate were reported to not change significantly after 40 minutes of treadmill running at 55% and 75% $\dot{V}O_{2max}$ in male and female joggers or after 90 minutes of treadmill running at the same intensities in male and female well-trained distance runners (marathoners and triathletes) (Mackinnon and Hooper 1994).

Training Effects on Mucosal Ig

Intense exercise training has been associated with decreased salivary IgA levels. In a study on male collegiate swimmers followed over a four-month season, absolute IgA concentration declined progressively with increasing training intensity (Tharp and Barnes 1990). Both resting and postexercise IgA concentrations decreased significantly by about 25% from early to late in the season as training intensity increased from "light" to "heavy." IgA concentration partially recovered during reduced training (taper) leading up to major competition, but still remained significantly lower than early in the season. These data are consistent with the suggestion by Tomasi et al. (1982) that cumulative suppression of IgA production may occur in athletes who train intensely over prolonged periods. A more recent study of male and female elite Australian swimmers also reported progressively decreasing resting and postexercise salivary IgA concentrations over a seven-month season (Gleeson et al. 1995) (figure 5.5). In contrast to the data of Tharp and Barnes described above, however, IgA concentration continued to decrease during the taper period, suggesting that the taper (of unspecified length, but usually two to three weeks) may have been insufficient to overcome the long-term suppression of mucosal immunity resulting from several months of intense training. Alternatively, immunosuppressive effects of psychological stress associated with the upcoming major competition may have obscured or prevented recovery of IgA levels with easing of the physical training stress. It was further noted in this study that salivary IgM was most often detected in samples obtained at the end of the season just before major competition, and it was suggested that increasing IgM concentration may partially compensate for decreasing IgA concentration. It was further suggested that future work should focus on

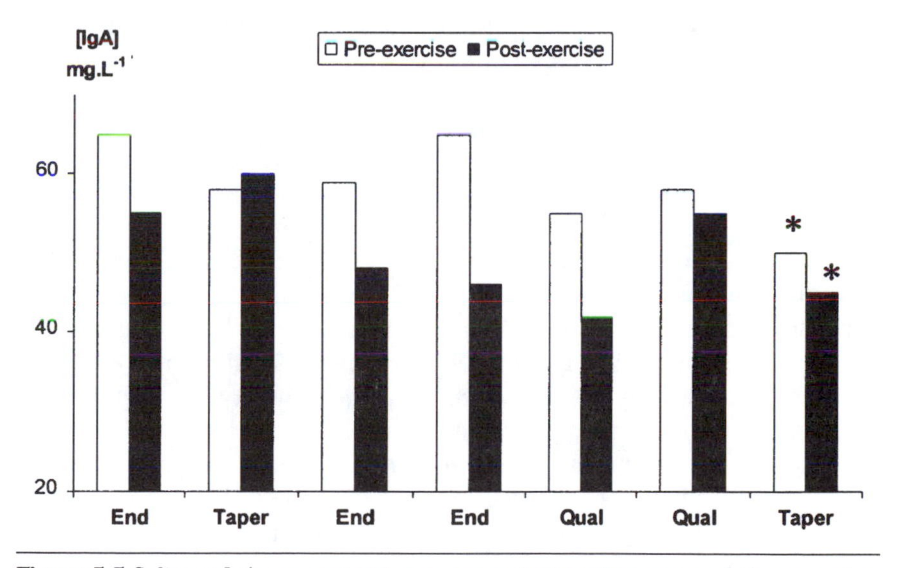

Figure 5.5 Salivary IgA concentration over a swim training season. IgA concentrations were measured before and after swimming training sessions over a seven-month training season in elite swimmers. See text for details. End = endurance training phase; taper = reduced training volume before competition; qual = quality training phase (low volume, high intensity). * = Significantly different from start of season.

Data from Gleeson, M., W.A. McDonald, A.W. Cripps, D.B. Pyne, R.L. Clancy, and P.A. Fricker. 1995. The effect on immunity of long term intensive training in elite swimmers. *Clinical and Experimental Immunology* 102: 210-216.

whether URTI is more frequent in those swimmers who do not exhibit this increase in IgM late in the competitive season.

In another study on elite Australian swimmers, the salivary IgA concentration relative to total protein was reported to be significantly lower, by 18% to 32%, throughout a six-month competition season in athletes who showed symptoms of "overtraining" (e.g., prolonged high fatigue levels, poor performance) compared with their well-trained (i.e., not overtrained) peers (Mackinnon and Hooper 1994). In contrast to data on other groups of swimmers followed over a competitive season as discussed above (Tharp and Barnes 1990; Gleeson et al. 1995), however, IgA levels did not change significantly over the six months in either group. The differences between studies may relate to the particular athletes involved and possibly to their training regimes. In the study by Mackinnon and Hooper (1994), swimmers began the season with high training volumes and intensity, and they may not have experienced the more gradual increase in training load as described in the other studies. The swimmers may also have been under considerable psychological stress from early in the season as they prepared for the upcoming World Titles held in Australia at the end of that season (January, 1991).

A cumulative suppressive effect of daily intense endurance exercise on salivary IgA levels has also been observed over the short term in runners (Mackinnon and Hooper 1994) and in hockey athletes (Mackinnon et al. 1994). In the study on runners (Mackinnon and Hooper 1994), male distance runners ran on a treadmill for 90 minutes at 75% $\dot{V}O_{2max}$ at the same time of day on three consecutive days; saliva was sampled before and after exercise on each day and at rest on day 4. IgA secretion rate declined significantly, by approximately 30% to 40%, after exercise on days 2 and 3, but not on day 1 (figure 5.6). Moreover, postexercise IgA secretion rate was significantly lower, by 27% to 37%, on days 2 and 3 compared with day 1, suggesting a cumulative effect of daily exercise. In the study on elite female hockey athletes (Mackinnon et al. 1991), pre- and postexercise IgA concentration relative to total protein declined progressively, by approximately 20%, during five days of major competition (national tournament).

Lower-volume exercise training does not appear to alter the IgA response to exercise. For example, an eight-week interval training program in active but previously untrained males did not alter the suppression of IgA secretion rate seen after intense interval exercise (Mackinnon and Jenkins 1993). Acute exercise involved five 60-second bouts of supramaximal cycle ergometry with 5 minutes of passive recovery between each bout; training consisted of the same exercise performed three times per week. McDowell et al. (1992b, 1992c) found no long-term

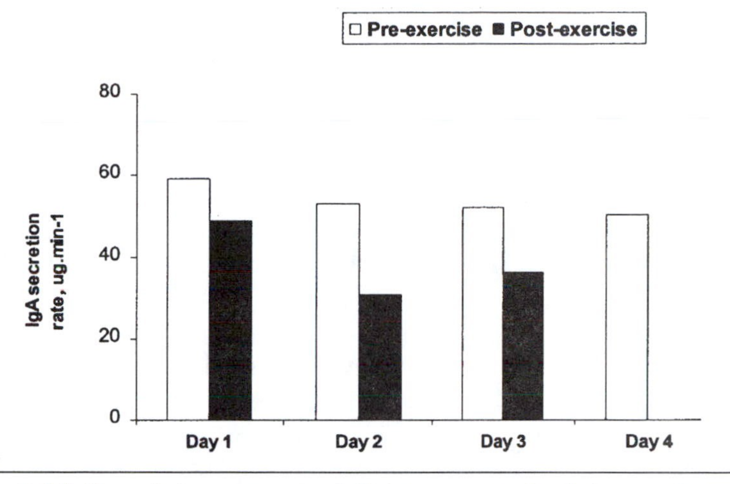

Figure 5.6 Salivary IgA response to daily intense exercise. IgA secretion rate was measured before and after 1.5 hours of treadmill running on three consecutive days and at rest on the fourth day. See text for details.

Data from Mackinnon, L.T., and S. Hooper. 1994. Mucosal (secretory) immune system responses to exercise of varying intensity and during overtraining. *International J Sports Medicine* 15, S179-S183.

effects of moderate running training or resistance (weight) training on pre- or postexercise IgA concentration ($\mu g \cdot ml^{-1}$) and flow rate ($\mu g \cdot min^{-1}$) in young adult males. In the weight training study, although absolute IgA concentration did not change significantly after each exercise session, IgA flow rate increased significantly to the same extent, by approximately 25%, after an identical resistance exercise session both before and after the 10-week training program (McDowell et al. 1992c).

Taken together, these training studies indicate that intense exercise training on a daily or more frequent basis, as performed by elite athletes, is associated with chronic and cumulative suppression of secretory IgA levels; the magnitude of decrease in IgA concentration and output appears to be related to exercise intensity. These data suggest that such suppression of IgA output may be one mechanism related to the high incidence of URTI among endurance athletes and may partially explain the perceived elevation of risk of infection during periods of intense training and major competition.

Are Exercise-Induced Changes in IgA Related to URTI?

As described above, several studies have shown acute decreases in secretory IgA secretion after intense exercise and chronic decreases during periods of intense training. Decreases in serum Ig and some specific antibodies have also been observed during prolonged periods of intense training. However, few studies have addressed the question of whether such changes are directly associated with an increase in the incidence of infectious illness among athletes. In a prospective study of elite hockey and squash athletes (Mackinnon et al. 1993b), exercise-induced decreases in the absolute concentration of salivary IgA ($\mu g \cdot ml^{-1}$) were temporally associated with subsequent appearance of URTI. Saliva samples were obtained in 19 male and female national team hockey athletes before and after normal training sessions at the same time each day during a nine-day training camp, and in 14 national team squash athletes before and after normal training sessions at the same time each day on the same day of the week over a 10-week period. Athletes completed daily logbooks detailing the presence, nature, duration, and severity of symptoms of all illnesses; team physicians also documented the severity, symptoms, duration, and cause of each episode of illness. Over the study periods, all but one episode of illness was due to viral URTI. Of the episodes of viral URTI, six of seven episodes in squash athletes, and all five episodes in hockey athletes were preceded, within two days, by a decrease in IgA concentration, averaging 20% to 25%, during normal training sessions (figure 5.7). In contrast, IgA concentration tended either to increase or to decrease only slightly (<10%) in athletes who did not develop URTI within two days of sampling. Although these data were obtained on a relatively small sample of only two types of elite athletes, they provide suggestive evidence of a link between exercise-induced decreases in salivary IgA and subsequent URTI. Whether such an association occurs in all endurance athletes and is directly related to susceptibility to URTI awaits further, larger studies on athletes from various sports.

Mechanisms Underlying Exercise-Induced Changes in Ig and Antibody

There are several possible mechanisms to explain exercise-induced changes in Ig and antibody. As depicted in figure 5.1, activation and regulation of Ig and antibody secretion are complex, involving several different types of cells (e.g., APCs, T cells, B cells) and messenger molecules (e.g., cytokines). Moreover, systemic and mucosal Ig and antibody secretion are mediated independently and by somewhat different mechanisms. Thus, exercise may elicit different responses in these two systems (e.g., suppression of mucosal yet maintenance of

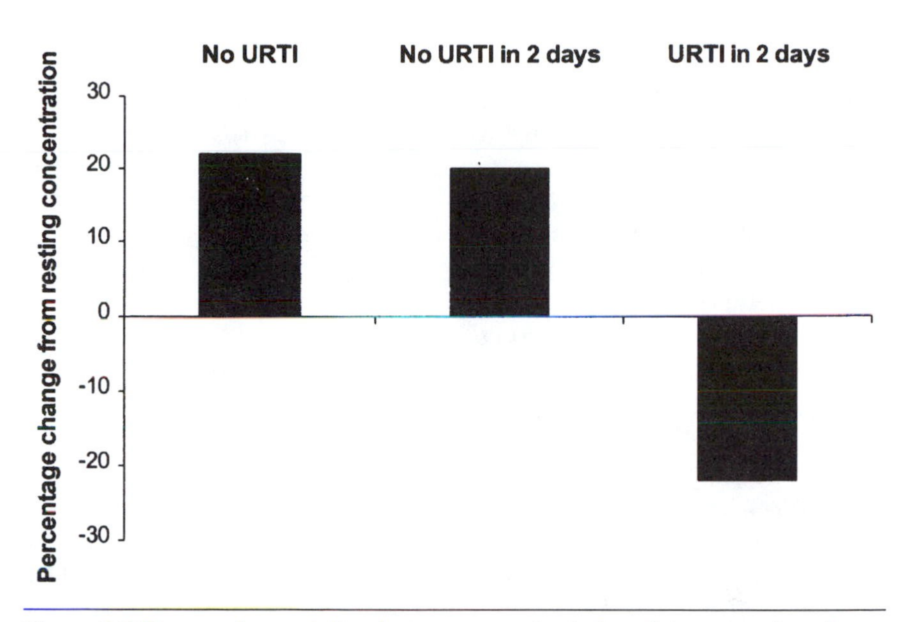

Figure 5.7 Temporal association between exercise-induced decreases in salivary IgA and appearance of URTI. Salivary IgA concentration was measured in elite squash athletes before and after a normal training session on a weekly basis. Viral URTI was documented by the team physician. For data analysis, athletes were first grouped according to whether the athlete developed URTI during the study or not. For athletes who developed URTI, data were further grouped by time of sample, i.e., whether URTI developed within or more than two days after saliva sampling. No URTI = IgA response in athletes who did not develop URTI; no URTI in 2 days = IgA response in athletes who developed URTI, but more than two days after saliva sample; URTI in 2 days = IgA response in athletes who developed URTI within two days after saliva sample. See text for discussion.

From Mackinnon, L.T., E. Ginn, and G.J. Seymour. 1993. Temporal relationship between exercise-induced decreases in salivary IgA and subsequent appearance of upper respiratory tract infection in elite athletes. *Australian Journal of Science and Medicine in Sport* 25: 94-99.

serum Ig levels). As described below, there are many possible steps in the pathway of regulation of Ig and antibody production that may be affected by exercise.

The Systemic Ig Response

As discussed above, the small increases (<20%) in serum concentrations of Ig observed after acute exercise can be accounted for mainly by changes in plasma volume. Those studies that have measured plasma volume and then adjusted Ig concentration accordingly report relatively minor or no acute changes in serum levels after exercise or differences between athletes and nonathletes (Sawka et al. 1989;

Nieman et al. 1992; Rocker et al. 1976; Nehlsen-Cannarella et al. 1991a, 1991b). Any remaining small increase in serum Ig concentration not resulting from hemoconcentration, usually less than 10%, can be explained by diurnal variations (Nieman and Nehlsen-Cannarella 1991) and by exchange from extravascular and lymphatic Ig pools to the circulation (Nieman and Nehlsen-Cannarella 1991; Poortmans 1971). Although there is evidence to suggest diurnal variation in serum Ig levels (Nehlsen-Cannarella et al. 1991a), there are few experimental data to support the suggestion that Ig is exchanged between various compartments. However, lymph flow is stimulated by exercise, and the enhanced lymph flow may increase the influx of various proteins into the circulation (Lindena et al. 1984).

More pronounced changes in both serum Ig and specific antibody concentrations observed after exercise training are likely to reflect a complex interaction of several factors involved in immune regulation, including acute effects after each exercise bout as well as long-term training adaptations. Many cells and soluble factors are involved in the regulation of production of Ig by B cells (see figure 5.1). Such factors may include alterations in the number and relative proportions of lymphoid cells both within the circulation and in lymphoid tissues; release of immunomodulatory factors such as cytokines and/or the number and sensitivity of lymphocyte receptors for these molecules; neuroendocrine changes such as circulating hormone levels and receptor sensitivity; and the effects of psychological stress. These factors need not be mutually exclusive and indeed may be synergistic. Moreover, the acute effects of a single exercise session may overlap or interact with long-term chronic effects of exercise training.

Most studies on athletes show clinically normal numbers and relative proportions of circulating lymphoid cells, which is consistent with the general finding of normal serum Ig levels in most athletes. Thus, it appears unlikely that transitory changes in the number and relative proportion of immune cells that occur acutely with exercise (Maisel et al. 1990; Nieman et al. 1992) influence the long-term Ig response to moderate exercise training in humans. In contrast, as described above, the enhanced secondary response of specific antibody after moderate exercise training in an animal model was attributed to increases in the proportion of B and CD4:CD8 lymphocytes in the spleen of trained animals (Kaufman et al. 1994). At present it is unclear whether such mechanisms also apply to humans, since human work is generally restricted to cells obtained from peripheral blood.

The apparent long-term suppression of certain classes of serum Ig noted in elite athletes undergoing prolonged periods of intense training (Garagiola et al. 1995; Gleeson et al. 1995) is more difficult to explain and most likely results from a combination of psychological and physiological responses to excessive training. Certainly, neuroendocrine factors play an important role in such responses, although it is not clear whether the primary stimulus is psychological or physiological, since training and competing at the elite level involve both physical and psychological stress. As with most aspects of immune function, regulation of Ig synthesis is complex and involves input from the neuroendocrine system (Calabrese et al. 1987). It is generally accepted that there is two-way communication between the neuroendocrine and immune systems (Roitt et al. 1993), with both systems capable of synthesizing and sharing many of the same messenger molecules (e.g., stress hormones, cytokines). The antibody response can be influenced by the neuroendocrine system both directly, via influencing B cell function, and indirectly, via actions on regulatory cells such as T lymphocytes and APCs.

Norepinephrine-releasing adrenergic nerve fibers innervate primary and secondary lymphoid organs, and nerve terminals are in direct contact with lymphoid cells within these organs. Lymphocytes express β-adrenergic receptors; the density of receptors varies between different lymphocyte subsets (e.g., high in NK cells, intermediate in B and CD8 cells and monocytes, low in CD4 cells) (Maisel et al. 1990). Thus, input from the sympathetic nervous system is one obvious mechanism by which exercise may exert both acute and chronic (training) effects on Ig synthesis. Acute exercise increases sympathetic input in a dose-dependent manner, with large increases in norepinephrine and lesser increases in epinephrine release in direct proportion to exercise intensity above a threshold of about 50-65% $\dot{V}O_{2max}$. Moreover, exercise training results in a damping of the catecholamine response to acute exercise at the same intensity (Wilmore and Costill 1994). In addition, lymphocyte β-adrenergic receptor density and sensitivity are also influenced by exercise (Maisel et al. 1990). For example, acute exercise appears to increase β-adrenergic receptor number, due to recruitment into the circulation of lymphocytes expressing this receptor with no change in receptor number per lymphocyte for all subsets except NK cells (Maisel et al. 1990). In contrast, exercise training appears to downregulate β-adrenergic receptor density and sensitivity.

B cells express β-adrenergic receptors, and it is generally accepted that adrenergic innervation influences antibody synthesis. For example, norepinephrine has been shown to enhance specific antibody synthesis in response to antigen by increasing the number of antigen-specific B cells that differentiate into antibody-secreting plasma cells (Sanders and Powell-Oliver 1992). Norepinephrine appears to mediate both suppression and stimulation of antibody synthesis, depending on the dosage and timing of administration in relation to antigen exposure (reviewed in Sanders and Munson 1985). Exposure to norepinephrine early in the antibody response appears to enhance, whereas later administration is associated with suppression of, antibody synthesis.

One possible scenario explaining the neuroendocrine control of the responses of Ig and antibody to both acute exercise and exercise training as described in this book is as follows. Acute exercise causes rapid and marked leukocytosis and lymphocytosis, presumably due to cells released from the spleen; this increase in cell number is related to the elevation in norepinephrine (Maisel et al. 1990). Despite the large increase in B cell number, however, these changes are transitory, and their time course is too short to influence serum Ig levels. However, in regular exercisers, repeated elevation in the number of lymphocytes, especially B cells, coupled with increases in norepinephrine, may lead to enhanced antibody synthesis over time, as shown in the training studies discussed above (Douglass 1974; Eskola et al. 1978; Kaufman et al. 1994; Liu and Wang 1986/87). However, exercise training also leads to downregulation of lymphocyte β-adrenergic receptors as well as a damping of the catecholamine response to exercise, indicating that some other mechanisms may be involved in the chronic effects of exercise training on antibody synthesis (e.g., cytokines). Excessive training (overtraining) has been associated with catecholamine depletion in endurance athletes. For example, in runners after a four-week period of intensified training, lower urinary excretion of norepinephrine, indicating reduced synthesis and release over time, and lower plasma norepinephrine concentration in response to maximal exercise have been observed (Lehmann et al. 1992). Given the importance of norepinephrine to antibody synthesis, catecholamine depletion may result in an impaired ability to mount an antibody response to antigenic challenge during periods of intense training in athletes. The selective reduction of some classes of Ig (e.g., IgG_2) suggests that certain types of lymphocytes are more susceptible to prolonged periods of intense exercise. Although speculative at this point, this hypothesis warrants further scrutiny in experimental studies.

Antibody synthesis is also regulated by a complex interaction between several cytokines that are released by lymphoid and other cells (in particular the interleukins IL-1 through 7) (Janeway and Travers 1996; Roitt et al. 1993). Different aspects of B cell function (e.g.,

B cell activation and proliferation, antibody synthesis) are targeted by specific cytokines. Although the topic has not been extensively studied, some cytokines appear to increase in response to acute exercise. For example, Sprenger et al. (1992) noted increased urinary excretion of IL-1 and IL-6 and the soluble IL-2 receptor after endurance exercise in distance runners (discussed in chapter 6). IL-6, and to a lesser extent IL-5, appear to markedly stimulate synthesis of certain Igs, such as secretory IgA (Beagley et al. 1989). At present, however, there are not enough data to indicate whether or not exercise-induced release of cytokines influences antibody synthesis.

The Secretory Ig Response

The mechanisms responsible for changes in output of secretory Ig (primarily salivary IgA) observed after intense exercise and during periods of intensive training are equally speculative at present. Possible mechanisms include alterations in the oral mucosal surface due to high ventilatory flow during exercise, suppression of IgA secretion and/or SC-mediated transport across the mucosal epithelium, and changes in the homing of IgA-secreting cells to the oral submucosal regions. In elite athletes, suppression of IgA levels noted during prolonged periods of intense training or associated with major competition may also involve psychological factors.

The observation that intense exercise selectively influences mucosal Ig levels (e.g., IgA and IgM but not IgG) suggests a specific effect on local production of secretory Ig. As noted above, IgA and IgM appearing in mucosal fluids are produced locally by plasma cells residing in the submucosa, and secretion requires transport by the polymeric Ig receptor, secretory component (SC). Although at present there are no data on the effects of exercise on SC-mediated transport of IgA, it is possible that this process is influenced by exercise, either by soluble factors or by mechanical or structural changes to the mucosal epithelium secondary to the high ventilation required during intense exercise.

The total amount of IgA available on the oral mucosal surface, as expressed by IgA secretion (or flow) rate ($\mu g \cdot min^{-1}$), decreases markedly after intense exercise (Mackinnon et al. 1993a; Mackinnon and Jenkins 1993). This decrease is due to a combination of decreases in both IgA concentration and saliva flow (Mackinnon and Jenkins 1993). Regulation of saliva flow is complex, involving input from the parasympathetic and sympathetic nervous systems. Increased sympathetic output reduces saliva flow or volume by limiting the water content of saliva and by vasoconstricting arterioles in the salivary gland. Input from higher CNS centers also appears to reduce saliva flow (e.g., during psychological stress). Saliva flow does not appear to be regulated by stress hormones such as the catecholamines. Sympathetic control may also influence migration of IgA-secreting B cells to the oral submucosa via vasoconstriction of blood vessels, effectively reducing the number of cells synthesizing and secreting IgA.

The stress hormone cortisol does not appear to mediate changes in IgA observed during either acute exercise or periods of intense training. For example, Tharp and Barnes (1990) failed to find a significant correlation between salivary IgA and cortisol concentrations over a four-month training season in swimmers. McDowell et al. (1992b) also found no significant correlation between salivary IgA and cortisol concentrations in response to maximal treadmill running before and after a 10-week running training program.

Psychological stress and certain personality attributes also influence mucosal IgA levels (Jemmott et al. 1983; McLelland et al. 1980). For example, in a prospective study on dental students, IgA secretion rate declined as perceived stress levels increased over the academic year (Jemmott et al. 1983). Moreover, IgA secretion rates were lower in students reporting higher compared with lower stress levels. Secretion rate was also lower in those characterized by a personality requiring a great need for power and high activity (similar to the so-called type A personality). In another study, male college students exhibiting this personality type also showed more frequent and severe illness, other signs of psychological stress, and lower IgA concentration (McLelland et al. 1980). These data suggest that psychological stress, the perception of stress, and personality attributes influence both IgA levels and resistance to illness. Intense exercise training is associated with disturbances in mood state, as indicated by the Profile of Mood States (POMS) questionnaire, as well as changes in certain stress hormones such as catecholamines (reviewed in Lehmann et al. 1993). Since susceptibility to illness and secretory IgA levels appear to be influenced by both physical and psychological factors, future work focusing on the relative contribution of each would greatly enhance our understanding of the immune response to intense exercise training in the context of competitive sport.

Summary and Conclusions

Serum Ig does not appear to change significantly with acute exercise or after normal exercise training. Small increases in serum Ig levels, usually less than 20%, have been observed in some studies, but these are generally explained by a combination of diurnal or seasonal effects and hemoconcentration occurring during exercise. Athletes tend to exhibit clinically normal serum Ig levels, although recent reports have shown low serum Ig levels, especially the subclass IgG_2, after prolonged periods (i.e., months) of intense exercise training in elite athletes. It is not currently known whether this apparent suppression of serum Ig levels is related to the purported high incidence of infectious illness among elite athletes. Despite clinically low levels of serum IgG and some IgG subclasses, the ability to mount an antibody response to bacterial antigenic challenge appears to be normal during intense training in elite athletes. Production of serum antibodies to specific antigens may be enhanced by moderate exercise training in animal models, although there are few data on humans.

Salivary IgA has been used as a marker of mucosal immune status. Salivary IgA levels decrease acutely with intense exercise and over prolonged periods of intense exercise training in elite athletes. The magnitude of decrease in IgA during acute exercise appears to be related to exercise intensity. Moderate exercise does not appear to alter salivary IgA levels. The apparent suppression of salivary IgA output resulting from intense exercise may be related to the high incidence of URTI in endurance athletes, especially elite athletes who train intensely on a daily basis. It is also likely that psychological stress associated with training and competing at the elite level contributes to the suppression of mucosal IgA levels in athletes.

It is currently unknown which mechanisms are responsible for the apparent suppression of both serum and mucosal Ig levels associated with intense exercise training in elite athletes. More research is needed to understand the relationship between physical and psychological stressors, the neuroendocrine responses to exercise training, and their combined (or synergistic) influences on immune function among athletes, as well as in the general populace who are encouraged to exercise for health benefits.

Research Findings

- Serum Ig level does not change after acute exercise or chronically after moderate exercise training.
- Clinically low serum Ig concentrations have been reported in some athletes during intense training periods. However, the ability to mount an appropriate antibody response to a specific antigen is not impaired in these athletes.
- Salivary IgA concentration and secretion rate decrease after intense brief and prolonged exercise and during periods of intense training, but not after more moderate exercise.

Possible Applications

- Moderate exercise training has little effect on serum or mucosal Ig and antibody, suggesting that such exercise does not compromise immunity to URTI.
- The decline in mucosal IgA concentration and secretion rate associated with intense exercise training may be related to the high incidence of URTI among endurance athletes.
- Monitoring mucosal IgA and IgM levels may provide a means to identify athletes susceptible to URTI and to assess changes in susceptibility as training intensifies over the season.

Yet to Be Explored

- Is the low serum Ig level observed in some endurance athletes of clinical relevance?
- Is mucosal IgM compensatory for decreasing IgA levels during intense training in athletes?
- Can changes in mucosal IgA predict (and thus be used to help prevent) appearance of URTI in various athletes?
- What is the relative contribution of physical and psychological stress to changes in mucosal immunity in athletes?
- Can training be modified to help maintain mucosal IgA levels and thus help prevent URTI in athletes?
- What mechanisms are responsible for the decrease in serum Ig and mucosal IgA concentrations in high-performance athletes?

Chapter 6

Exercise and Cytokines

© Caroline Woodham

Soluble factors such as cytokines are important initiators and regulators of the immune response, influencing virtually all immune functions. As described in chapter 2, cytokines are growth factors, produced by leukocytes and other cells, that stimulate proliferation and differentiation of various immune cells; cytokines may also influence the activity of other cells and tissues (e.g., fibroblasts, neural cells). That some cytokines, in particular the pro-inflammatory cytokines IL-1, IL-6, and TNFα, are released during exercise is expected for a number of reasons. These cytokines play an important role in inflammation, and exercise may cause damage to and inflammation within skeletal muscle. Some cytokines such as IL-1 have also been implicated in muscle proteolysis and repair following injury/inflammation, which is known to occur as a result of exercise. Moreover, the pro-inflammatory cytokines are also mediators of the acute phase response, and prolonged exercise induces release of some acute phase reactants; it has been suggested that exercise-induced elevation of cytokines may precede or underlie such acute phase response. In addition, some cytokines such as IL-1 and TNFα are released in response to high levels of stress hormones, such as catecholamines and corticosteroids (Kunkel and Remick 1992), the levels of which rise dramatically during physical stress including exercise.

The cytokine response to exercise has received relatively less attention than the response of immune cells, possibly due to methodological considerations relating to measurement of cytokines, such as the extremely low concentrations normally present in healthy individuals and rapid clearance of cytokines from the blood and other body compartments (Shephard et al. 1994a, 1994b). Earlier studies often relied on bioassays to quantify cytokine levels, but the specificity of such assays can be retrospectively questioned, considering the myriad and overlapping actions of many cytokines and other growth factors present in serum or other tissue fluids (Weight et al. 1991). Moreover, biological activity of some cytokines may be masked by endogenous inhibitors, carrier proteins, or soluble receptors. Over the past several years, however, development of sensitive and highly specific commercially available immunoassays has facilitated a recent focus on this topic.

Although recent work suggests that some cytokines, especially the pro-inflammatory cytokines IL-1, IL-6, and TNFα, are released during and after exercise, responses appear to be subtle and have not been consistently observed. Moreover, because of the myriad and diverse actions of cytokines, the biological significance of such changes in response to exercise are often difficult to interpret. For example, because cytokines are rapidly cleared from the blood via receptors located on immune and other cells, blood levels may not necessarily reflect changes in production and activity of a particular cytokine. In addition, cytokines may be produced or may act locally within a small area of tissue (e.g., damaged skeletal muscle), and accessing such sites after exercise is not always feasible, especially in human studies. A summary of the cytokine responses to exercise is provided in tables 6.1 and 6.2. This chapter will first focus on the effects of exercise on each cytokine; a later section will deal with the significance of such changes. While reading this chapter, it is helpful to bear in mind the complexity of biological effects of cytokines on diverse cells and tissues, the difficulties in accurately assessing changes in cytokine levels in various compartments within the body, the minute amounts

Table 6.1 Effects of Exercise on IL-1, IL-2, and IL-6

Cytokine	Variable measured	Main finding	Representative references
IL-1	Plasma	Higher at rest in end athletes	Evans et al. 1986; Sprenger et al. 1992
		No difference at rest in end athletes	Smith et al. 1992
		↑ after mod-int ex	Cannon and Kluger 1983; Evans et al. 1986; Cannon et al. 1986
		No change after mod-int ex	Cannon et al. 1991; Drenth et al. 1995; Smith et al. 1992; Sprenger et al. 1992; Ullum et al. 1994a
	Urinary excretion	↑ 2-fold after end ex	Sprenger et al. 1992
	In skeletal muscle	↑ 2-fold up to 5 d after eccentric ex	Cannon et al. 1989, 1991, 1995; Fielding et al. 1993
	In vitro production	↑ after ≤ 1 h ex ↓ after 4 h ex	Cannon et al. 1991; Haahr et al. 1991; Drenth et al. 1995
IL-2	Plasma	No change after ex	Mackinnon et al. 1988; Nosaka and Clarkson 1996; Sprenger et al. 1992
	IL-2R	↑ in plasma after end ex	Dufaux and Order 1989; Shek et al. 1995; Sprenger et al. 1992
	In vitro production	↓ 20-80% after ex	Baj et al. 1994; Lewicki et al. 1988; Haahr et al. 1991; Tvede et al. 1993
IL-6	Plasma	↑ at rest in end athletes	Sprenger et al. 1992
		↑ 3-fold after end ex	Drenth et al. 1995; Sprenger et al. 1992; Ullum et al. 1994a

(continued)

Table 6.1 (continued)

Cytokine	Variable measured	Main finding	Representative references
IL-6 *(continued)*	Plasma mod ex	No change after	Smith et al. 1992
	Urinary excretion	↑ 3-fold after end ex	Sprenger et al. 1992
	In vitro production	↑ after end ex	Haahr et al. 1990

Abbreviations: mod = moderate; int = intense; end = endurance; ex = exercise, ↑ = increase; ↓ = decrease.

of cytokines present at any given time, and the fact that changes in concentration do not necessarily mean changes in biological activity.

Interleukin-1

IL-1 is a multifunctional cytokine that in the human appears in two forms (IL-1α and IL-1β) with similar function. Although many cell types may produce IL-1, it is mainly secreted by circulating monocytes and tissue macrophages, although B, T, and NK cells may also produce this cytokine (Scales 1992). IL-1 serves many functions, most notably induction of fever, alterations in trace metal levels in the blood, stimulation of the acute phase response, and activation of lymphocytes via stimulating release of IL-2 and expression of its receptor (Janeway and Travers 1996; Scales 1992).

Plasma IL-1

Plasma IL-1 was the first cytokine to be studied in the exercise literature (Cannon and Kluger 1983; Cannon et al. 1986) and has received more attention than the other cytokines (table 6.1). Although early studies suggested that plasma IL-1 concentration increases during and after exercise, later studies using more sensitive and specific assays suggest otherwise. As described below, these discrepancies may be attributed to different assay techniques.

IL-1 activity increased above resting levels in plasma obtained immediately and 3 hours after 1 hour of cycling at 60% $\dot{V}O_{2max}$ in

Table 6.2 Effects of Exercise on IFN and TNFα

Cytokine	Variable measured	Main finding	Representative references
IFN	Plasma	↑ IFNα after end ex No change IFNγ after end ex	Viti et al. 1985 Sprenger et al. 1992
	Urinary excretion	↑ IFNγ after end ex	Sprenger et al. 1992
TNFα	Plasma	At rest in end athletes No change after end ex	Sprenger et al. 1992 Cannon et al. 1991; Sprenger et al. 1992; Ullum et al. 1994a
	Urinary excretion	↑ 2-fold after end ex	Sprenger et al. 1992
	In vitro production	No change after brief ex	Haahr et al. 1991; Rivier et al. 1994
		↓ after long or exhaustive ex	Bagby et al. 1994; Drenth et al. 1996; Kvernmo et al. 1992; Natelson et al. 1996
		↑ after eccentric ex	Cannon et al. 1991

See table 6.1 for abbreviations.

untrained males (Cannon and Kluger 1983). A later study using a more sensitive bioassay reported that IL-1 activity did not increase immediately after exercise, but was elevated by 50% for 3 to 6 hours after, returning to preexercise values by 9 hours (Cannon et al. 1986). The increase in IL-1 activity was inhibited by addition of anti-IL-1 antibody to the bioassay, suggesting a specific effect of IL-1 and not other cytokines.

Resting IL-1 levels have been reported to be higher in distance runners than nonathletes (Evans et al. 1986; Sprenger et al. 1992). Sprenger et al. (1992) noted elevation of resting plasma IL-1 concentration, measured by ELISA, in distance runners; in contrast, IL-1 was not detected in plasma from matched nonathlete control subjects.

The IL-1 response to exercise may also differ between athletes and nonathletes (Evans et al. 1986). For example, in untrained subjects, IL-1 activity increased after 3 hours moderate eccentric exercise (27-42% $\dot{V}O_{2max}$) sufficient to cause muscle soreness; in contrast, IL-1 activity increased slightly or decreased in four runners after the same exercise. For both groups, IL-1 activity returned to preexercise levels by 24 hours after exercise. Because the runners had refrained from training for two days before testing, the authors noted that it was unclear whether the higher resting levels reflected a training effect or simply a long-lasting acute effect from the last exercise session (Evans et al. 1986). In this study, plasma IL-1 level was restored to resting values by 9-24 hours after exercise (Cannon et al. 1986; Evans et al. 1986), arguing against a long-lasting acute effect of prior exercise. However, at the time this paper was published, it was not known how much time was required for restoration of plasma IL-1 activity to resting levels after prolonged intense exercise such as these athletes would have performed. A later study (Sprenger et al. 1992) showed elevation of urinary excretion of IL-1β for at least 24 hours after distance running, suggesting that IL-1 may be released for some time after a single session of exercise. An elevated resting IL-1 level in athletes such as runners who participate in prolonged weight-bearing exercise is consistent with high resting concentrations of other markers of inflammation, such as C-reactive protein (see chapter 4; Dufaux et al. 1984) and creatine kinase (Evans et al. 1986); these changes may reflect chronic inflammation or muscle damage due to intense daily exercise. It was suggested that high resting IL-1 levels in runners may reflect muscle proteolysis and repair resulting from intense regular exercise (Evans et al. 1986). In contrast, Smith et al. (1992) failed to show significant differences in resting plasma IL-1β concentration between untrained subjects, triathletes, elite distance runners, and elite swimmers. IL-1β concentration was nonsignificantly higher in swimmers than in the other groups, which argues against the concept that persistently elevated plasma IL-1 would result from damage induced by repetitive load-bearing exercise such as running.

In contrast to these studies, later work using more sensitive and specific immunoassays has not shown consistent effects of exercise on IL-1 concentration in plasma. For example, using a commercially available enzyme-linked immunoassay (ELISA), Sprenger et al. (1992) could not detect IL-1 in plasma from 22 distance runners before and 2, 3, 7, and 24 hours after a 20 km run, despite evidence for elevation of IL-1 production (see below). Similarly, using an ELISA, Ullum et al. (1994a) reported undetectable levels of plasma IL-1α and IL-1β both before and after 1 hour of cycling at 75% $\dot{V}O_{2max}$ in untrained males. Smith et al. (1992) found no change in plasma IL-1β concentration, measured by immunoradiometric assay, 3 and 6 hours after moderate exercise (1 hour of cycling at 60% $\dot{V}O_{2max}$) in endurance-trained and untrained subjects beyond that which could be attributed to normal diurnal variation. Northoff et al. (1994) also reported no significant change in plasma IL-1 levels up to 48 hours after marathon running. Cannon et al. (1991) reported undetectable plasma IL-1 concentrations, measured by radioimmunosassay, in most untrained subjects immediately, 3, 6, and 24 hours after 45 minutes of downhill running. Although there was a trend for IL-1β level to increase 6 hours after exercise, IL-1β was detectable in plasma from no more than 50% of the subjects. Drenth et al. (1995) found no increase in plasma IL-1β concentration immediately after a 65 km (6 hours) run in endurance athletes. Nosaka and Clarkson (1996) could not detect IL-1α or IL-1β in plasma of subjects before and up to five days after maximal eccentric exercise of the elbow flexors that induced clear indicators of muscle cell damage. Simpson and Hoffman-Goetz (1991) reported unchanged IL-1 concentration immediately after cycle ergometry of various durations and intensities (30-75% $\dot{V}O_{2max}$ and 30-120 minutes) in subjects of mixed fitness levels.

Differences between earlier studies suggesting elevation of plasma IL-1 by exercise and later studies that failed to find such increases, or indeed reported undetectable levels of plasma IL-1, most likely relate to the assay methods. Earlier studies used bioassays in which serum, plasma, or an extract from these fluids is cultured with a cell line that requires IL-1 for growth. Growth of these cells is then quantified and IL-1 activity estimated from this augmentation of growth. Thus, although these assays estimate the biological activity of IL-1, it is not always clear whether this activity relates only to IL-1 or also to other substances contained in plasma/serum. In contrast, recently developed immunoassays use highly specific antibodies to IL-1 to measure the concentration, rather than biological activity, of IL-1. Not all IL-1 appearing in plasma is biologically active, however, since endogenous inhibitors may block its action (Liles and Van Voorhis 1995). Thus, although the concentration of IL-1 may be measured, it is unclear whether all the IL-1 is biologically active. Although most recent studies are consistent in showing a lack of response of plasma IL-1 concentration to exercise, the following discussion presents evidence that IL-1 is released during and for some time after exercise.

Urinary Excretion of IL-1 After Exercise

In contrast to the lack of clear response to exercise of plasma IL-1, one recent study has shown increased urinary excretion of IL-1 (figure 6.1)

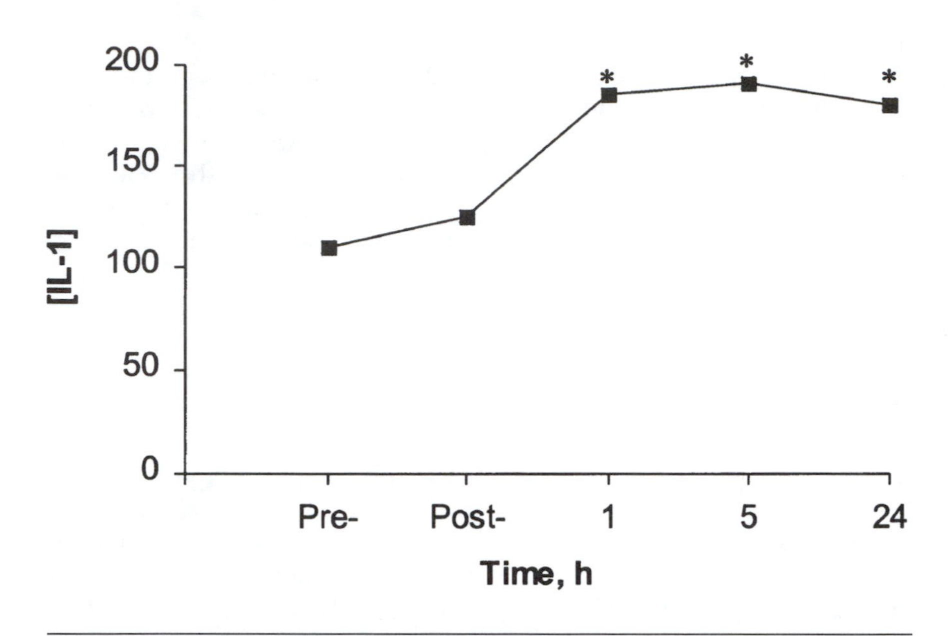

Figure 6.1 Urinary excretion of IL-1β in response to endurance exercise. Urine was collected from distance runners pre-, immediately post-, and 1, 5, and 24 hours after a 20 km run. Units are pg/ml × kg/osmol. * = Significantly higher compared with resting level.

Adapted from Sprenger, H., C. Jacobs, M. Nain, A.M. Gressner, H. Prinz, W. Wesemann, and D. Gemsa. 1992. Enhanced release of cytokines, interleukin-2 receptors, and neopterin after long-distance running. *Clinical Immunology and Immunopathology* 53: 188-195.

(Sprenger et al. 1992). Appearance of IL-1 in urine suggests that IL-1 is secreted but then rapidly cleared from the blood, during and for some time after exercise. Sprenger et al. (1992) were the first to note a disparity between the plasma and urinary IL-1 response to exercise. Twenty-two distance runners (8 women, 14 men) ran 20 km, and urinary IL-1β concentration was measured by ELISA and corrected for changes in urine osmolality. Despite undetectable levels in plasma, urinary IL-1β concentration did not change immediately after exercise, but was nearly double resting levels between 3 and 24 hours after exercise.

IL-1 in Skeletal Muscle After Exercise

IL-1 appears to stimulate skeletal muscle proteolysis, and it has been proposed that IL-1 may play a role in inflammation and repair of skeletal muscle after intense exercise that causes damage to muscle cells (Evans et al. 1986). Exercise with a large eccentric bias, such as downhill running, induces significant damage to muscle cells that is accompanied by an acute phase response and infiltration of neutrophils and mononuclear cells into damaged muscle. Using a model of eccentric exercise (downhill running), Cannon and associates have demonstrated the appearance of IL-1β in skeletal muscle for up to five days after exercise (figure 6.2) (Cannon et al. 1989, 1991, 1995; Fielding et al. 1993). Untrained subjects ran for a total of 45 minutes at 75% $\dot{V}O_{2max}$ at –10% treadmill incline in three intervals (two 15 minutes and one 10 minutes). Muscle biopsies were obtained before and then at various times up to five days after exercise; these biopsies showed clear evidence of damage to skeletal muscle cells after exercise. Immunohistochemical staining of muscle samples showed increases of 135% in IL-1β concentration 45 minutes after and 250% five days

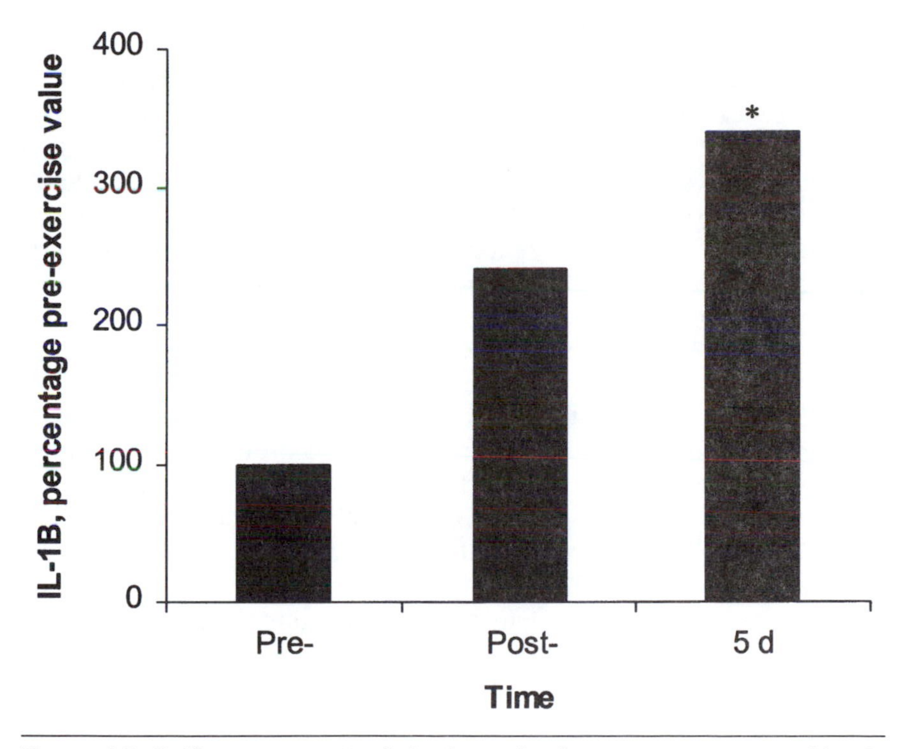

Figure 6.2 IL-1β appearance in skeletal muscle after eccentric exercise. Muscle biopsies were obtained pre-, post-, and five days after 45 minutes downhill running at –16% grade at 70% APMHR. Muscle IL-1β was quantified by immunohistochemical staining with antibodies to IL-1β and image analysis; data are expressed relative to preexercise levels. * = Significantly higher compared with preexercise value.

Adapted from Fielding, R.A., T.J. Manfredi, W. Ding, M.A. Fiatarone, W.J. Evans, and J.G. Cannon. 1993. Acute phase response in exercise III. Neutrophil and IL-1β accumulation in skeletal muscle. *American Journal of Physiology (Regulatory Integrative Comparative Physiology* 34): R166-R172.

after exercise (Fielding et al. 1993). Moreover, the presence of IL-1β was significantly correlated with neutrophil accumulation in the muscle samples. Since IL-1β is chemotaxic for neutrophils, possibly by promoting endothelial adhesion, it was suggested (Fielding et al. 1993) that IL-1β accumulation may mediate neutrophil infiltration of damaged skeletal muscle after eccentric exercise. It was further suggested that preventing accumulation of IL-1β may help limit or prevent skeletal muscle damage after eccentric exercise (see chapter 8; Cannon et al. 1989).

Effects of Exercise on Lymphocyte IL-1 Production

In vitro production of IL-1 also appears to be elevated following exercise (Cannon and Kluger 1983; Cannon et al. 1991; Haahr et al. 1991; Lewicki et al. 1988). To assess cytokine production in vitro, mononuclear cells are isolated from blood samples and incubated for up to three days with various mitogens or other stimulants, similar to the assay for lymphocyte proliferation described in chapters 2 and 3. IL-1 (or other cytokines) secreted into the culture medium (cell supernatant) is then measured by standard assays.

In adherent cells (primarily monocytes) isolated from blood samples taken from trained cyclists immediately after exercise (progressive cycle test to volitional fatigue, about 20 minutes' duration), a 78% increase in IL-1 production, measured by bioassay, was observed (Lewicki et al. 1988). A further increase in IL-1 production was noted in cells obtained 2 hours after exercise. Because a constant number of adherent cells was used in the in vitro cultures, it is likely that the increase in IL-1 concentration reflects enhanced IL-1 production by each cell. IL-1-like activity was also demonstrated in cells obtained immediately after, but not before, 1 hour of exercise at 60% $\dot{V}O_{2max}$ in untrained subjects (Cannon and Kluger 1983).

Haahr et al. (1991) reported significant twofold increases in IL-1α and IL-1β production by mononuclear cells isolated 2 hours after exercise (1 hour of cycling at 75% $\dot{V}O_{2max}$ in untrained subjects); in contrast, IL-1 production was not stimulated above resting levels immediately or 24 hours after exercise. Mononuclear cells were obtained immediately before and after, and 2 and 24 hours after, exercise and incubated with bacterial lipopolysaccharide (LPS); IL-1 concentration was measured in the supernatants. Increases in IL-1α and IL-1β concentration, each measured by ELISA, and IL-1 activity, assessed by bioassay, were of similar magnitude. Since cell cultures used a fixed number of mononuclear cells, the increase in IL-1α and IL-1β production 2 hours after exercise was attributed to an increase in the proportion of monocytes in the circulation at this time.

In contrast, Drenth et al. (1995) reported a significant decrease in LPS-stimulated production of IL-1β in cells immediately after a 65 km (6 hours) run in endurance athletes. To control for possible diurnal variation, blood samples were obtained at rest 24 hours before the projected end of the run (i.e., at the same time of day as the immediately postexercise sample). It was suggested that the postexercise suppression of IL-1β production may reflect selective downregulation of pro-inflammatory cytokine production (discussed below).

Taken together, these studies on the IL-1 response to exercise suggest that IL-1 is released during prolonged weight-bearing exercise such as distance running. Although plasma levels do not change appreciably, if at all, IL-1 has been detected in urine up to 24 hours postexercise and in damaged skeletal muscle up to five days postexercise. Monocyte production of IL-1 appears to increase after moderate endurance exercise, but may be suppressed by very vigorous or prolonged exercise (several hours). Whether changes in IL-1 mediate any of the immune cell changes observed during and after exercise (e.g., leukocyte trafficking via effects on adhesion molecules, migration of cells to damaged tissue) has not been elucidated and is discussed further below.

Interleukin-2

IL-2 exerts regulatory effects on most cells in the body, especially immune cells. Actions of IL-2 include stimulation of proliferation and differentiation of B and T cells, enhancement of lymphocyte cytotoxicity by both NK and cytotoxic T cells, activation of monocytes/macrophages, and release of other cytokines such as TNFα and IFNγ (Anderson 1992; Janeway and Travers 1996).

IL-2 in Plasma

IL-2 concentration in plasma may exhibit a complex pattern of change in response to exercise (table 6.1) (Espersen et al. 1990; Lewicki et al. 1988). For example, in trained runners, plasma IL-2 concentration, measured by radioimmunoassay, decreased 50% immediately after a 5 km race, returned to baseline levels by 2 hours after exercise, and then increased by 50% 24 hours after the race (Espersen et al. 1990). It was suggested that the transitory decrease in IL-2 concentration immediately after exercise may reflect a concomitant increase in the number of activated lymphocytes expressing the IL-2 receptor; that is, more cells with more receptors would quickly remove IL-2 from the circulation. This concept is supported by two observations: first, the large increase in CD16+ (NK) cell number immediately after exercise in this study, since NK cells exhibit the highest density of IL-2 receptor of all lymphocytes (Shek et al. 1995); and second, a rapid increase in IL-2 receptor appearance in urine 1 hour after distance running (Sprenger et al. 1992; see below). In the study by Espersen et al. (1990), the delayed increase in IL-2 level observed 24 hours postexercise could not be accounted for by changes in lymphocyte number or distribution and may reflect post-race inflammation, which may stimulate release of cytokines such as IL-1 and IL-2. In contrast to the data from Espersen et al. (1990), IL-2, measured by ELISA, was not detectable in plasma at any time before or up to 24 hours after 20 km running in distance runners (Sprenger et al. 1992), up to 48 hours after 2-hour cycling at 70-75% $\dot{V}O_{2max}$ in trained cyclists (Mackinnon et al. 1988), or up to five days after maximal eccentric contractions of the elbow extensors (Nosaka and Clarkson 1996).

Effects of Exercise on Lymphocyte IL-2 Production

IL-2 production can also be measured in vitro, by similar methods to those described for IL-1 above. Changes in IL-2 concentration in plasma or IL-2 receptor number (discussed above and below, respectively) may be reflected in changes in IL-2 production. Although not extensively studied, IL-2 production appears to be suppressed after endurance exercise. While acute changes in CD4 T cell number and proportion may account for some of the observed decrease in in vitro IL-2 production after exercise, they cannot fully explain changes observed in all conditions, as discussed further below.

In trained cyclists, in vitro IL-2 production was reduced 27% in mononuclear cells obtained immediately after a 20-minute incremental cycling test to volitional fatigue (Lewicki et al. 1988); IL-2 production was 40% lower 2 hours after exercise compared with baseline levels. The decrease in IL-2 production may have been related to a 30% decrease in CD4 cell number, with a corresponding decline in the CD4:CD8 ratio. This decline in IL-2 production after brief maximal exercise may indicate a reduced ability of lymphocytes to respond to immunological challenge and may be causally related to the compromised lymphocyte proliferative responses observed after intense exercise, as described in chapter 3.

Haahr et al. (1991) reported no change in IL-2 production by PHA-stimulated cells obtained immediately after 1 hour of cycling at 75%

$\dot{V}O_{2max}$ in untrained subjects. There was, however, a significant 20% decrease in IL-2 production measured by ELISA, but not when measured by bioassay, in cells obtained 2 hours postexercise; IL-2 production was restored by 24 hours after exercise. Similarly, Tvede et al. (1993) reported a decline in PHA-stimulated IL-2 production in mononuclear cells obtained during and 2 hours after more intense, but not moderate, exercise (1 hour of cycling at 75% vs. 50% $\dot{V}O_{2max}$, respectively) in untrained males. The decline in IL-2 production was partially attributed to a decrease in $CD4^{+}$ cell number, suggesting fewer IL-2-producing T cells in samples obtained after exercise at the higher intensity. The postexercise suppression of IL-2 production was also partially attributed to prostaglandins, since in vitro addition of the prostaglandin inhibitor, indomethacin, restored IL-2 production in cells obtained after exercise at 75% $\dot{V}O_{2max}$.

Baj et al. (1994) found significantly lower, by 50-80%, PHA-stimulated in vitro production of IL-2 by lymphocytes isolated at rest from experienced male cyclists compared with matched nonathletes. Moreover, in these cyclists, in vitro IL-2 production was reduced by more than 50% after, compared with before, a six-month period of intense training. Training consisted of more than 500 km per week, more than triple the preseason training volume. Importantly, in vitro IL-2 production remained unchanged in cells obtained at the same times from nonathletes, indicating an effect of training and not seasonal variation. Although there were slight differences in total ($CD3^{+}$) and CD4 T cell numbers between cyclists and nonathletes, and within athletes over the six months, the lower IL-2 production in athletes could not be completely attributed to differences in cell counts: cell counts differed by no more than 10%, yet differences in IL-2 production were far greater, about 50-80%.

IL-2 Receptor

A cross-sectional comparison reported significantly higher expression of the high-affinity IL-2 receptor β (IL-2Rβ) subunit on lymphocytes in endurance-trained runners compared with nonathletes (Rhind et al. 1994). Resting lymphocytes do not generally express the high-affinity receptor, which is upregulated during activation of lymphocytes. In athletes compared with nonathletes, the proportion of NK cells expressing the IL-2Rβ subunit was 10% and 26% higher for $CD16^{+}$ and $CD56^{+}$ cells, respectively, and there was a strong correlation between aerobic power and IL-2Rβ expression. Moreover, as discussed in the next chapter, runners exhibited significantly higher circulating NK cell number. Thus, the greater expression of the IL-2Rβ subunit in these runners could be explained by the higher number of NK cells in the circulation, with each cell expressing more of the higher-affinity receptor. These data suggest that endurance training may enhance the ability of NK cells to respond to IL-2.

In contrast to the equivocal data on the response to exercise of plasma IL-2 concentration and IL-2 production, there is clear, although not abundant, evidence of exercise-induced changes in IL-2R expression (Dufaux and Order 1989a; Sprenger et al. 1992; Shek et al. 1995). Although plasma concentration of soluble IL-2R did not change immediately and for several hours after a 2.5-hour run in distance runners, levels were significantly higher 24-48 hours postexercise (Dufaux and Order 1989b). Urinary output of soluble IL-2R (adjusted for changes in urine osmolality) doubled immediately after a 20 km run in 22 distance runners (Sprenger et al. 1992); normal levels were restored by 1 hour and remained low for 24 hours postexercise, indicating a brief, transitory increase in IL-2R excretion. Increased IL-2R expression occurs in activated T and NK cells. The increased excretion immediately after exercise may reflect activation of cells, increased recruitment of NK cells into the circulation, enhanced clearance of IL-2, or some combination of these factors. Since soluble IL-2R in the blood can still bind IL-2 (Anderson 1992), increased release of the receptor may explain the inability to detect IL-2, or declining plasma IL-2 levels after exercise (discussed above).

Interleukin-6

IL-6 may be produced by many types of cells, primarily by stimulated monocytes/macrophages and activated T and B cells. IL-6 exerts a wide range of activities, including stimulation of growth and differentiation of lymphocytes and antibody production (Cox and Gauldie 1992; Janeway and Travers 1996). Released in response to elevated IL-1 and TNFα levels, IL-6 is the major cytokine responsible for release of acute phase proteins from liver (Cox and Gauldie 1992).

Plasma IL-6

Of the various cytokines measured in plasma, IL-6 appears to be the only one for which an increase has been consistently noted after prolonged exercise (table 6.1) (Drenth et al. 1995; Sprenger et al. 1992). Sprenger et al. (1992) reported higher resting plasma IL-6 concentration, measured by ELISA, in 22 distance runners compared with matched nonathletes who did not exhibit detectable levels of IL-6 at rest. A sustained elevation of plasma IL-6 was noted after a 20 km run in the runners; values peaked immediately after exercise, and were markedly elevated for at least 5 hours after exercise (figure 6.3). A significant 80% increase in IL-6, measured by ELISA, was also reported near the end of 1 hour of cycling at 75% $\dot{V}O_{2max}$ in moderately fit males, with return of IL-6 to baseline levels by 2 hours postexercise (Ullum et al. 1994a). Drenth et al. (1995) observed a significant threefold increase in plasma IL-6 immediately after a 65 km (6 hours) run in 19 endurance athletes.

In contrast, no change in plasma IL-6 concentration, measured by immunoradiometric assay, was observed immediately, 3, 6, and 24 hours after 1 hour of cycling at 60% $\dot{V}O_{2max}$ in nonathletes (Smith et al. 1992); however, only 50% of subjects had detectable IL-6 levels in plasma. Similarly, IL-6 was not detected in plasma before or up to five days after maximal eccentric exercise of the elbow flexors despite evidence of muscle cell damage. These data suggest that prolonged (e.g., > 1 hour), intense (e.g., > 60% $\dot{V}O_{2max}$) exercise using large muscle groups may be needed to increase plasma IL-6 concentration. Cannon et al. (1991) noted methodological problems relating to consistent

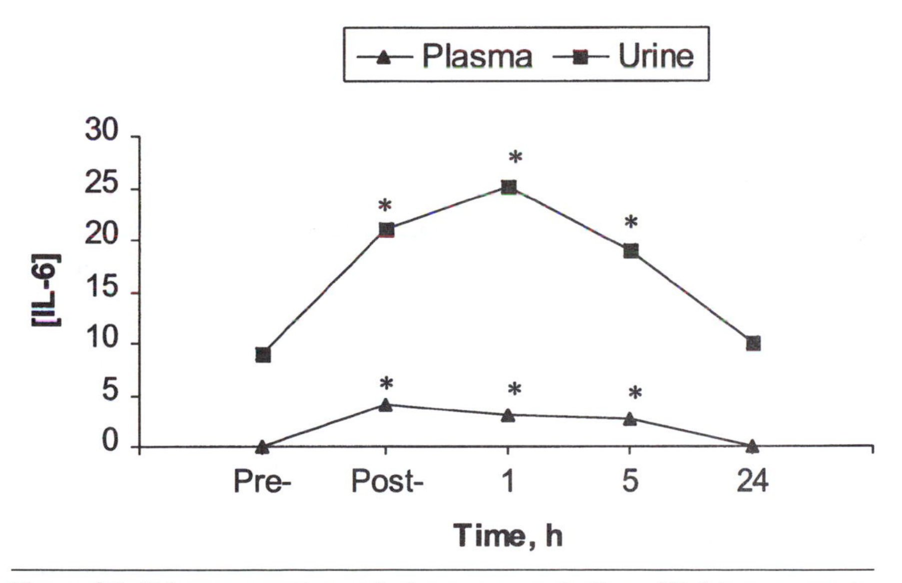

Figure 6.3 Urinary excretion and plasma concentration of IL-6 in response to endurance exercise. Urine was collected from distance runners pre-, immediately post-, and 1, 5, and 24 hours after a 20 km run. Units are pg/ml × kg/osmol for urine and pg/ml for plasma IL-6. * = Significantly higher compared with resting level.

Adapted from Sprenger, H., C. Jacobs, M. Nain, A.M. Gressner, H. Prinz, W. Wesemann, and D. Gemsa. 1992. Enhanced release of cytokines, interleukin-2 receptors, and neopterin after long-distance running. *Clinical Immunology and Immunopathology* 53: 188-195.

measurement of IL-6, which may account for differences between studies; these methodological problems may have since been resolved with more recently developed immunoassays for measurement of IL-6.

Urinary Excretion of IL-6 After Exercise

As with plasma IL-6, marked urinary excretion of IL-6 occurs with exercise (figure 6.3). Sprenger et al. (1992) reported a more than twofold increase in IL-6 concentration in urine (measured by ELISA and corrected for osmolality changes) immediately after a 20 km run, peaking at nearly three times resting level by 1 hour postexercise; IL-6 excretion was still elevated 5 hours postexercise and gradually declined over the next 24 h. At each time point, urinary IL-6 levels were markedly (i.e., 10 times) higher than in plasma. The large increase in urinary output and relatively smaller increase in plasma level of IL-6 strongly suggest that IL-6 is released during and after prolonged exercise and that the rate of release may exceed that of removal for several hours after exercise.

Effects of Exercise on IL-6 Production

Exercise may also stimulate IL-6 production, as assessed in in vitro studies (Haahr et al. 1991). IL-6 production by LPS-stimulated mononuclear cells increased about 65% in cells obtained 2 hours after exercise (1 hour at 75% $\dot{V}O_{2max}$ in untrained subjects) (Haahr et al. 1991); IL-6 production was unchanged immediately and 24 hours after exercise, suggesting a transitory effect. The authors attributed this increase in IL-6 production 2 hours postexercise to an increase in the number of monocytes in the circulation, resulting in more IL-6-producing cells in the cultures. A small but significant increase in in vitro IL-6 production by monocytes was also observed immediately after a cycling test to $\dot{V}O_{2max}$ in young and masters cyclists (ages 21-32 and 56-70 years, respectively) (Rivier et al. 1994); at all time points, IL-6 production was higher in the younger compared with masters cyclists.

Although IL-6 may be released during exercise, as evidenced by increases in plasma and urinary IL-6 concentration and IL-6 production by mononuclear cells, this increase does not appear to occur via de novo synthesis of the cytokine during exercise. For example, Ullum et al. (1994a) reported no change in pre-mRNA for IL-6 in mononuclear cells isolated during and for up to 4 hours after exercise (1 hour of cycling at 75% $\dot{V}O_{2max}$), despite a significant increase in plasma IL-6 level during exercise. It was concluded that other processes, such as release by cells other than mononuclear cells (e.g., endothelial cells, fibroblasts), may be responsible for the observed release of IL-6 during and after exercise.

Interferon

IFNs are produced by T and NK cells and other leukocytes in response to viral infection. IFNs exert antiviral activity by inhibiting viral protein translation and degrading viral mRNA (Janeway and Travers 1996). IFNs also stimulate macrophage antigen presentation and cytotoxicity, as well as NK cytotoxic activity against virally infected cells.

Plasma IFN

There are only a few studies that have investigated the plasma IFN response to exercise (table 6.2). IFN activity, as measured in a bioassay using virally infected cells and compared with IFN standards, more than doubled immediately and 1 hour after 1 hour of cycling at 70% $\dot{V}O_{2max}$, but returned to preexercise values by 2 hours postexercise (Viti et al. 1985). This increase in IFN activity was attributed to IFNα, since it was heat labile and inhibited by anti-IFNα neutralizing antibodies; IFNβ and IFNγ levels did not appear to increase during exercise. Sprenger et al. (1992) reported no detectable IFNγ before or up to 24 hours after 20 km running in distance runners.

Urinary Excretion of IFN After Exercise

As with other cytokines, there is evidence for release of IFN during exercise despite no detectable levels in plasma. Sprenger et al. (1992) reported a marked, more than fivefold, increase in urinary excretion of IFNγ levels, measured by ELISA and adjusted for urinary osmolality changes, immediately after a 20 km run in distance runners (figure 6.4). IFNγ concentration was nonsignificantly elevated 1 hour postexercise, but returned to preexercise values by 5 hours after exercise.

Effects of Exercise on IFN Production

Few studies have looked at the effects of exercise on IFN production. Haahr et al. (1991) reported no change in IFNγ production, measured by immunoradiometric assay, in mononuclear cells obtained during, 2 hours after, and 24 hours after exercise (1 hour of cycling at 75% $\dot{V}O_{2max}$).

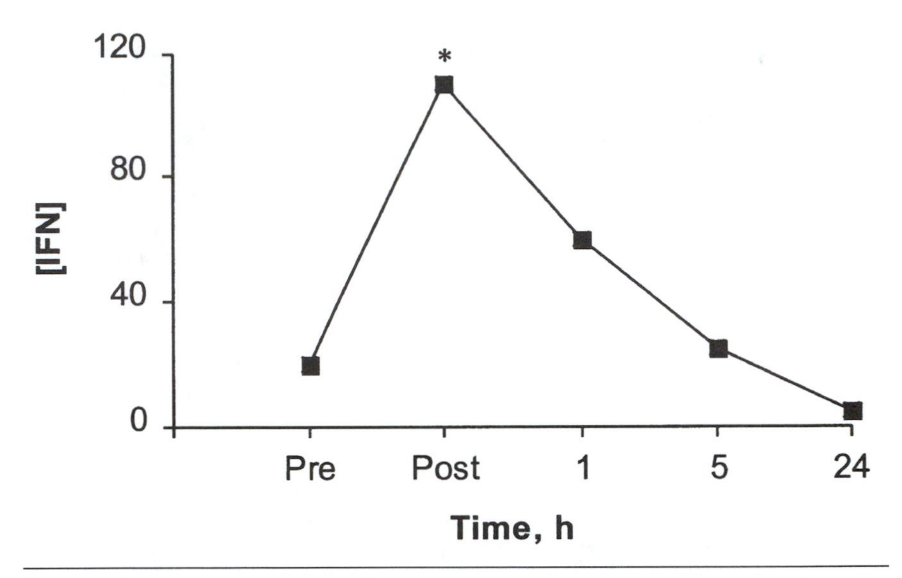

Figure 6.4 Urinary excretion of IFNγ in response to endurance exercise. Urine was collected from distance runners pre-, immediately post-, and 1, 5, and 24 hours after a 20 km run. Units are pg/ml × kg/osmol. * = Significantly higher compared with resting level.

Adapted from Sprenger, H., C. Jacobs, M. Nain, A.M. Gressner, H. Prinz, W. Wesemann, and D. Gemsa. 1992. Enhanced release of cytokines, interleukin-2 receptors, and neopterin after long-distance running. *Clinical Immunology and Immunopathology* 53: 188-195.

Neopterin

Neopterin is a metabolite of IFNγ. Since IFNγ is degraded rapidly after release, some studies have used levels of the more stable neopterin as an indirect indicator of IFNγ release during and after exercise (Smith et al. 1992; Sprenger et al. 1992). Smith et al. (1992) reported no change in plasma neopterin concentration, measured by radioimmunoassay, immediately and 3 hours after exercise (1 hour of cycling at 60% $\dot{V}O_{2max}$ in endurance athletes and nonathletes); a nonsignificant increase was observed 6 hours postexercise. Sprenger et al. (1992) could not detect neopterin in the plasma for up to 24 hours after a 20 km run in trained runners, but did note a significant 30% elevation of urinary excretion of neopterin beginning 1 hour and continuing for at least 24 hours after exercise. Dufaux and Order (1989b) reported a small (about 15%) but significant elevation of plasma neopterin concentration 1-24 hours after a 2.5-hour run in moderately trained runners; neopterin levels were not elevated immediately after the run, and returned to baseline by 48 hours postexercise. These data are consistent with observations that IFNγ is released after prolonged load-bearing exercise such as distance running; whether such changes occur after shorter or less strenuous exercise is unclear at present.

Tumor Necrosis Factor α

TNFα released from monocytes/macrophages, NK cells, and other cells exerts a wide range of effects on many cell types including localized inflammation, expression of cell surface receptors on lymphocytes, stimulation of cytotoxicity, lymphocyte proliferation, release of acute phase reactants, protein catabolism, and increase in core temperature (Janeway and Travers 1996; Liles and Van Voorhis 1995; Tsuji and Torti 1992).

Plasma TNFα

Elevated resting TNFα concentration, measured by ELISA, was reported in 22 distance runners compared with nonathlete control subjects who did not exhibit measurable levels of this cytokine (table 6.2) (Sprenger et al. 1992). A significant correlation (R = 0.83) was reported between plasma TNFα and IL-1β concentration in endurance-trained individuals and nonathlete controls (Smith et al. 1992), suggesting that IL-1 and TNFα may be released in response to the same types of physical stress, such as distance running.

As with other cytokines, plasma levels of TNFα may not change during exercise, despite evidence for release of this cytokine, as discussed below. Plasma TNFα was either not detectable or remained unchanged at various points after exercise, including up to 6 hours after 1-hour cycling at 75% $\dot{V}O_{2max}$ in moderately trained men (measured by ELISA; Ullum et al. 1994a); up to 24 hours after 1 hour of cycling at 60% $\dot{V}O_{2max}$ in both endurance-trained and untrained subjects (immunoradiometric assay; Smith et al. 1992); up to 24 hours after a 20 km (ELISA; Sprenger et al. 1992); after 65 km (6 hours) running in distance runners (RIA; Drenth et al. 1995); up to five days after maximal eccentric exercise of the elbow flexors (ELISA; Nosaka and Clarkson 1996); and up to 24 hours after 45 minutes downhill treadmill running in untrained subjects (ELISA; Cannon et al. 1991). In contrast, a significant 40% increase in plasma TNFα concentration, measured by radioimmunoassay, was reported 2 hours but not immediately after a 5 km run in distance runners (Espersen et al. 1991), although values were still within the clinically normal range; and a 60% increase was observed 1 hour after a 2.5-hour run in endurance athletes (ELISA; Dufaux and Order 1989b). Thus, there are inconsistent data on the plasma TNFα response to endurance exercise. Most studies show no changes in plasma concentration. When observed, increases in plasma TNFα levels appear to peak 1 to 2 hours postexercise, but values may still appear to be within clinically normal ranges, which may explain the general lack of response reported in most studies. This apparent lack of effect on plasma TNFα is consistent with unchanged or decreased TNFα production, as described below.

Urinary Excretion of TNFα After Exercise

There is only one published report of urinary excretion of TNFα after exercise. TNFα concentration in urine, measured by ELISA and corrected for osmolality, increased by about twofold immediately and 1 hour after a 20 km run in distance runners (Sprenger et al. 1992); resting level was restored by 5 hours postexercise (figure 6.5). These data suggest that if TNFα is released during exercise, this occurs only during the first few hours afterward. These data are also consistent with a general lack of response of plasma levels and in vitro production of TNFα to most types of exercise, except for very intense or prolonged exercise.

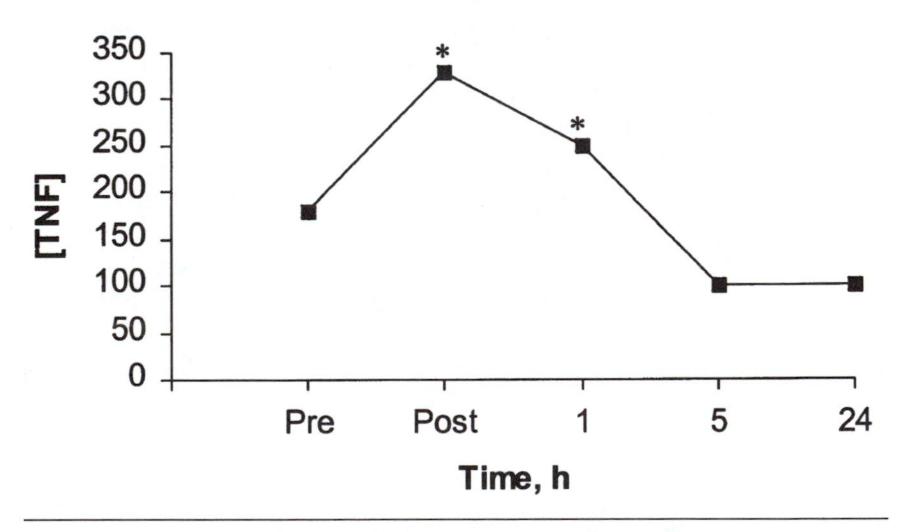

Figure 6.5 Urinary excretion of TNFα in response to endurance exercise. Urine was collected from distance runners pre-, immediately post-, and 1, 5, and 24 hours after a 20 km run. Units are pg/ml × kg/osmol. * = Significantly higher compared with resting level.

Adapted from Sprenger, H., C. Jacobs, M. Nain, A.M. Gressner, H. Prinz, W. Wesemann, and D. Gemsa. 1992. Enhanced release of cytokines, interleukin-2 receptors, and neopterin after long-distance running. *Clinical Immunology and Immunopathology* 53: 188-195.

Effects of Exercise on TNFα Production

TNFα production by mononuclear cells was reported to be unchanged after 1 hour of cycling at 75% $\dot{V}O_{2max}$ (Haahr et al. 1991). Cells were isolated from blood obtained during the last minute of exercise, and then 2 and 24 hours postexercise, and cultured with bacterial LPS; TNFα released into the supernatant was measured by ELISA. Using a similar protocol, a subsequent study from the same laboratory showed no change in pre-mRNA for TNFα up to 4 hours after exercise, suggesting no short-term stimulation of synthesis of TNFα by exercise (Ullum et al. 1994a). Spontaneous (nonstimulated) release of TNFα by blood monocytes did not change significantly immediately after a cycle ergometer test to $\dot{V}O_{2max}$ in young adult and masters cyclists (Rivier et al. 1994).

In contrast, TNFα production by mononuclear cells was reported to decline significantly after exercise in 25 athletes of varying competitive level (Kvernmo et al. 1992). Cells were isolated from blood obtained before, immediately after, and up to 24 hours after exercise and stimulated in vitro for 2 hours with bacterial LPS. Athletes were classified by extent of training: highly trained cross-country skiers or runners who trained at least daily; moderately trained who exercised 3-7 times per week; and less trained who exercised 4-5 times per month. Resting in vitro IL-6 production was inversely related to training volume, with the lowest level observed in the most highly trained, intermediate production in the moderately trained, and highest level in the least active group; IL-6 production in the most highly trained was about 67% lower than in the least active. LPS-stimulated TNFα production declined immediately after exercise in all groups (1.5-2.0 hours cross-country skiing or running in two most active groups, unspecified in least active group). IL-6 production was restored by 2 hours postexercise in the moderately trained group and by 6 hours in the highly trained athletes, suggesting a transient suppressive effect.

TNFα production by LPS-stimulated mononuclear cells increased significantly by about 60% in cells obtained within the first 24 hours after 45 minutes downhill running that induced clear indicators of muscle cell damage, such as IL-1β appearance within damaged skeletal muscle (discussed above) in untrained subjects (Cannon et al. 1991). Taken together, data from these studies suggest that the effects of exercise on mononuclear cell production of TNFα may depend on exercise mode, training volume, and possibly the extent of tissue damage.

A recent study using an experimental animal model suggests that exhaustive exercise may reduce the ability of immune cells to produce TNFα in response to immunological challenge (Bagby et al. 1994).

Male rats were first familiarized with treadmill running by brief (10 minutes) daily exposure to the treadmill. Animals were then exercised to exhaustion and bacterial LPS was injected immediately, 30 minutes, and 2, 6, and 24 hours after exercise; matched control rats were exposed to identical procedures except for sitting on top of the treadmill rather than running. LPS induced a large increase (more than 10-fold) in plasma TNFα in nonexercised animals (figure 6.6). In contrast, the TNFα was significantly attenuated, by 83%, in LPS-challenged exercised animals. Attenuation of the TNFα response occurred in rats injected with LPS up to 6 but not 24 hours after exercise, suggesting short-term suppression of TNFα production by exhaustive exercise. It was suggested that such suppression of cytokine production may increase susceptibility to infection following exhaustive exercise; alternatively, it was also postulated that such response may be beneficial by limiting inflammatory responses to intense exercise (discussed further below).

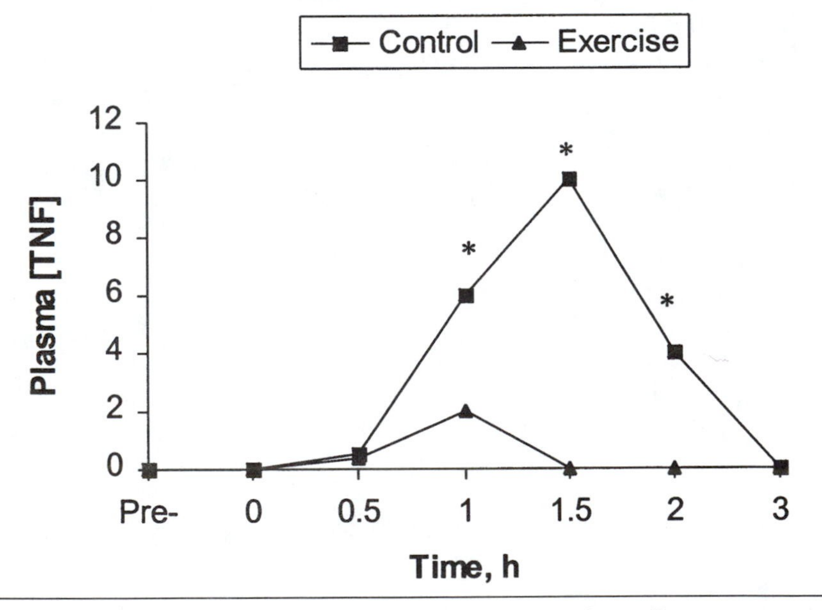

Figure 6.6 Plasma in vivo TNFα production in exercised rats. Rats were exercised to near exhaustion before intravenous infusion of LPS; control rats received the LPS challenge but were not exercised. Blood was collected before exercise (pre-), immediately after exercise (0), and then at various times up to 3 hours postexercise. * = Significantly higher in vivo TNFα production in control compared with exercised animals.

Adapted from Bagby, G.J., D.E. Sawaya, L.D. Crouch, and R.E. Shepherd. 1994. Prior exercise suppresses the plasma tumor necrosis factor response to bacterial lipopolysaccharide. *Journal of Applied Physiology* 77: 1542-1547.

This latter suggestion is supported by a study showing an association between low TNFα production and enhanced recovery after infection in exercised animals (Chao et al. 1992). Mice were inoculated with a relatively avirulent strain of the intracellular parasite, *Toxoplasma gondii,* and trained by daily swimming for 45 minutes on the day of inoculation and then for 25 days after. Exercise had no effect on survival (none of the mice died). Serum TNFα concentration, measured by bioassay, increased sixfold by day 11 in infected, nonexercised mice. In exercised infected mice, however, serum TNFα concentration was less than half the level observed in infected nonexercised mice; exercise had no effect on serum TNFα levels in noninfected animals. This attenuation of the TNFα response to infection was not associated with impairment of immune cell response to the infection but was linked to a faster recovery of appetite and body mass in the exercised mice, suggesting a beneficial effect of the partial suppression of TNFα during this type of infection. Whether such suppression of TNFα occurs in other types of infections and in other animal species, and is beneficial to host defense, is not known at present.

Although the data are far from conclusive, it would appear that TNFα production and gene expression are unchanged by moderate exercise (Haahr et al. 1991; Ullum et al. 1994a), transiently suppressed following more intense or prolonged exercise (Babgy et al. 1994; Chao et al. 1992; Drenth et al. 1995; Kvernmo et al. 1992; Natelson et al. 1996), and increased after exercise with an eccentric bias that induces muscle cell damage (Cannon et al. 1991).

Mechanisms Underlying the Cytokine Response to Exercise

The mechanisms responsible for release of cytokines during exercise are not currently known, and are likely to be complex. It has been suggested that release of the pro-inflammatory cytokines during and after exercise represents an inflammatory reaction, presumably initiated by damage to and perhaps inflammation within skeletal muscle (Cannon et al. 1991; Drenth et al. 1995; Dufaux and Order 1989a, 1989b; Smith et al. 1992; reviewed in Tidball 1995). The exact mechanisms mediating communication between skeletal muscle and cells releasing cytokines during and after exercise have not been studied as yet. It has been proposed that proteolytic fragments released by damaged skeletal muscle, presumably through a leaky cell membrane, interact with leukocytes and other cells (e.g., fibroblasts), initiating release of cytokines (Tidball 1995). This would explain the observation that only intense prolonged exercise or that which induces muscle cell damage (e.g., eccentric exercise) is associated with elevated cytokine concentration.

It is not clear at present whether circulating leukocytes are the major source of cytokines released during and after prolonged exercise. For example, studies of cytokine gene expression in cells isolated during and after exercise do not show clear evidence for de novo production of cytokines (Natelson et al. 1996; Ullum et al. 1994a). As assessed by a reverse transcriptase polymerase chain reaction, peripheral blood leukocyte gene expression of various cytokines (IL-1, IL-2, IL-4, IL-6, IL-10, and TNFα) did not increase immediately after maximal treadmill exercise to volitional fatigue (average 45 minutes duration) (Natelson et al. 1996). Moreover, expression of TNFα declined after exercise, although such decline was not deemed of biological significance. Mononuclear cell pre-mRNA for IL-1α, IL-1β, and IL-6 content did not change during or up to 4 hours after exercise (1 hour of cycling at 75% $\dot{V}O_{2max}$) in untrained subjects (Ullum et al. 1994a). Since pre-mRNA for these cytokines is stimulated by short-term in vitro exposure of mononuclear cells to mitogens such as LPS (Ullum et al. 1994a), these data suggest that de novo production is not actually stimulated during exercise. However, the level of exercise imposed in both of these studies (60 minutes of cycling at 75% $\dot{V}O_{2max}$ [Ullum et al. 1994a] and 45-minute treadmill running to volitional fatigue [Natelson et al. 1996]) does not appear to be sufficient to induce release of cytokines, and further work is needed to determine whether cytokine gene expression is altered during prolonged exercise.

Release of stress hormones does not appear to contribute to the release of IL-1 during exercise. For example, physiological concentrations of epinephrine or hydrocortisone did not increase in vitro human monocyte IL-1 production, and suppression of IL-1 release was observed in cells exposed to both hormones together (Cannon et al. 1986). Exercise-induced hyperthermia does not appear to directly mediate release of cytokines during exercise. For example, in vitro release of IL-1β, IFNγ, and IL-6 declined, and release of IL-1α and TNFα were unchanged, in peripheral blood mononuclear cell cultures exposed to temperatures up to 39.5 °C (Kappel et al. 1991a). However, in vivo hyperthermia may cause release of hormones such as catecholamines (Kappel et al. 1991b), which may indirectly contribute to release of cytokines during or after exercise by recruitment and activation of immune cells.

Significance of the Cytokine Response to Exercise: A Role in Tissue Repair?

Many cytokines have similar and overlapping functions and may also act synergistically; IL-1 and TNFα may stimulate the release of one another as well as IL-6 (Anderson 1992; Baumann and Gauldie 1994; Cox and Gauldie 1992; Janeway and Travers 1996; Liles and van Voorhis 1995; Moldawer 1992; Scales 1992; Tsuji and Torti 1992). Moreover, because of the extensive overlap in function, it is often difficult to isolate a specific action of a particular cytokine. The physiological effects of many cytokines are similar to many of the changes occurring during and after exercise, including increased body temperature (IL-1, IL-6, TNFα), increased NK cytotoxic activity (IL-1, IL-6, TNFα; discussed in the next chapter), activation of macrophages (IL-1, IL-6, TNFα), recruitment and priming/activation of neutrophils (TNFα, IL-8), release of acute phase reactants (IL-1, IL-6, TNFα) and prostaglandins (TNFα), and enhanced antibody production (IL-6). IL-1 and TNFα may be involved in muscle degradation and repair following injury, as often occurs during intense physical activity, especially that involving eccentric lengthening of skeletal muscle. Both cytokines stimulate protein catabolism, fibroblast proliferation, and collagen synthesis (Liles and Van Voorhis 1995; Moldawer 1992). Some cytokines (e.g., IL-1 and TNFα) are involved in leukocyte trafficking via their ability to increase expression of adhesion molecules on leukocytes and endothelial cells (Scales 1992). As discussed in chapter 3, there are some data showing that exercise enhances expression of leukocyte adhesion molecules (Baum et al. 1994; Gabriel et al. 1994a). Thus, these cytokines may be involved in release of leukocytes into the circulation during exercise and subsequent removal from the blood after exercise.

The chemokine (chemotactic cytokine) IL-8 also appears to mediate inflammation via its ability to attract and activate neutrophils to damaged tissue (Baggiolini 1993). IL-8 may arise in any tissue and is produced in response to tissue damage inducing elevated IL-1 and TNFα levels. IL-8 is slowly cleared from damaged tissue, and is one possible mediator of the accumulation of neutrophils in skeletal muscle cells damaged by prolonged load-bearing and eccentric exercise (Fielding et al. 1993) (although at present there do not seem to be any reports on IL-8 and exercise).

As discussed above, IL-1, IL-6, and TNFα do not always change concomitantly during and after all types of exercise, although there is evidence that all three are released at some point in the days after prolonged exercise such as distance running (Sprenger et al. 1992). Moreover, endogenous cytokine inhibitors, such as the IL-1 receptor antagonist (IL-1ra), may also change in response to exercise. For example, IL-1ra concentration increased more than four times after a 65 km (6 hours) run in endurance athletes, despite no change in plasma IL-1β levels (although urinary excretion and tissue concentration of IL-1β were not reported) (Drenth et al. 1995). Thus, an elevation in the concentration of a single cytokine does not necessarily reflect a change in biological activity, since a larger increase in its endogenous inhibitor(s) may negate such changes.

The actions of cytokines appear to be of a dual nature; that is, cytokines may be associated with both beneficial and adverse responses, depending on the level produced, endogenous inhibitors and regulators, and interaction with other cytokines. A particular cytokine rarely acts alone, and the term "cytokine network" is used to refer to the complex interaction of the various cytokines. Although elevation of some cytokines may be associated with adverse physiological responses (e.g., TNFα and tissue wasting), it has been proposed that release of the pro-inflammatory cytokines (IL-1, IL-6, TNFα) may be beneficial by stimulating nonspecific host defense early in infection and contributing to activation of cellular immunity (table 6.3) (Moldawer 1992). As discussed throughout this book, there is ample evidence for suppression of various aspects of immune func-

Table 6.3 Possible Role of Cytokines in Mediating Exercise Induced Changes in Immune Parameters

Immune parameter and exercise effect	Possibly related to cytokine action
Demargination of leukocytes during exercise and remargination after exercise via altered adhesion molecule expression	IL-1, TNFα
Decreased lymphocyte proliferation after intense exercise	IL-2
Enhanced NK cell cytotoxic activity	IL-1, IL-2, IL-6, TNFα, IFN
Enhanced monocyte phagocytic and cytotoxic activity	IL-1, IL-6, TNFα, IFN
Enhanced neutrophil priming	TNFα, IL-8
Neutrophil chemotaxis	IL-8
Release of acute phase proteins	IL-1, IL-6, TNFα
Antibody production	IL-1, IL-2, IL-6

tion during and after intense prolonged exercise. Thus, it may be speculated that release of cytokines during and after exercise may be a protective mechanism to counteract the general suppression of some aspects of immunity during the postexercise period. In addition, many of the acute phase proteins released in response to elevated cytokine levels are protease inhibitors or free radical scavengers that may help to limit tissue damage associated with release of toxic molecules and free radicals from activated neutrophils (Moldawer 1992) (recall from chapter 4 that neutrophils are activated by most types of exercise). It should be noted that direct experimental evidence is at present lacking to either support or refute such speculation.

Summary and Conclusions

The cytokine response to exercise appears to be complex and to be related to exercise parameters, previous training, the site of cytokine measurement (e.g., tissue, blood, urine), and perhaps the method of measurement. Cytokines do not normally appear in detectable levels in the blood of healthy individuals, and the observation of detectable resting cytokine concentration, in particular the pro-inflammatory cytokines IL-1β, IL-6, and TNFα, in endurance athletes suggests either chronic inflammation or a long-lasting effect of prior exercise.

Although earlier studies using bioassays suggested that plasma levels of cytokines may increase after moderate exercise, more recent studies using sensitive and highly specific immunoassays generally show few changes in plasma cytokines after exercise, with the exception of IL-6, which appears to be elevated after distance running. In contrast, however, urinary excretion of cytokines including IL-1β, IL-6, IFNγ, and TNFα has been noted after prolonged exercise. The time course of appearance of these cytokines differs between cytokines, however. IFNγ and TNFα increase immediately after exercise but IL-1β displays a delayed response; urinary cytokine excretion may persist for more than 24 hours after prolonged exercise. The low levels of cytokines in plasma despite high levels in the urine during and after exercise suggest that cytokines are released during exercise, but are quickly taken up and excreted. IL-1β has been localized within damaged skeletal muscle after eccentric exercise, suggesting a role for this cytokine in inflammation within and repair of damaged skeletal muscle.

Research Findings

- Much of the inconsistency in the research literature on cytokines and exercise may be related to methodological issues such as the very low concentrations, assay sensitivity and specificity, and the source of samples.

- In general, cytokines do not appear in plasma during or after exercise, with the exception of IL-6, which is elevated in plasma after prolonged weight-bearing exercise such as distance running.
- Because cytokines act locally and are rapidly metabolized, analysis of urinary cytokine excretion gives a better indication of cytokine release. Urinary excretion of IL-1β, IL-6, TNFα, and IFNγ increases after prolonged exercise, but the time course of these increases and subsequent return to baseline levels vary between cytokines.
- IL-1β concentration increases for up to five days in skeletal muscle damaged by high-force muscle lengthening (eccentric exercise), suggesting a possible role for this cytokine in muscle damage and repair.
- Exercise does not appear to alter cytokine gene expression in peripheral blood mononuclear cells obtained after moderate exercise, suggesting either that other cells are the source of this cytokine or that exercise influences posttranscriptional mechanisms. (However, it cannot yet be discounted that intense or prolonged exercise influences cytokine gene expression.)

Possible Applications

- Cytokine release after intense exercise has been implicated as one mediator of skeletal muscle damage and repair. High resting cytokine levels observed in endurance athletes suggest chronic inflammation. Manipulation of cytokine release may influence inflammatory processes, possibly helping to limit damage and/or speed repair following damage after exercise.
- Cytokines appear to be released only after prolonged exercise or exercise causing skeletal muscle damage. If release of some cytokines is shown to induce adverse responses (e.g., infiltration of inflammatory cells into skeletal muscle), it may be possible to manipulate training programs to avoid or minimize exercise that induces cytokine release.

Yet to Be Explored

- Does cytokine release influence immune function or the immune response to exercise in athletes? Are cytokines involved in the apparent "downregulation" of certain aspects of immune function (e.g., neutrophil activation) in athletes?
- Is cytokine release after prolonged exercise generally beneficial or harmful to immune function in athletes?
- What is the role, if any, of the various cytokines in the processes of inflammation and repair in skeletal muscle after prolonged exercise?

Chapter 7

Exercise and Cytotoxic Cells

© Human Kinetics

Cytotoxic (killing) activity is exhibited by several types of immune cells, in particular cytotoxic T lymphocytes (CTL; $CD3^+/CD8^+$), natural killer (NK; $CD3^-/CD16^+/CD56^+$) cells, and monocytes/macrophages. As described in chapter 2, CTL, NK cells, and monocytes are major effectors of host defense against tumor growth and virally infected cells.

Much attention has focused on the effects of exercise on cytotoxic cells for several reasons (Mackinnon 1989; Pedersen and Ullum 1994; Pedersen 1997; Shephard et al. 1994a, 1994b). Exercise appears to influence host defense against both viral infection and cancer (discussed further in chapters 1 and 8). Exercise also causes the release of several cytokines involved in resistance to tumors and viral infection, such as IFN, TNFα, IL-1, and IL-6, which may also influence the activity of cytotoxic cells (discussed in chapter 6). Moreover, stress influences resistance to tumor growth and viral infection, and some stress hormones released during exercise, such as corticosteroids or catecholamines, can modulate the ability of immune cells to kill tumor cells. Thus, it can be postulated that exercise may influence host defense against tumor growth and viral infection via directly or indirectly modulating the activity of cytotoxic cells. Most research in this area has focused on NK cells, with relatively less attention to the effects of exercise on CTL and monocyte cytotoxicity.

Exercise and NK Cells

NK cells are a distinct but heterogeneous subset of lymphocytes capable of recognizing and killing virally infected cells, certain tumor cells, and some microorganisms without prior exposure (thus the name *natural* immunity). NK cells have been defined as large granular lymphocytes (LGLs) that do not express T cell surface antigens (e.g., CD3), but that do express antigens CD16 and CD 56 (in humans) and that mediate cytotoxicity without prior exposure to, and without expression of MHC antigen on, target cells (Janeway and Travers 1996; Trinchieri 1989). The exact lineage of NK cells is still debated (Roitt et al. 1993; Trinchieri 1989), but it is generally accepted that NK cells descend from a distinct progenitor cell in the bone marrow and that they represent a unique leukocyte subset. In the exercise literature, NK cells are operationally identified by flow cytometry as $CD3^-/CD16^+/CD56^+$ cells. NK cells exert a variety of functions including cytotoxicity against various targets that provides natural resistance to some viral and bacterial infection as well as some types of tumors; production of some cytokines; and regulation of hematopoiesis (Trinchieri 1989).

Exercise exerts a profound effect on both circulating NK cell number and cytotoxic activity. Changes in NK cell number and activity may occur very early during exercise, may persist for up to days after a single bout of exercise, and may also persist for months during intense exercise training. Of all the immune cells (e.g., B and T lymphocytes, neutrophils, monocytes) studied in the exercise literature, NK cell number and function appear to change the most dramatically in response to a single bout of exercise. Moreover, there is some evidence that endurance exercise training induces changes in resting NK cell number and function.

As discussed in chapter 3, NK cells are quickly recruited into the circulation at the onset, and leave the circulation rapidly at the end, of a single bout of exercise; such trafficking may be mediated by adhesion molecules found on the surface of NK cells, which may facilitate binding of NK cells to the vascular endothelium and thus their exiting from the circulation. Total NK activity generally increases during and immediately after exercise, whether brief or prolonged, moderate or intense. For brief exercise lasting 30 minutes or less, NK activity is generally restored to resting levels by 1 hour postexercise. In contrast, intense or prolonged exercise lasting more than 30 minutes may decrease total NK activity for up to 6 hours after cessation of exercise.

Resting NK Cell Number and Activity in Athletes

Resting NK cell number and percentage of lymphocytes appear to be normal in athletes compared with nonathletes, although there are some reports of higher NK cell number in athletes (table 7.1). Cross-sectional comparisons of circulating NK cell number and percentages have generally shown no significant differences between nonathlete control subjects and endurance-trained cyclists (Baj et al. 1994; Nieman et al. 1995b, 1995c; Tvede et al. 1991) or runners (Nieman et al. 1993a, 1995b, 1995c).

In contrast, a recent study on a small group of distance runners compared with matched nonathletes (N = 7 and 6, respectively) reported twofold higher NK cell ($CD16^+$ and $CD\ 56^+$) number and percentage in the runners compared with control subjects (Rhind et al. 1994). NK cells from athletes also exhibited significant elevation of expression of the β subunit of the IL-2 receptor, suggesting a more highly activated NK cell (discussed below). Although these athletes were rested for 36 hours before blood sampling, resting leukocyte profiles are suggestive of a long-lasting acute effect of the last exercise session, as shown by significantly higher total leukocyte, granulocyte, and NK cell number and lower lymphocyte number. Whether such differences in NK cell number and other parameters are due to the small sample size reported in this study, to a long-lasting acute effect of exercise, or to training-induced enhancement of NK cell number and activation awaits further study.

Two other recent reports suggest the possibility that, in athletes, resting NK cell number may differ from that for nonathletes depending on the intensity of training at the particular time sampled. For example, compared with matched nonathlete control subjects, in 26 elite swimmers, NK cell number was 55% higher and NK cell percentage 61% higher during low-intensity training at the onset of a seven-month season (Gleeson et al. 1995). At the end of the season after several months higher-intensity training prior to national competition, however, NK cell number was 33% lower and NK cell percentage 22% lower in

Table 7.1 Effects of Training on Resting NKCA

Subjects	Exercise	Main results	Reference
UTr	15 wk running 40-50 min 6 × wk	25% ↓ resting NKCA	Watson et al. 1986
7 moderately trained, 7 UTr 72 yr	16 wk moderate walking 60 min, 3 × wk	Total NKCA higher in trained	Crist et al. 1989
27 elite cyclists, 15 matched nonathletes	Normal training	Total NKCA higher in cyclists; correlated with $\dot{V}O_{2max}$	Pedersen et al. 1989
36 obese UTr	15 wk walking 45 min 5 × wk at 60% HRR	↑ total NKCA after 6 but not after 15 wk	Nieman et al. 1990b

29 elite cyclists, 15 matched nonathletes	Low- and high-intensity training	Total NKCA higher in cyclists than controls; higher during high- vs. low-intensity training	Tvede et al. 1991
62 Japanese men	Normal activities	NKCA higher in those reporting regular physical activity (> 1 per wk)	Kusaka et al. 1992
7 runners, 6 UTr	Normal training	More NK cells expressing CD122 in runners	Rhind et al. 1994
10 runners, 8 cyclists, 11 matched nonathletes	Normal training	No difference in total NKCA or NKCA per cell; no relationship with $\dot{V}O_{2max}$	Nieman et al. 1995b
22 runners, 18 matched nonathletes	Normal training	57% higher in runners; inversely correlated with body fat level	Nieman et al. 1995c

Abbreviations: ↑, ↓ = increase and decrease, respectively, compared with prestudy values; Tr = trained; UTr = untrained; NKCA = NK cytotoxic activity; NKCA/cell = NKCA per cell; LAK = lymphokine-activated killer activity; APMHR = age-predicted maximum heart rate; HRR = heart rate reserve; VF = volitional fatigue; ND = no data reported; Ep = epinephrine.

swimmers compared with nonathletes. Tvede et al. (1991) also reported significantly higher, by about 50%, NK cell percentage in 29 male road cyclists during low-intensity training compared with nonathlete control subjects; NK cell percentage did not differ between groups during high-intensity training. These data suggest that, in general, athletes exhibit normal NK cell number and percentage but that these variables may be influenced by prolonged periods of intense exercise training (as discussed below).

In athletes, resting and postexercise NK cytotoxic activity (NKCA) also appear to be within the normal range (Brahmi et al. 1985; Mackinnon et al. 1988; Nieman et al. 1995b; Pedersen et al. 1989) or possibly higher in trained compared with untrained subjects (Crist et al. 1989; Nieman et al. 1993a, 1995c; Pedersen et al. 1989). For example, both Pedersen et al. (1989) and Nieman et al. (1993a, 1995c) reported significantly higher resting NKCA in athletes compared with nonathletes. In the study by Pedersen et al., resting NK cell number was 50% higher and NKCA 20% higher in 27 male cyclists compared with age-matched nonathletes. In the study by Nieman et al. (1995c), resting NKCA and NK cell numbers were compared in 22 male distance runners and 18 nonathletes matched for age and height but not body mass or composition. Resting NKCA was 57% higher in runners despite no difference in NK cell ($CD3^-/CD16^+/CD56^+$) number, indicating higher NKCA per cell. Higher NKCA per cell suggests the presence in blood of a more activated cell, and is consistent with the observation of increased expression of the high-affinity IL-2 receptor on NK cells from distance runners (Rhind et al. 1994; discussed below). In the study by Nieman et al. (1995c), NKCA was inversely correlated with percent body fat in all subjects and in the runners ($r = -0.48$ and -0.49, respectively). Although the runners were rested for 36 hours before blood sampling, the possibility of long-lasting stimulation of NKCA from last exercise session cannot be ruled out. Alternatively, it is possible that the higher NKCA per cell may be partially related to differences in diet, since endurance athletes generally consume a high carbohydrate and low fat diet; a high fat diet has been associated with lower NKCA (Hebert et al. 1990).

Higher resting NKCA among endurance-trained individuals has also been noted in older persons (Crist et al. 1989; Nieman et al. 1993a). Crist et al. (1989) reported higher resting NKCA in elderly women (mean age 72 years) who had participated in a moderate exercise training program (60 minutes moderate exercise, three times per week for 16 weeks) compared with control women who did not exercise. Similarly, Nieman et al. (1993a) reported a 50% higher resting NKCA in highly conditioned elderly women runners compared with sedentary older women. As in the younger male runners discussed above, leanness (low percent body fat) was the strongest predictor of NKCA in these elderly women.

In contrast, differences between endurance athletes and nonathletes were not confirmed by a later study from the same laboratory comparing endurance-trained runners and cyclists with matched nonathletes (Nieman et al. 1995b). Neither NK cell number nor NKCA was significantly different between groups, nor was there a significant relationship between NKCA and $\dot{V}O_{2max}$. Subjects had refrained from exercise for at least 15 hours, and only moderate exercise was performed 15-24 hours before blood sampling, ruling out any acute suppression of NKCA by prior exercise in the trained subjects. There is no ready explanation for the apparent discrepancies between studies, even from the same laboratory. At present, it seems that an increase in resting NKCA is an inconsistent observation among endurance-trained athletes that may relate to small sample size or to variations by season or training intensity/volume. The majority of studies report normal NK cell number and cytotoxic activity in various athletes studied to date.

Acute NK Cell Response to Exercise

The mode, intensity, and duration of exercise, as well as the fitness level of the subject, may all influence the NK cell response to a single bout of exercise.

NK Cell Number

NK cell number rises early in exercise and generally remains elevated throughout exercise regardless of duration (see table 3.7). After brief or moderate exercise, NK cell number returns to resting levels soon after cessation of exercise (figure 7.1); however, NK cell number may decline below baseline levels for at least 1 hour and possibly up to several hours or even days after intense prolonged exercise before returning to normal values.

It has been suggested (Nielsen et al. 1996a, 1996b) that the magnitude of changes in NK cell number is more a function of exercise intensity than duration. For example, NK cell ($CD16^+$) number increased immediately after 1-hour cycle ergometry at 25%, 50%, and 75% $\dot{V}O_{2max}$; the magnitude of changes in NK cell number during exercise, however, was directly related to exercise intensity (Tvede et al. 1993). NK cell number declined below resting levels 2 hours after the more intense exercise. In male distance runners, NK cell ($CD3^-$/$CD16^+$/$CD56^+$) number was greater immediately after 45 minutes intense compared with moderate exercise (treadmill running at 80% and walking at 50% $\dot{V}O_{2max}$, respectively); NK cell number increased about

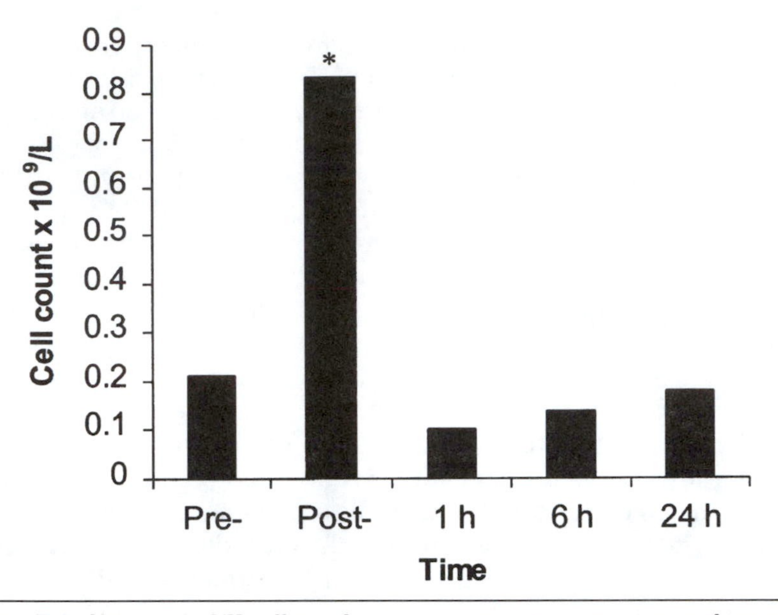

Figure 7.1 Changes in NK cell number in response to intense interval exercise. Peripheral blood was obtained pre-, immediately post-, and at various times up to 24 hours postexercise; exercise consisted of repeated 1-minute maximal treadmill runs to exhaustion (mean of 16 intervals completed). NK cells were identified as $CD3^-$, $CD16^+/CD56^+$ cells. * = Significantly different from preexercise value.

Data from Gray, A.B. R.D. Telford, M. Collins, and M.J. Weidemann. 1993b. The response of leukocyte subsets and plasma hormones to interval exercise. *Medicine and Science in Sport and Exercise* 25: 1252-1258.

fourfold after high-intensity but only about 50% after moderate-intensity exercise (Nieman et al. 1993b). However, NK cell number declined below resting level to the same extent up to 3.5 hours after both exercise sessions. These data suggest that the extent of NK cell mobilization into the circulation is sensitive to exercise intensity but that the postexercise decline in cell number may relate more to exercise duration. NK cell number generally returns to baseline levels within 2 hours after brief submaximal or maximal exercise lasting less than 30 minutes (Brahmi et al. 1985; Field et al. 1991; Nielsen et al. 1996a). In contrast, circulating NK cell number declines below baseline values between 1 and 2 hours after prolonged (> 30 minutes) moderate and intense exercise, including 1-2 hours of cycling at 50-80% $\dot{V}O_{2max}$ (Mackinnon et al. 1988; Pedersen et al. 1988, 1990; Tvede et al. 1993), 45 minutes of treadmill running at 80% $\dot{V}O_{2max}$ (Nieman et al. 1993b), 2-hour treadmill running at 65% $\dot{V}O_{2max}$ (Shek et al. 1995), 3 hours of treadmill running at marathon race pace (Berk et al. 1990), repeated maximal sprint running (figure 7.1) (Gray et al. 1993b), and repeated weight lifting to exhaustion (Nieman et al. 1995e).

NK cell number may remain depressed below preexercise values for several hours or days after a single session of prolonged intense exercise. For example, in male distance runners, a 3-hour simulated marathon run on a treadmill increased NK cell ($CD16^+/CD56^+$) number by 55% after 1 hour of exercise (figure 7.2) (Berk et al. 1990). NK cell number returned to preexercise levels immediately postexercise, then declined by 30-60% between 1 and 21 hours after exercise. The lowest level was observed 6 hours postexercise, and NK cell number was still 43% below baseline level at 21 hours postexercise.

In another study, six male subjects ran on a treadmill at 65% $\dot{V}O_{2max}$ for 2 hours (or until core temperature reached 40 °C) (Shek et al. 1995). NK cell ($CD16^+$) number increased by 50% during the first 30 minutes and a further 20% during the last hour of exercise. NK cell number then declined to 60% of preexercise levels by 60 minutes postexercise and

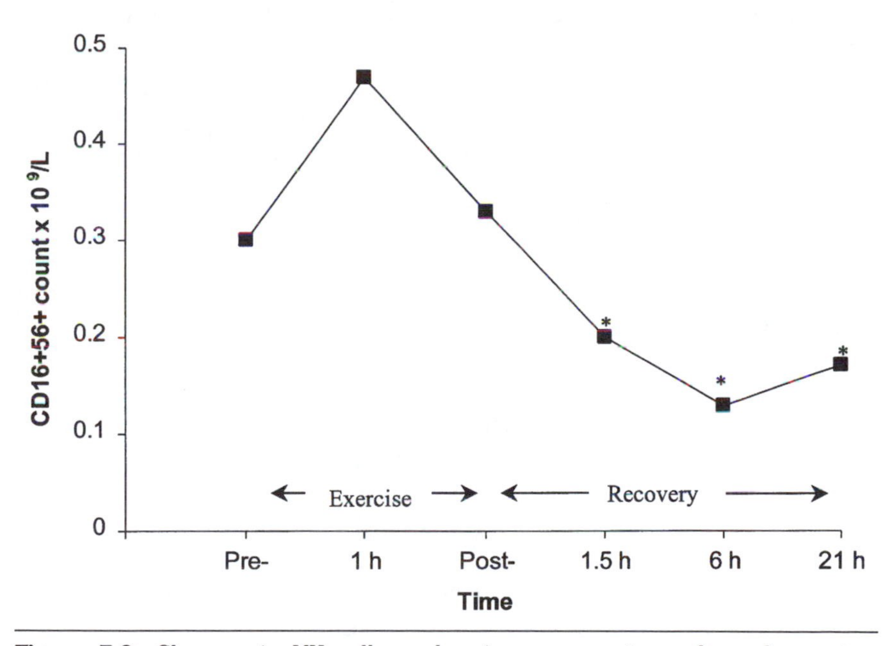

Figure 7.2 Changes in NK cell number in response to prolonged exercise. Peripheral blood was obtained pre-, immediately post-, and at various times up to 21 hours postexercise; exercise consisted of 3 hours treadmill running at marathon pace. NK cells were identified as $CD16^+/CD56^+$ cells. * = Significantly different from preexercise value.

Data from Berk, L.S., Nieman, D.C., Youngberg, W.S., Arabatzis, K., Simpson-Westerberg, M., Lee, J.W., Tan, S.A., and Eby, W.C. 1990. The effect of long endurance running on natural killer cells in marathoners. *Medicine and Science in Sport and Exercise* 22: 207-212.

remained at this level at one and seven days postexercise. This prolonged suppression of NK cell number contrasts with a previous study on cycling and running of similar duration but of higher intensity in well-trained cyclists (e.g., 70-75% $\dot{V}O_{2max}$; Mackinnon et al. 1988), in which NK cell number was restored by 24-48 hours postexercise. These differences may relate to different demands of activity, especially the larger eccentric component in running compared with cycling. It is possible that the decline in circulating NK cell number for seven days after running reflects homing of NK cells to damaged tissues (i.e., skeletal muscle) during the days following exercise, and that such movement of NK cells may not occur after cycling with a negligible eccentric component, thus little muscle cell damage (discussed further below).

Most studies of the NK cell response to exercise have included endurance exercise, although recent reports indicate that NK cells are also mobilized into the circulation after shorter-duration maximal exercise, sprinting, or power/strength activities. For example, in male triathletes, NK cell ($CD3^-/CD16^+/CD56^+$) number increased 2.5 to 4 times immediately after intense interval treadmill running; exercise consisted of repeated 1-minute sprints at 100% $\dot{V}O_{2max}$ to volitional fatigue with 1-minute active rest between (average 15 1-minute sprints) (Gray et al. 1993b). NK cell number decreased to 50% below preexercise levels 1 hour and remained low for 6 hours postexercise, although these decreases were not significant. Six minutes of maximal exercise on a rowing ergometer was also reported to increase circulating NK cell ($CD16^+$) number in elite rowers (Nielsen et al. 1996a); NK cell number increased about fourfold immediately after exercise, but returned to preexercise levels by 2 hours postexercise. NK ($CD16^+$) cell number also increased dramatically (nearly 10 times) after cycle ergometry at 100% $\dot{V}O_{2max}$ to exhaustion (mean 13 minutes' duration) in untrained subjects (Field et al. 1991). The magnitude of increase in NK cell number during exercise was similar after a second identical bout of exercise 1 hour later.

A recent study indicates that NK cell number may also exhibit a biphasic response to intense resistance exercise (Nieman et al. 1995e). Ten experienced male weight lifters performed repeated sets of leg squats until volitional fatigue, each set consisting of 10 repetitions at 65% 1 RM, and mean time to fatigue was 37 minutes. NK cell ($CD56^+$) number increased threefold immediately after exercise but decreased to 46% below resting values 2 hours postexercise. This pattern is similar to that seen after intense endurance exercise, as discussed above.

NK Mobilization During Exercise

Recruitment of NK cells into the circulation does not appear to involve the spleen. For example, 20 minutes of cycle ergometry at 80% age-predicted maximal heart rate induced similar increases in NK cell (CD undesignated) number in splenectomized subjects as in untrained subjects or swimmers (Grazzi et al. 1993). Resting NK cell number was also similar in splenectomized and the other subjects. Thus, NK cells must be recruited into the circulation from compartments other than the spleen during brief, moderate exercise, although a role for the spleen during more intense or prolonged exercise cannot be discounted.

Exercise mode and possibly muscle group involvement may influence mobilization of NK cells during exercise. In a crossover study of 10 male distance runners, changes in circulating NK cell number were compared between 1-hour level and downhill (–10% grade) treadmill running at equal metabolic cost (70% $\dot{V}O_{2max}$) (Pizza et al. 1995). NK cell ($CD3^-/CD56^+$ and $CD16^+$) number in the circulation increased dramatically during exercise and then decreased below preexercise levels 1.5 hours postexercise. However, NK cell number increased significantly more during downhill compared with level running; NK cell number increased about three times during downhill running, but only about twofold during level running despite equal metabolic cost in the two exercise bouts. NK cell numbers were similar 1.5 hours after exercise in both conditions. As expected, serum creatine kinase activity was significantly elevated from 1.5 to 48 hours after downhill but not level running, indicative of muscle cell leakage and possibly cellular damage during downhill running. These data suggest that eccentrically biased downhill running is associated with greater mobilization of NK cells into the circulation during exercise and that these NK cells rapidly exit the circulation after exercise. No muscle biopsies were performed in this study, and it is not possible to determine if NK cells were selectively recruited to damaged skeletal muscle cells. However, NK cells do not generally infiltrate damaged skeletal muscle (reviewed in Tidball 1995).

NKCA

The literature is fairly consistent in showing that NKCA increases acutely during exercise, even of brief duration, and remains elevated throughout exercise, regardless of exercise duration. In contrast, the postexercise pattern of NKCA is complex, and is related to exercise intensity and duration as well as time of blood sample.

The magnitude of increase in NKCA during and immediately after exercise is related to exercise intensity (tables 7.2 and 7.3). For example, total NKCA doubled immediately after 45 minutes running at 80% $\dot{V}O_{2max}$ in 10 distance runners, but increased only marginally (and nonsignificantly) after moderate exercise (50% $\dot{V}O_{2max}$) in the same subjects (Nieman et al. 1993b). Marked increases in NKCA, ranging from 50% to 400%, have been noted after brief high-intensity exercise, such as 6 minutes of all-out rowing in competitive rowers (Nielsen et al. 1996a), or 12-18 minutes of progressive exercise testing to volitional fatigue in untrained young adult and older subjects (Brahmi et al. 1985; Crist et al. 1989; Fiatarone et al. 1988, 1989; Field et al. 1991). Brief, moderate exercise is associated with much smaller increases (e.g., 20-50%) in NKCA (Klokker et al. 1995; Kotani et al. 1987; Targan et al. 1981). NKCA generally returns to baseline levels within 1 hour after brief moderate exercise (Field et al. 1991; Nielsen et al. 1996a; discussed below).

Many studies have documented marked increases in NKCA during and immediately after longer-duration (>30 minutes) exercise, including a progressive rise in NKCA during, and up to two times resting values after, 2 hours of treadmill running at 65% $\dot{V}O_{2max}$ in male distance runners (Shek et al. 1995); 25% increase after 3 hours of treadmill running at marathon pace in male distance runners (Berk et al. 1990); 100% increase after 45-minute treadmill running at 80% $\dot{V}O_{2max}$ in male distance runners (Nieman et al. 1993b); 50% increases after 1 hours of cycling at 60-80% $\dot{V}O_{2max}$ in untrained subjects (Kappel et al. 1991c; Pedersen et al. 1988, 1990); and 20% rise after 2 hours of cycling at 70-75% $\dot{V}O_{2max}$ in well-trained cyclists (Mackinnon et al. 1988). As with brief exercise, the NKCA response to longer-duration exercise is dependent on exercise intensity. For example, in six untrained males who cycled for 1 hour at 25%, 50%, and 75% $\dot{V}O_{2max}$, the magnitude of increase in NKCA was directly related to exercise intensity (Tvede et al. 1993). In contrast, high-intensity resistance exercise to fatigue (mean duration 37 minutes total) was not reported to change NKCA immediately after exercise (Nieman et al. 1995e).

A number of studies have documented a postexercise suppression of NKCA below preexercise values, lasting at least 1-2 hours, after prolonged exercise (table 7.3) (Berk et al. 1990; Kappel et al. 1991c; Mackinnon et al. 1988; Nieman et al. 1993b; Pedersen et al. 1988, 1990; Tvede et al. 1991, 1993; Shek et al. 1995). NKCA has been reported to decrease by about 50% 2 hours after 60 minutes of cycling at 60-80% $\dot{V}O_{2max}$ in untrained subjects (Kappel et al. 1991c; Pedersen et al. 1988, 1990; Tvede et al. 1991, 1993); 25% 1 hour after 2 hours of cycling at 70-80% $\dot{V}O_{2max}$ in competitive cyclists (Mackinnon et al. 1988); 30% 1 hour after 45-minute treadmill running at 80% $\dot{V}O_{2max}$ in distance runners

Table 7.2 Effects of Brief Exercise on NKCA

Subjects	Exercise	Main results*		Reference
		Duringpost	≥ 1 h post	
4 UTr	5 min moderate (unspecified)	30% ↑; augmented by IFN	ND	Targan et al. 1981
5 UTr	5 min stair climbing	40% ↑	ND	Edwards et al. 1984
5 Tr, 10 UTr	Graded cycling to VF	50% ↑; similar response in both groups	50% ↓ 2 h post; similar response in both groups	Brahmi et al. 1985
8 UTr	Graded cycling to VF (~ 75% APMHR)	100% ↑; prevented with naloxone	ND	Fiatarone et al. 1988
8 UTr 30 yr; 9 UTr, 71 yr	Graded cycling to VF (~ 75% APMHR)	100% ↑; no age difference	ND	Fiatarone et al. 1989
7 UTr, 7 Tr, 72 yr	Graded walking to VF	30% ↑ UTr, 50% ↑ Tr	ND	Crist et al. 1989
12 UTr	13 min cycle ergometry at 100% $\dot{V}O_{2max}$	4 × ↑ NKCA; 100% ↑ NKCA/cell	Normal NKCA and NKCA/cell by 1 h post	Field et al. 1991

(continued)

Table 7.2 *(continued)*

Subjects	Exercise	Main results*		Reference
		Duringpost	≥ 1 h post	
7 UTr	2 × 20 min cycling at 60%$\dot{V}O_{2max}$ with 60 min rest between	20-40% ↑ NKCA and LAK	ND	Klokker et al. 1995
10 weight lifters	37 min weight lifting to exhaustion	No change NKCA; 40% ↓ NKCA/cell	60% ↓ NKCA and % ↓ NKCA/cell 2 h post	Nieman et al. 1995e
8 UTr	4 × 5 min 1-legged eccentric cycling	60% ↑ total NKCA; no change NKCA/cell	No change 4 h post total NKCA and NKCA/cell	Palmo et al. 1995
8 elite rowers	3 × 6 min max rowing on 2 consecutive days	50-100% ↑ NKCA	↑ at rest 24 h after 2nd day of exercise	Nielsen et al. 1996a
10 elite powers	6 min max rowing	20-100% ↑ NKCA and LAK	Normal by 2 h post	Nielsen et al. 1996b

Abbreviations: ↑, ↓ = increase and decrease, respectively, compared with preexercise or resting values; see table 7.1 for other abbreviations. * Data refer to total NKCA unless otherwise indicated (e.g., LAK or NKCA per cell).

Table 7.3 Effects of Prolonger Exercise on NKCA

Subjects	Exercise	Main results*		Reference
		Duringpost	≥ 1 h post	
8 cyclists	2 h cycle ergometry at 70-75% $\dot{V}O_{2max}$	No change; 40% ↑ NKCA/cell	25% ↓ NKCA 1 h post; 40% ↑ NKCA/cell 1 h post	Mackinnon et al. 1988
6 UTr	60 min cycle ergometry at 80% $\dot{V}O_{2max}$	50% ↑	50% ↓ 2 h post; abolished by indomethacin	Pederson et al. 1988
10 runners	3 h running	↑ 25%	↓ 15% 1.5 h post	Berk et al. 1990
15 UTr	60 min cycle egometry at 75% $\dot{V}O_{2max}$	50% ↑	50% ↓ 2 h post; prevented with indomethacin	Pederson et al. 1990
UTr	60 min cycle ergometry at 75% $\dot{V}O_{2max}$	↑ 100% mimicked by Ep infusion	Normal by 2 h post; mimicked by Ep infusion	Kappel et al. 1991c; Tvede et al. 1994
10 runners	45 min running at 50% (low) and 80% (high) $\dot{V}O_{2max}$	50% ↑ after low, 100% ↑ after high; no change NKCA/cell after low and high	Similar 40% ↓ 1 h post; 60% ↑ NKCA/cell 2 h post for high only	Nieman et al. 1993b

(continued)

Table 7.3 *(Continued)*

Subjects	Exercise	Main results* Duringpost	≥ 1 h post	Reference
6 UTr	60 min cycle ergometry at 25, 50, 75% $\dot{V}O_{2max}$	50-70% ↑ NKCA and LAK	25% ↓ 2 h postonly after 75% $\dot{V}O_{2max}$ prevented with indomethacin	Tvede et al. 1993
22 runners	2.5 run at 75%$\dot{V}O_{2max}$	No change total NKCA and NKCA/cell	50-60% ↓ 1.5-6 h post total NKCA; not prevented with indomethacin; no change NKCA/cell	Nieman et al. 1995a
6 Tr	2 h run at 65%$\dot{V}O_{2max}$	100% ↑ NKCA	25-40% ↓ 30 min-7 d post	Shek et al. 1995
8 triathletes	Triathlon (swim, cycle, run)	No change in NKCA and LAK after swim; 40% ↓ after cycle, no change after run	40% ↓ 2 h post both NKCA and LAK	Rohde et al. 1996

See table 7.1 and 7.2 for abbreviations. * = Data refer to total NKCA unless otherwise indicated (e.g., LAK or NKCA per cell).

(Nieman et al. 1993b); 10-30% 1.5 to 6 hours after 3 hours of treadmill running in distance runners (Berk et al. 1990); 60% 2 hours after 37 minutes of repetitive weight lifting to volitional fatigue (Nieman et al. 1995e); and 25-40% up to seven days after 2 hours of treadmill running at 65% $\dot{V}O_{2max}$ in distance runners (Shek et al. 1995).

The latter two studies are interesting in showing a long-lasting suppression of NKCA after exercise, although the duration of suppression varies between studies. In the study by Berk et al. (1990), although NKCA was suppressed for up to 6 hours after a 3-hour simulated marathon run on a treadmill in male distance runners, NKCA appeared to be restored by 21 hours postexercise. In contrast, in the study by Shek et al. (1995), NKCA remained 25-40% below resting values starting 30 minutes postexercise and lasting seven days after a 2-hour run at 65% $\dot{V}O_{2max}$ (or up to a core temperature of 40 °C). Since both studies included well-trained distance runners (as evidenced by high $\dot{V}O_{2max}$ values), and the latter study appeared to have involved less rigorous exercise (2 hours vs. 3 hours), there is no ready explanation for these differences in results; and quantifying the duration of suppression of NKCA after exercise awaits confirmation by further study. This prolonged suppression of NK cell number reported by Shek et al. (1995) also contrasts with previous reports showing restoration of NKCA 24-48 hours after exercise (e.g., cycling for 2 hours at higher intensity [70-75% $\dot{V}O_{2max}$] in well-trained cyclists) (Mackinnon et al. 1988). As noted above, these differences may relate to different demands of activity, especially the larger eccentric component in running compared with cycling. In the study by Shek et al. (1995), the extended low NKCA may be a function of low circulating NK cell number, possibly reflecting homing of NK cells to damaged tissues (i.e., skeletal muscle) during the days following exercise, and that such movement of NK cells may not occur after cycling with a negligible eccentric component. As discussed below, mechanisms responsible for the enhancement of NKCA during, and delayed suppression of NKCA after, exercise have been attributed to a complex interaction of changes in NK cell number in the circulation and effects of various hormones and other factors such as cytokines.

In contrast to prolonged exercise, brief, even intense, exercise does not appear to cause the delayed postexercise suppression of NKCA (Nielsen et al. 1996a; Field et al. 1991). For example, in 10 elite male rowers, 6 minutes all-out rowing increased NKCA by about 30% during exercise, but NKCA returned to preexercise values by 2 hours postexercise (Nielsen et al. 1996a); NK cell ($CD16^+$) number increased to a much greater extent (fourfold increase) during, but also returned to baseline levels after, exercise. Similarly, Field et al. (1991) reported normal NKCA 1 hour after exercise in untrained males (cycling to volitional fatigue at 100% $\dot{V}O_{2max}$, mean duration 13 minutes) despite a fourfold increase in NKCA immediately postexercise. Nielsen et al. (1996a) suggested that the increase in NK cell number is determined by exercise intensity, whereas changes in NKCA during and after exercise are more a function of exercise duration.

NK Cell Response to Exercise Training

As described above, resting NK cell number and cytotoxic activity are normal in most athletes, suggesting no chronic effect of exercise training on these variables. However, a few recent reports suggest a possible decline in NK cell number and function during periods of intense training in athletes such as competitive swimmers and runners. In contrast, moderate exercise training may have little effect, or may enhance NK cell function in recreationally active individuals. Whether such changes are clinically relevant remains to be seen. It may be speculated that because NK cells are important to early defense against some viral infections, reduced NK cell function may contribute to impaired immunity in athletes. As discussed below, however, few studies have directly addressed this question.

Intense Exercise Training

There has been one report of a decrease in resting NKCA after intense endurance training in previously untrained young adult subjects (Watson et al. 1986). Thirty-six males trained at 70-85% $\dot{V}O_{2max}$ for 40-50 minutes per day, five times per week for 15 weeks. NKCA declined by 25% after 15 weeks, suggesting suppression of resting NKCA; NK cell number was not reported. However, a nonexercising group was not included to control for possible seasonal variation (see below). In addition, it was not stated how long subjects were rested before blood samples were obtained, to rule out any possible acute effect of the last exercise session before blood sampling. As discussed above, suppression of NK cell number or function may persist for up to 21 hours (Berk et al. 1990) and possibly days (Shek et al. 1995) after a single bout of exercise.

This observation of lower NKCA after a period of intense training in previously sedentary subjects (Watson et al. 1986) is consistent with data from recent studies showing a decline in NK cell number after intense training in elite or well-trained athletes (Fry et al. 1992a, 1994; Gleeson et al. 1995). For example, Fry et al. (1992a, 1994) reported progressively declining NK cell number ($CD56^+$) over 10 days of very intense sprint training in five well-trained military personnel (figure 7.3). Training consisted of 15 1-minute all-out sprints on the treadmill each morning followed by 10 such sprints each evening for 10 consecutive days. NK cell number declined 47% from day 0 to 6 and was 40% below prestudy levels at day 10; moreover, NK cell number remained 50% below prestudy values even after five days recovery training consisting of 20 minutes very light jogging per day. These data suggest that NK cell number responds rapidly to large increases in training intensity and that several days may be needed for full recovery after 10 days intense training.

Gleeson et al. (1995) also reported significantly lower NK cell ($CD16^+/56^+$) number after seven months intense swim training in elite (Australian National Team) swimmers. In the swimmers, NK cell number declined by 43% and NK cell percentage by 32% from the start to end of the season and was 33% and 22% lower, respectively, than in nonathlete control subjects by the end of the seven months. These changes were unique to NK cells, since the number of other cells (e.g., $CD3^+$, $CD4^+$, $CD8^+$) did not change significantly over the seven months.

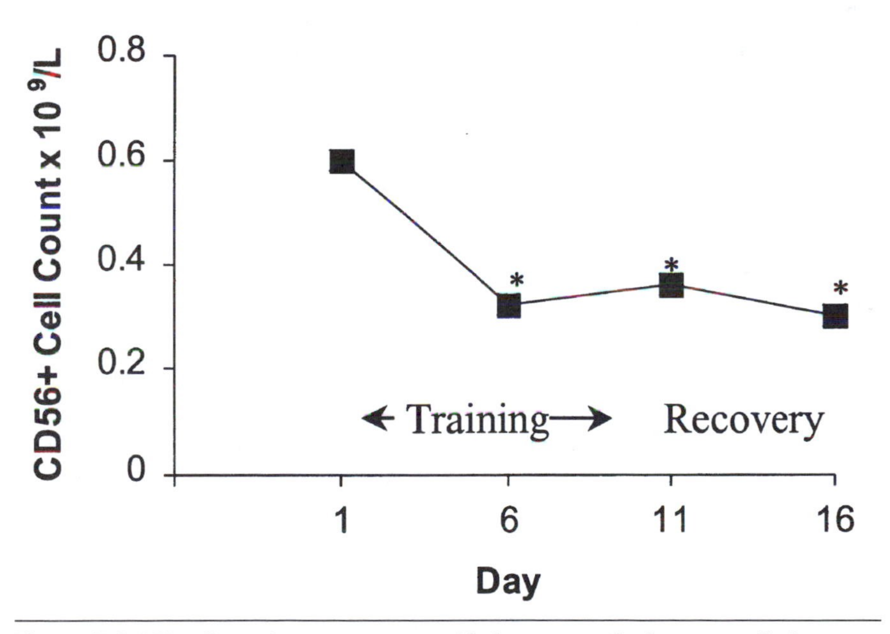

Figure 7.3 NK cell number response to 10 days intensified training. Endurance-trained military personnel performed intense interval training twice daily for 10 days followed by very moderate recovery training for 5 days. Peripheral blood was sampled at rest on days 1, 6, 11, and 16. NK cells were identified as $CD56^+$. * = Significantly different from day 1.

Data from Fry, R.W., A.R. Morton, P. Garcia-Webb, G.P.M. Crawford, and D. Keast. 1992. Biological responses to overload training in endurance sports. *European Journal of Applied Physiology* 64: 335-344.

In contrast, some studies have reported no significant changes in NK cell number or percentage over a training season. Baj et al. (1994) and Tvede et al. (1991) found no changes in $CD16^+/56^+$ cell number or percentage after six months of intense training in endurance cyclists. However, in the study by Tvede et al. (1991), NK cell number and percentage increased by nearly 50% over the same time period in untrained control subjects, suggesting a seasonal change that did not occur in the athletes. Taken together, these data suggest that NK cell number may decline and remain low throughout periods of intense exercise training in some elite athletes; recovery of NK cell number may require at least several days after cessation of intense training. Whether such suppression of circulating NK cell number during intense training is related to changes in immune function is not currently known.

Moderate Exercise Training

Moderate exercise training may enhance or have no effect on resting NKCA in previously sedentary individuals (Nieman et al. 1990b). In an exercise intervention study, 36 mildly obese women were randomized into either an exercise or nonexercise control group (Nieman et al. 1990b). Exercisers participated in a 15-week program of moderate exercise consisting of five 45-minute sessions per week of brisk walking at 60% heart rate reserve. NKCA increased significantly by 55% within the first 6 weeks in the exercisers while remaining unchanged in the control subjects. However, NKCA did not increase further from 6 to 15 weeks in the exercisers, whereas in control subjects, NKCA increased to the level of the exercisers between 6 and 15 weeks. NK cell number ($CD16^+$, $CD56^+$) remained unchanged in both groups throughout the study, suggesting that increased NKCA occurred via stimulation of cytotoxic activity per NK cell. In exercisers, body mass and composition did not change significantly over the 15 weeks, thus the increase in NKCA occurred independently of changes in body fat levels. Improvement in submaximal heart rate during a standardized exercise test (indicative of improvement in cardiorespiratory fitness) was significantly correlated ($r = 0.35$) with the change in NKCA from week 0 to 6, but not between week 6 and 15. These data suggest that enhancement of resting NKCA occurs during moderate training; the data are consistent with some cross-sectional studies comparing endurance-trained athletes and nonathletes, as discussed above. However, the increase in NKCA from week 6 to 15 in the nonexercisers is problematic, and no firm conclusion can be drawn from this study.

In another study from the same laboratory on the NK response to moderate exercise in older women (Nieman et al. 1993a), NK cell number and NKCA were compared in a group of well-trained distance runners and in previously untrained women before and after moderate endurance training. As mentioned above, resting NKCA was 50% higher in the endurance-trained older women compared with sedentary women of the same age at the start of the study (65-84 years). The sedentary women were randomly allocated to either a walking or calisthenic group, both of which met for 30-40 minutes per day, five days per week for 12 weeks. The walking group trained at 60% heart rate reserve, and the calisthenic group performed easy range-of-motion and flexibility exercises while maintaining heart rate near resting levels. There were no changes in NK cell number or NKCA after the 12 weeks in either group, indicating no effect of 12 weeks of moderate exercise training on NK cells in elderly women. Taken together, these studies suggest little, if any, effect of short-term moderate exercise training on NK cell function in humans. Some cross-sectional studies have reported higher NKCA in endurance-trained compared with sedentary individuals, suggesting that more vigorous training over the longer term (i.e., years) may enhance resting NKCA; the higher NKCA occurs despite no differences in NK cell number, suggesting activation of NK cells. However, this difference is not always a consistent finding, and it is still unclear what effect, if any, long-term exercise training has on NK cell function.

In an experimental animal model, 30 minutes of daily moderate exercise over nine weeks was associated with enhanced splenic NKCA (MacNeil and Hoffman-Goetz 1993a). Mice were randomly assigned to one of four groups: sedentary control; voluntary exercisers allowed free access to running wheels; treadmill control, which were given very brief daily exposure to the treadmill that was insufficient to induce a training effect; and treadmill exercisers trained for 30 minutes per day, equivalent to the amount of exercise performed by voluntary exercisers. After the nine weeks, both exercised groups (treadmill and voluntary) exhibited significantly higher in vitro splenic NKCA compared with the two control groups (sedentary and treadmill control). Animals had been rested for at least 20 hours since the last training session, indicating a chronic rather than transient effect of exercise. Moreover, compared with control groups, both exercise-trained groups exhibited lower retention of injected lung tumor cells, suggesting a role for enhanced natural immunity in the in vivo defense against tumors (discussed in chapter 8).

Exercise-Induced Changes in NK Activity: Cell Distribution or Changes in Activity Per Cell?

Total NKCA, usually expressed as percent specific lysis or in lytic units, is a function of both the killing activity of each cell and the number of cells in the assay system. Since exercise may influence each of these parameters independently, both must be considered as possible contributors to exercise-induced changes in NKCA (Mackinnon 1989; Nieman et al. 1993b). The in vitro assay system commonly used to quantify total cytotoxic activity system that uses either a fixed number of lymphocytes or volume of blood, both of which contain a mixture of different subsets of lymphocytes. Since NK cells compose only about 15% of total lymphocytes, any change in the relative proportion of lymphocytes, which often occurs during exercise (see chapter 3), may thus influence total cytotoxic activity.

There has been debate over the past several years as to whether changes in NK cell total cytotoxic activity reflect changes in cellular distribution, which would influence the number of NK cells in the assay, or changes in the activity of each cell (Mackinnon 1989; Berk et al. 1990; Nieman et al. 1993b, 1995a, 1995e). Nieman et al. (1993b) concluded that redistribution of NK cells accounted for a large part of the changes in total NKCA observed during and after moderate exercise, and for the delayed suppression of NKCA observed 1-2 hours after intense exercise (discussed further below).

In some instances, it appears that changes in circulating NK cell number may account for most of the change in total NKCA, especially during and immediately after exercise (Kappel et al. 1991c; Mackinnon 1989; Nieman et al. 1993b, 1995a, 1995e; Palmo et al. 1995). For example, the exercise-induced increase in total NKCA after 1 hour of moderate exercise was accompanied by a similar increase in the number of CD16$^+$ (NK) cells (Kappel et al. 1991c). However, not all of the effect of exercise on total NKCA can be attributed solely to changes in NK cell number, especially changes that persist for several hours after cessation of exercise (Berk et al. 1990; Pedersen et al. 1988, 1990). For example, identification of NK cell subsets via flow cytometry failed to show a correlation between changes in total NKCA and any specific NK cell subset during or after intense prolonged exercise (3 hours of simulated marathon run on a treadmill in male distance runners) (Berk et al. 1990). However, during and 1.5 hours after exercise, the changes in total NKCA paralleled changes in the number of CD16$^+$/CD56$^+$. During exercise, NKCA increased 24%; taking data from Berk et al. (1990), it can be calculated that the number of CD16$^+$/CD56$^+$ cells in the assay system increased by a similar amount, 21%. However, these comparisons do not hold up for each time point studied: for example, at 1.5 hours postexercise, the decrease in NK cell number exceeded that observed for NKCA, and NKCA returned to preexercise values by 6 hours despite a 50% decline in NK cell number. Obviously, there is a complex relationship between exercise-induced changes in NK cell number and NKCA; it is likely that, at any given time after exercise, changes in total NKCA reflect both changes in NK cell number in the circulation and cytotoxic activity per cell, both of which may be influenced by myriad factors such as hormones and cytokines released during exercise.

Single-cell assays measure cytotoxic activity at the cellular level. Only one study has measured NKCA at the single-cell level after exercise (Targan et al. 1981). Following brief moderate exercise (5 minutes unspecified exercise), NK cells bound tumor cells at the same rate as before exercise. However, because total NKCA increased, it was reasoned that the rate of killing must also have increased after exercise, indicating that each NK cell was killing more targets. It was suggested that exercise may activate previously inactive cells, possibly by enhancing their sensitivity to cytokines or hormones that stimulate NKCA. Alternatively, it is possible that the influx of NK cells into the circulation during exercise (as evidenced by the large but transitory increases in NK cell number) brings into the circulation a more active subset of NK cell.

A few studies have attempted to address this issue by calculating NKCA on a per-cell basis (Mackinnon et al. 1988; Nieman et al. 1993b, 1995a, 1995e). NKCA is calculated by dividing total NKCA by the number of NK cells in each assay well, for assays using either whole blood or isolated peripheral blood mononuclear cells. NK cell number is quantified by flow cytometry to obtain the percentage of lymphocytes staining positively with NK cell markers (e.g., $CD16^+$ and/or $CD56^+$). NK cell percentage is then used to calculate NK cell number per volume of blood. By knowing the total activity per well and the number of NK cells per well, NKCA per cell can be calculated. This method, of course, assumes a constant activity per NK cell (i.e., does not distinguish between different subsets of NK cells).

Mackinnon et al. (1988) reported an apparent stimulation of calculated NKCA per cell immediately and 1 hour after 2 hours cycling at 70-80% $\dot{V}O_{2max}$ in well-trained male cyclists. Compared with resting values, NKCA per cell increased 40% immediately and 1 hour after exercise, returning to preexercise values by 24 hours postexercise. Nieman et al. (1993b) reported no change in NKCA per cell immediately and 1 hour postexercise compared with preexercise values; exercise consisted of 45 minutes running at 50% and 80% $\dot{V}O_{2max}$ in 10 distance runners. The higher-intensity (80% $\dot{V}O_{2max}$) running was associated with a significant twofold increase in NKCA per cell 2 hours postexercise, which returned to preexercise values by 3.5 hours after exercise, whereas NKCA per cell was unchanged at all times after the lower-intensity exercise (figure 7.4).

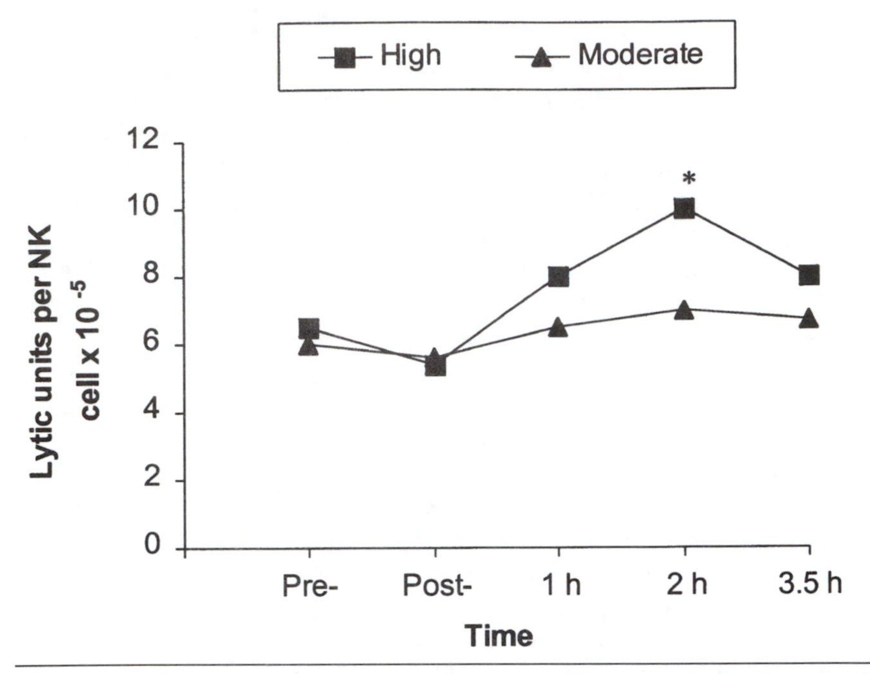

Figure 7.4 Effects of exercise at two intensities on NK cytotoxic activity per cell. Peripheral blood was sampled pre-, immediately post-, and up to 3.5 hours postexercise. Each subject completed two 45-minute exercise sessions, a high-intensity run at 80% $\dot{V}O_{2max}$ and a moderate-intensity walk at 50% $\dot{V}O_{2max}$. NK cytotoxic activity (NKCA) was expressed as lytic units per NK cell to adjust for changes in NK cell number in peripheral blood. * = Significantly different from preexercise value for high-intensity exercise only.

Adapted from Nieman, D.C., A.R. Miller, D.A. Henson, B.J. Warren, G. Gusewitch, R.L. Johnson, J.M. Davis, D.E. Butterworth, and S.L. Nehlsen-Cannarella. 1993b. Effects of high- vs moderate-intensity exercise on natural killer activity. *Medicine and Science in Sport and Exercise* 25: 1126-1134.

Another study from the same laboratory (Nieman et al. 1995) reported no change in NKCA per cell for up to 6 hours after 2.5 hours of intense exercise (treadmill running at 75% $\dot{V}O_{2max}$) in distance runners. NKCA and NK cell number were also shown to increase proportionately after shorter-duration one-legged cycling exercise with an eccentric bias (4 × 5 minutes) (Palmo et al. 1996). In this study, NKCA per cell was unchanged immediately and 4 hours postexercise after adjustment of NKCA to account for changes in NK cell number.

Total NKCA represents a complex interaction of changes in both NK cell distribution and killing activity during and after exercise, as well as possible training effects. It is possible that all changes occur simultaneously, but to varying extents from one time to another. The literature is consistent in showing enhanced cytotoxic activity during and immediately after most forms of exercise studied to date, mainly running and cycling but also interval training (figure 7.5). A delayed suppression of total NKCA observed between 1 and 6 hours postexercise appears to be at least partially related to redistribution of NK cells between the circulation and other body compartments; that is, NK cells exit from the circulation, reducing the proportion (hence number) of NK cells in the in vitro assay. In some circumstances, this delayed suppression may also be attributed to a true suppression of cellular function possibly related to release of certain hormones (see discussion below). However, at present, the relative contribution of changes in cell number and suppression of killing activity appears to vary between exercise conditions, subject fitness level, and possibly between laboratories. Further work is needed on purified NK cell fractions, or with more sophisticated quantification of NK cell subsets, to fully understand the NK cell response to exercise.

One recent study suggested that, at least in an animal model, different mechanisms may underlie changes in NKCA after acute exercise and exercise training (Hoffman-Goetz 1995). Untrained mice performed an acute bout of exercise (30 minutes of treadmill running) or nine weeks of run training (45 minutes per day). NKCA and serine esterase activity were measured in splenocytes obtained immediately after the acute exercise bout or in rested trained animals. A family of lytic enzymes found in cytotoxic cells (e.g., NK cells, LAK cells, and CTL), release of serine proteases mediates target cell killing. In this study (Hoffman-Goetz 1995), acute exercise induced a parallel increase in splenocyte NKCA and serine esterase activity, without any change in NK cell number. In contrast, splenocytes from resting trained mice exhibited significantly higher NKCA and NK cell number, without any change in serine esterase activity. These data suggest that the increase in NKCA after acute exercise resulted from enhanced release of lytic enzymes such as serine esterases, but that changes in lytic enzyme activity were not responsible for the higher splenocyte NKCA observed at rest.

Other Methodological Considerations

NKCA appears to exhibit some diurnal and seasonal variations, although this has not been studied extensively. NKCA appears to be highest early in the morning, with a second smaller peak of activity in the late afternoon (reviewed in Trinchieri 1989). In addition, there may be seasonal changes in NK cell function. For example, in a randomized 15-week exercise intervention study of NK cell function in previously

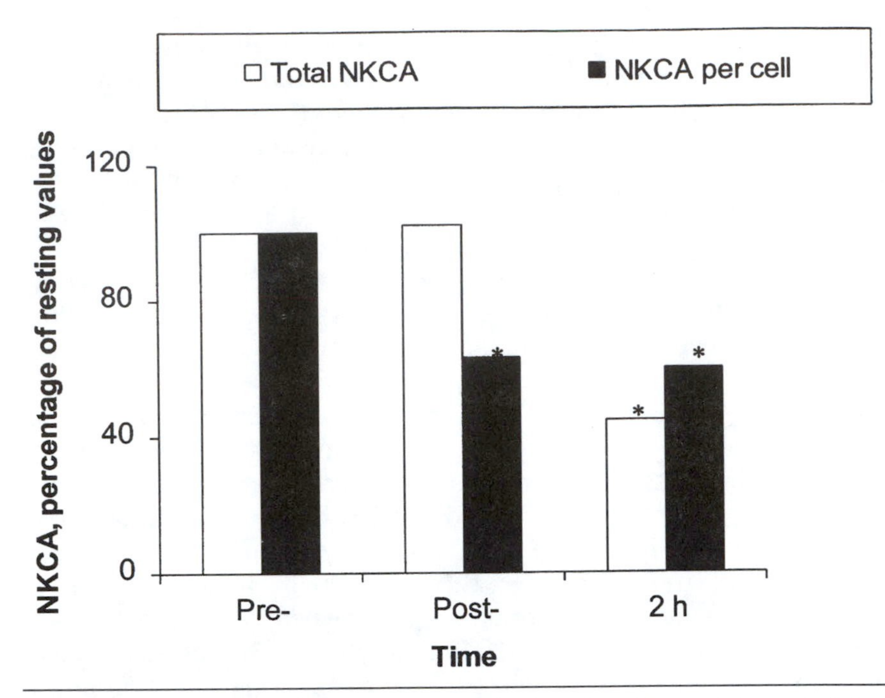

Figure 7.5 Responses of NK cytotoxic activity to exhaustive resistance training. Peripheral blood was sampled from experienced weight lifters pre-, immediately post-, and 2 hours postexercise; exercise consisted of repeated weight lifting to exhaustion (mean time 37 minutes). NK cytotoxic activity (NKCA) is expressed two ways: total NKCA and NKCA expressed per NK cell to adjust for changes in NK cell number in the circulation; data are presented relative to preexercise values. * = Significantly different from preexercise values.

Adapted from Nieman, D.C., D.A. Henson, C.S. Sampson, J.L. Herring, J. Stulles, M. Conley, M.H. Stone, D.E. Butterworth, and J.M. Davis. 1995e. The acute immune response to exhaustive resistance exercise. *International Journal of Sports Medicine* 16: 322-328.

sedentary obese women, NKCA increased significantly between weeks 6 and 15 in the nonexercising group, suggesting some seasonal change in NKCA (Nieman et al. 1990b). These data suggest that experiments designed to study exercise-induced changes in NK cell function should include appropriate nonexercising subjects to control for possible diurnal and seasonal variations.

Mechanisms Affecting NK Cell Number and NKCA During and After Exercise

Several mechanisms have been proposed to explain the changes in NK cell number and cytotoxic activity during and after exercise. These mechanisms include changes in circulating NK cell number, recruitment of activated cells into the circulation, and modulation of NK killing activity of each cell mediated by an increase in core temperature and/or various soluble factors released during exercise such as cytokines, prostaglandins, catecholamines, opioid peptides, and growth hormone (table 7.4) (Brahmi et al. 1985; Crist et al. 1989; Fiatarone et al. 1988; Kappel et al. 1991c, 1994; Mackinnon et al. 1988; Nieman et al. 1993b, 1995a, 1995c, 1995e; Palmo et al. 1996; Pedersen et al. 1988, 1990; Targan et al. 1981; Tvede et al. 1991, 1994). It is likely that the different responses of NK cell number and cytotoxic activity during and after exercise (e.g., stimulation vs. suppression; suppression despite normal cell number) are mediated by a complex interaction of more than one of these factors. Moreover, it also is likely that changes in NK cell number and cytotoxic activity occur independently of one another and may be independently regulated. In addition, the relative contribution to total NKCA attributed to redistribution of NK cells and activation/suppression of killing activity may differ according to the time of sampling after exercise.

Catecholamines

Several studies noting large increases in NK cell number and/or NKCA during or immediately after exercise have also reported significant rises in catecholamine concentration (Mackinnon et al. 1988; Nieman et al. 1993b, 1995e; Kappel et al. 1991c, 1992; Palmo et al. 1995; Tvede et al. 1994).

Acute exercise-induced changes in NK activity and cell distribution are similar to those induced by administration of epinephrine to physiological levels observed during and after moderate exercise. For example, NK activity increased 40% during 1 hours of cycling (75% $\dot{V}O_{2max}$) and then decreased 30% below preexercise levels 2 hours after exercise (Kappel et al. 1991c; Tvede et al. 1994). Similar changes in NK activity were noted with infusion of epinephrine at doses that increased plasma epinephrine concentration to the same level observed during exercise. Moreover, exercise and epinephrine infusion induced similar changes in $CD16^+$ cell number (i.e., increased during, and normal by 2 hours postexercise) (figure 7.6) (Tvede et al. 1994). These data indicate that epinephrine mediates at least part of the acute change in NK cell distribution and total cytotoxic activity during and after exercise, although there is strong evidence to implicate other factors as well (discussed below). It is likely that the main effect on NKCA of catecholamines is mediated via redistribution of NK cells and not via direct action on NK cell-killing activity. Catecholamines do

Table 7.4 Possible Substances Mediating Exercise-Induced Changes in NKCA

Substance	Effect on NKCA	Possible role mediating exercise-induced changes in NKCA	Representative references
Catecholamines	Recruits NK cells into circulation	↑ NKCA via ↑ NK cell number during exercise	Kappel et al. 1991c; Mackinnon et al. 1988; Nieman et al. 1993b
PGE_2	Suppresses cytotoxic acivity	↓ NKCA 1-6 h postexercise in untrained, no apparent effect in trained	Pedersen et al. 1988, 1990; Nieman et al. 1995a
Growth hormone	Stimulates cytotoxic activity	No likely acute role; possible chronic role	Kappel et al. 1993; Crist et al. 1988
β-endorphin	Stimulates cytotoxic activity	↑ NKCA without altering cell number during exercise; possibly no role	Fiatarone et al. 1988; Klokker et al. 1995
IL-2	Stimulates cytotoxic and LAK activity	↑ NKCA without altering cell number during exercise	Nielsen et al. 1996a, 1996b; Pedersen et al. 1988; Tvede et al. 1993
IFN	Stimulates cytotoxic activity	Synergistically ↑ NKCA during exercise	Nielsen et al. 1996b; Targan et al. 1981

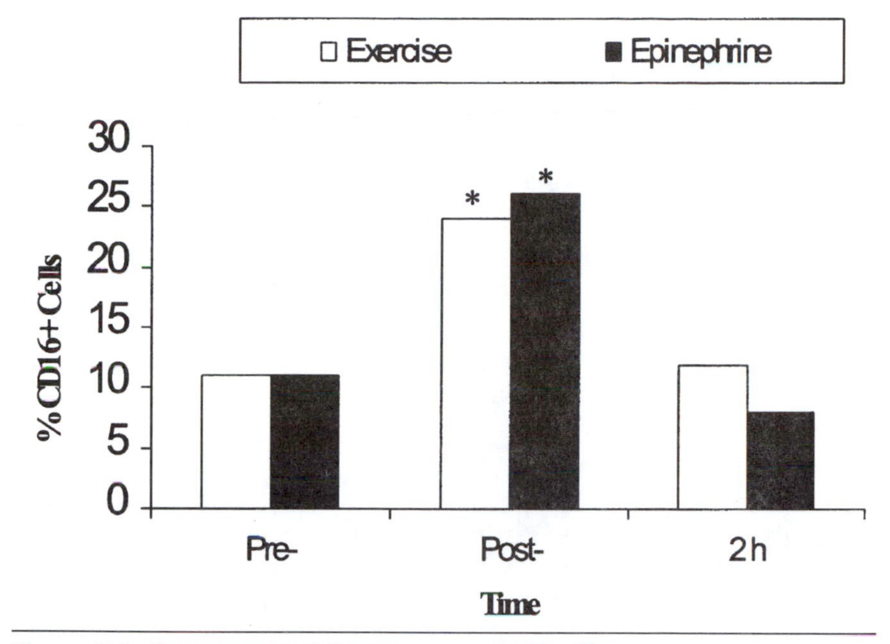

Figure 7.6 Comparison of the response of NK cell number to exercise and epinephrine infusion. Subjects performed 1-hour cycle ergometry at 75% $\dot{V}O_{2max}$ on one day, and on another were infused with epinephrine to match blood levels observed during exercise. Peripheral blood was sampled pre-, immediately post-, and 2 hours postexercise or infusion. NK cells were identified as CD16+. * = Significantly different from preexercise or infusion levels.

Adapted from: Tvede, N., M. Kappel, K. Klarlund, S. Duhn, J. Halkjaer-Kristensen, M. Kjaer, H. Galbo, and B.K. Pedersen. 1994. Evidence that the effect of bicycle exercise on blood mononuclear cell proliferative responses and subsets is mediated by epinephrine. *International Journal of Sports Medicine* 15: 100-104.

not appear to be involved in training-induced changes in NKCA, however. For example, in a double-blind study using β-adrenergic blockade by atenolol or propranolol in previously untrained males, 15 weeks intense endurance exercise training resulted in a similar decrease in resting NKCA in subjects taking placebo and β-blockers (Watson et al. 1986).

It appears that, during exercise, recruitment of NK cells into the circulation from the marginated pool is mediated, at least partially, by epinephrine-induced changes in NK cell adhesion to the vascular endothelium. For example, using an in vitro model, Benschop et al. (1993) showed a dose-response relationship between epinephrine concentration and NK-endothelial cell adhesion. Enriched NK cell fractions and endothelial cells isolated from human umbilical cord vein were incubated together, and binding between cells was assessed using monoclonal antibodies and flow cytometry. Five minutes of incubation at physiological concentrations of epinephrine was sufficient to elicit a significant reduction in NK-endothelial cell adhesion. Moreover, detachment of NK cells was shown to be selective for NK cells and to be mediated by activation of adenylate cyclase via β_2-adrenergic receptors. These data suggest that recruitment of NK cells into the circulation during exercise is mediated via epinephrine-induced alterations in NK cell adhesion to the vascular endothelium; whether a reversal of this process mediates the rapid removal of NK cells from the circulation after exercise is not currently known.

Prostaglandins

Suppression of total NKCA 1 to 2 hours after prolonged (e.g., >1 hour) exercise has been attributed, at least partially, to prostaglandins released from monocytes during exercise. Several prostaglandins (e.g., PGE_1, PGE_2, PGA_1, PGA_2) are known inhibitors of NKCA, and PGE_2 production by monocytes increases after exercise (Pedersen et al. 1988, 1990).

In untrained males, total NKCA was stimulated in cells obtained near the end of exercise (1 hour of cycling at 80% $\dot{V}O_{2max}$), but suppressed below preexercise values in cells obtained 2 hours postexercise (Kappel et al. 1991c; Pedersen et al. 1988, 1990). Oral administration of the prostaglandin inhibitor indomethacin before exercise did not alter the stimulation of NKCA during exercise, but did prevent the decrease in NKCA after exercise (figure 7.7) (Pedersen et al. 1990). Similar results were obtained when indomethacin was added to the in vitro assay of NKCA (without oral administration of indomethacin). Moreover, the postexercise suppression of NKCA was correlated to the increase in circulating monocyte number during exercise. Monocytes are a source of prostaglandins released during exercise, and depletion of monocytes from blood samples before the start of the in vitro NKCA assay restored cytotoxic activity in cells obtained 2 hours postexercise (Pedersen et al. 1990). These data suggest that a monocyte-derived factor inhibited by indomethacin, such as prostaglandins, is released during exercise and that this factor suppresses NKCA after but not during exercise. Since NKCA is generally restored by 6 hours postexercise (except in prolonged, more intense exercise), prostaglandin-induced suppression of NKCA appears to be transitory.

In contrast, work from another laboratory has failed to confirm a relationship between prostaglandin release during and suppression of NKCA after prolonged exercise (Nieman et al. 1995a). Twenty-two

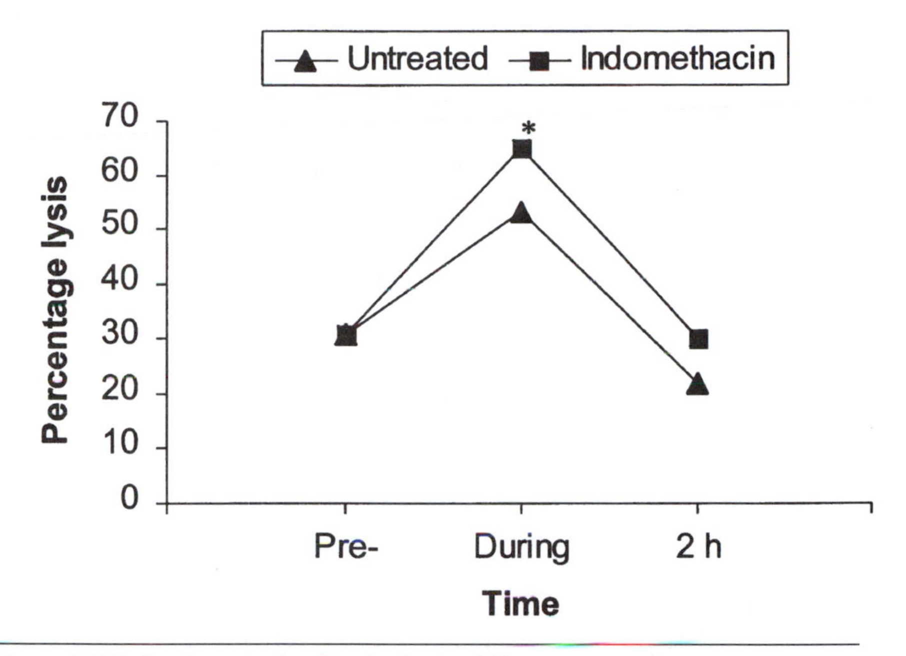

Figure 7.7 Role of prostaglandins in altering NK cytotoxic activity in response to exercise. NKCA was measured in peripheral blood sampled pre-, during, and 2 hours after 1 hour of cycle ergometry at 75% $\dot{V}O_{2max}$. Subjects exercised after oral administration of indomethacin or without (untreated). * = Significantly different between treatment conditions.

Adapted from Pedersen, B.K., N. Tvede, K. Klarlund, L.D. Christensen, F.R. Hansen, H. Galbo, and A. Kharazmi. 1990. Indomethacin in vitro and in vivo abolishes post-exercise suppression of natural killer cell activity in peripheral blood. *International Journal of Sports Medicine* 11: 127-131.

male distance runners ran on a treadmill for 2.5 hours at 75% $\dot{V}O_{2max}$. Blood samples were obtained before, immediately after, and then 1.5, 3, and 6 hours postexercise; to control for diurnal variation, blood samples were also obtained at the same times from nonexercising control subjects. Total NKCA per fixed number of mononuclear cells did not change immediately after exercise, but declined by 50-60% 1.5 hours postexercise and remained at this low level up to 6 hours postexercise. Indomethacin added to the NK assay had no effect on NKCA at any time point in both runners and nonexercising controls. Moreover, when NKCA was adjusted for changes in NK cell number and plasma volume, NKCA per cell was unchanged at any time after exercise. These data suggest that changes in total NKCA during and up to 6 hours after prolonged intense exercise are due primarily to changes in circulating NK cell number and not to any changes in killing activity.

Differences between studies may relate to the mode and duration of exercise, fitness level of the subjects, and the hormonal responses to the different exercises. For example, in the studies by Pedersen's group (Pedersen et al. 1988, 1990; Kappel et al. 1991c; Tvede et al. 1994), subjects were untrained and shorter-duration cycling exercise was used (e.g., 1 hour at 60-80% $\dot{V}O_{2max}$). In contrast, the study by Nieman et al. (1995a) included trained distance runners and longer-duration treadmill running (e.g., 2.5 hours at 75% $\dot{V}O_{2max}$). It is possible that exercise training reduces prostaglandin release, or raises the threshold of exercise intensity/duration required to induce prostaglandin release, during exercise. Alternatively, longer-duration exercise may induce release of other factors (e.g., cytokines) that counteract the suppressive effect of prostaglandins on NKCA observed after shorter (i.e., 1 hour) exercise. It has also been suggested that the hormonal response to 1 hour of cycling differs significantly from the response to 2.5 hours running: 1 hour of exercise is associated with high epinephrine and relatively low cortisol levels, whereas longer exercise (>2 hours) is associated with moderately elevated epinephrine but much higher cortisol concentrations (Nieman et al. 1995a). It is possible that prostaglandin-induced suppression of NKCA occurs only after specific durations and/or intensities of exercise (e.g., after 1 but not 2.5 hours, as discussed above), or only in particular subjects (e.g., untrained but not trained). Clarification of the role, if any, of prostaglandins in suppressing NKCA after intense exercise awaits further study directly comparing the effects of various exercise durations and intensities.

β-Endorphin

There is some evidence that β-endorphin may be involved in stimulation of NK activity during exercise. For example, in young women who exercised to 75% $\dot{V}O_{2max}$ in a graded cycle test (approximate duration 10 minutes), both NK cell number and NKCA increased immediately after exercise (Fiatarone et al. 1988). Administration of the opioid antagonist naloxone before exercise blunted the increase in NKCA observed during exercise with placebo. In contrast, naloxone did not alter the increase in NK cell number during exercise, indicating an effect on cytotoxic activity of NK cells and not an indirect effect via modulating circulating NK cell number. Moreover, the postexercise rise in NKCA was blunted by addition of β-endorphin to the in vitro assay. It was suggested that NK cells were already maximally stimulated by β-endorphin released during exercise, or there may be a refractory period after exercise or previous exposure to β-endorphin during which time NK cells are unresponsive to further stimulation by opioid peptides.

In contrast, data from a subsequent study using sensory nerve blockade (SNB) via epidural anesthesia (Klokker et al. 1995) do not support a role of β-endorphin in stimulation of NKCA during exercise. In this study, SNB was used to block release of β-endorphin during exercise. On two different occasions, seven untrained males cycled in a recumbent position at 60% $\dot{V}O_{2max}$ for two 20-minute sessions separated by 120 minutes rest. The first exercise of each session was performed under normoxic and the second exercise under hypoxic (11.5% O_2) conditions. NK cell number ($CD16^+$ and $CD56^+$) and NKCA were assessed in blood sampled before and after each 20-minute exercise. As expected, NK cell number and cytotoxic activity increased significantly during normoxic exercise. The increase in NK cell number was greater during hypoxic compared with normoxic exercise, although hypoxia had no effect on the increase in NKCA during exercise; hypoxia is known to recruit lymphocytes into the circulation. SNB had little or no further effect on the increases in NK cell number or NKCA during exercise under either condition, which was contrary to the expectation, based on previous work by Fiatarone et al. (1988), that preventing the rise in β-endorphin release would abrogate the exercise-induced increase in NKCA. One explanation offered by Klokker et al. (1995) to account for the differences between studies was the prior suggestion by Fiatarone et al. (1988) that inhibition of the exercise-induced increase in NKCA by in vivo administration of naloxone could occur via a direct effect on NK cells independent of β-endorphin release during exercise. That is, the earlier observation of lower NKCA after exercise with naloxone (Fiatarone et al. 1988) may have been due to a direct inhibition of NKCA by naloxone and not via inhibition of release of β-endorphin. If confirmed by further study, these data argue for little involvement of β-endorphin in exercise-induced stimulation of NKCA.

Other Hormones

A role of other hormones in contributing to exercise-induced alterations in NK cell number and NKCA is based on known effects of hormones such as growth hormone or cortisol on NK cell function (Pedersen 1997). It is unlikely, however, that cortisol and growth hormone contribute significantly to the acute effects of exercise on NK cell number and NKCA for the following reasons: plasma cortisol levels generally increase only after intense exercise (Wilmore and Costill 1994), but, as described above, NK cell number and activity have been observed to change dramatically after brief and prolonged moderate exercise in which plasma cortisol levels are not substantially elevated (Berk et al. 1990; Mackinnon et al. 1988; Nieman et al. 1993b, 1995e; Pedersen et al. 1988).

A bolus infusion of human growth hormone to physiological levels observed during exercise had no effect on lymphocyte subset distribution, NK cell number, NKCA, or cytokine production, suggesting no acute effect of growth hormone on NK cell function (Kappel et al. 1993). However, it is possible that growth hormone may contribute to long-term changes in NK cell function. For example, growth hormone deficiency is associated with lower NKCA, and administration of recombinant human growth hormone over six weeks resulted in a significant increase in NKCA in growth hormone-deficient women (Crist et al. 1988) and in healthy adults (Crist and Kraner 1990).

Cytokines

Cytokines such as IFNα, IL-2, and TNFα are potent stimulators of NK activity, and IL-1 may also indirectly enhance NKCA by stimulating release of IL-2 and TNFα (Janeway and Travers 1996; Liles and van Voorhis 1995; Trinchieri 1989). A more detailed discussion of cytokine release during and after exercise was presented in the previous chapter.

Concomitant and proportional increases in IL-1 and total NK activity were reported after brief maximal exercise testing (Lewicki et al. 1988). In this study, NK cell percentage was unchanged by exercise, suggesting possible activation of killing by IL-1 released during exercise, although an indirect effect of IL-1 causing release of other factors cannot be discounted. In contrast, however, a delayed rise in IL-1 appearance in urine observed 1 hour after 20 km running (Sprenger et al. 1992) argues against a role for IL-1 in stimulating NKCA during exercise. Moreover, mild endurance exercise, such as 1 hour of cycling at 60% $\dot{V}O_{2max}$, has been shown to increase NKCA (Tvede et al. 1993), but IL-1 does not appear to be released during such exercise (Smith et al. 1992). Thus, any involvement of IL-1 in exercise-induced stimulation of NKCA is likely to occur some time after exercise, and only after prolonged, vigorous exercise (e.g., distance running). It is possible that the delayed release of IL-1 is related to restoration of NKCA despite low NK cell numbers 1-6 hours postexercise (Berk et al. 1990; Mackinnon et al. 1988; Nieman et al. 1993b, 1995a).

TNFα may be released during prolonged exercise (Dufaux and Order 1989b; Sprenger et al. 1992), which may possibly stimulate NKCA. However, it is unclear at present whether changes in TNFα are related to enhanced NKCA during and after exercise.

It is unclear whether IL-2 is involved in exercise-induced stimulation of NKCA. Circulating levels of IL-2 may decrease during brief high-intensity exercise (Lewicki et al. 1988), although there is evidence of IL-2 release during prolonged exercise such as distance running (Sprenger et al. 1992). Addition of IL-2 to cells obtained during exercise further augments the exercise-induced increase in NKCA (Nielsen et al. 1996b; Pedersen et al. 1988). For example, addition of IL-2 to the in vitro NKCA assay using cells obtained pre-, immediately post-, and 2 hours postexercise (6 minutes maximal rowing) increased NKCA compared with that in control (no IL-2) assays of cells obtained at the same time points (figure 7.8) (Nielsen et al. 1996b). Thus, NK cells are capable of responding to IL-2 stimulation during and after exercise. However, IL-2 cannot overcome the delayed suppression of NKCA below baseline 2 hours after prolonged exercise (Pedersen et al. 1988). IL-2 was also reported to induce greater stimulation of NKCA compared with IFNα (Nielsen et al. 1996b; Pedersen et al. 1988), suggesting that exercise enhances NK cell sensitivity to IL-2, or that a more responsive NK cell is recruited into the circulation during exercise, or some combination of the two.

In contrast, in another study, NK cells were found to be less responsive to IL-2 after shorter-duration exercise (Fiatarone et al. 1989). Untrained younger and older women (mean ages 30 and 71 years, respectively) completed a 10-12-minute progressive cycle ergometer test to approximately 75% $\dot{V}O_{2max}$. NKCA was assessed before and after exercise, with and without recombinant IL-2 added to the assay system. Compared with preexercise levels, NKCA doubled immediately after exercise in control (no added IL-2) assays. Addition of recombinant IL-2 stimulated NKCA by 29% above control levels in cells obtained before exercise, but by only 7% in cells obtained immediately after exercise. Moreover, there was a significant strong negative correlation ($r = -0.954$) between the magnitude of IL-2 stimulation in NK cells obtained before and after exercise. There were no differences between younger and older subjects on any NK variable measured. These data suggest that, after exercise, NK cells are less responsive to IL-2 stimulation, possibly due to maximal stimulation of NKCA during exercise.

Exercise training may increase expression of the high-affinity IL-2β subunit (Rhind et al. 1994), which may increase sensitivity of NK cells to IL-2. In a cross-sectional comparison of seven male distance runners and six untrained control subjects, runners exhibited significantly more T ($CD3^+/CD8^+$) and NK ($CD16^+$ and $CD56^+$) cells expressing the β subunit (CD122); the β subunit was expressed on far more NK compared with T cells (70-90% of NK cells compared with 20-37% of

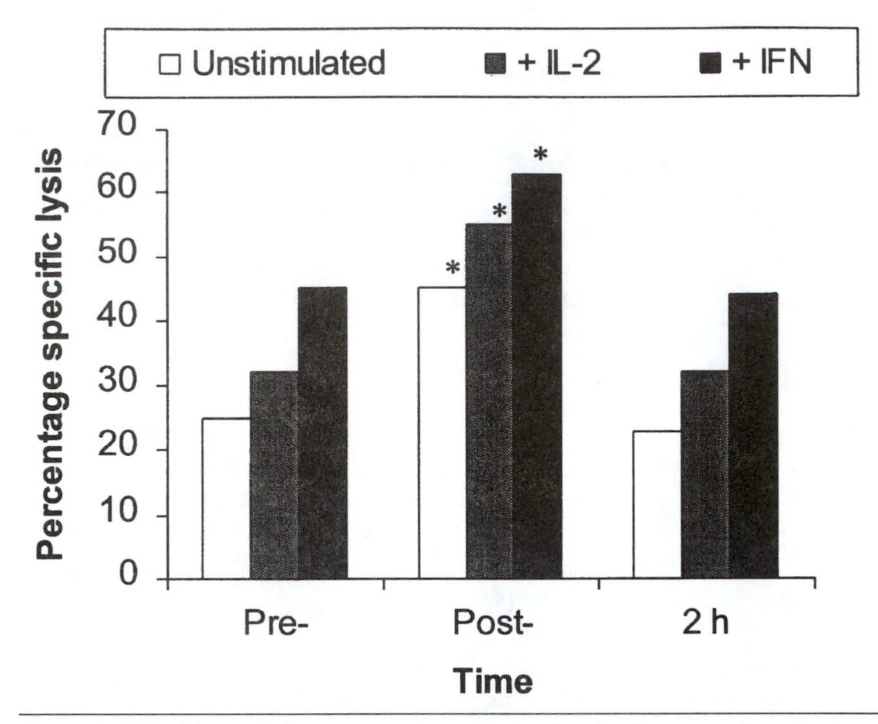

Figure 7.8 NKCA response to exercise in unstimulated cells or in cells stimulated with IL-1 or IFNα. Trained rowers completed a maximal 6-minute rowing ergometer test and peripheral blood was sampled pre-, immediately post-, and 2 hours postexercise. Mononuclear cells were incubated with medium (unstimulated) IL-2 or IFNα for 1 hour before the standard ^{51}Cr release NK assay. * = Significantly different from unstimulated cells.

Adapted from Nielsen, H.B., N.H. Secher, M. Kappel, B. Hanel, and B.K. Pedersen. 1996b. Lymphocyte, NK and LAK cell responses to maximal exercise. *International Journal of Sports Medicine* 17: 60-65.

$CD8^+$ T cells). β subunit expression was evident on 10% to 26% more NK cells in trained compared with untrained subjects. Moreover, there was a significant and strong correlation (r ~ 0.90) between $\dot{V}O_{2max}$ and expression of the IL-2 receptor β subunit for pooled data from both groups as well as within the group of runners. These data suggest that endurance training induces expression of the higher-affinity IL-2 receptor on NK cells, which may indicate a more active NK cell in the circulation in athletes; increased sensitivity to IL-2 may be related to stimulation of NKCA during exercise or as a result of exercise training. It should be noted that, although these athletes were rested for 36 hours before blood sampling, a long-lasting effect of the last training session on IL-2β expression cannot be ruled out. However, since well-trained athletes exercise daily, these data provide evidence for continued activation of NK cells by exercise training (regardless of whether it is due to a chronic training or long-lasting acute effect).

IFNα and IFNβ enhance NK activity when added to cells obtained either before or after exercise (Brahmi et al. 1985; Nielsen et al. 1996b; Pedersen et al. 1988; Targan et al. 1981). Exercise-induced stimulation of NK activity is further augmented by addition of IFNα to cells obtained after exercise (Nielsen et al. 1996b; Pedersen et al. 1988; Targan et al. 1981), and exercise and IFNα appear to have a synergistic effect on NK activity, at least after brief exercise (Targan et al. 1981).

Other Factors

NK cells are sensitive to temperature, both in vitro and in vivo. Core temperature may increase to 39-40 °C during prolonged exercise, and the magnitude of this increase is related to exercise intensity. Increased core temperature has been postulated as one mechanism underlying the increases in NKCA observed during exercise (Kappel et al. 1991b; Pedersen and Ullum 1994). For example, immersion of subjects in hot water to achieve rectal temperature of 39.5 °C resulted in 50-100% increases in NK cell number and 20-40% increases in NKCA, compared with immersion in thermoneutral water that did not increase core temperature (Kappel et al. 1991b); these increases in NK cell number and activity are similar to those observed during prolonged exercise. However, the increase in core temperature during hot water immersion was associated with only small increases in catecholamine levels, suggesting that catecholamines account for only part of the hyperthermia-induced enhancement of NKCA.

Neural input may also be involved in recruitment of NK cells into the circulation, although neural factors cannot entirely account for the increase in NK cell number and NKCA during exercise. For example, Nash (1994) reported 50% lower resting NKCA and NK ($CD3^-$/$CD56^+$) cell number in eight healthy quadriplegic patients compared with able-bodied control subjects. However, in these patients, NKCA and NK cell number increased more than twofold (i.e., as would be expected in an able-bodied population) after 30 minutes cycling exercise using computer-sequenced electrically stimulated contractions of the muscles used in cycling. The authors suggested that these increases in NK cell number and activity may have resulted from release of factors such as endogenous opioids despite the lack of sympathetic neural input. As described above, sensory nerve blockade by epidural anesthesia at a level lower than the splenic nerve did not alter the response of NK cells to normoxic and hypoxic exercise, suggesting that afferent nerve impulses from working muscles are not necessary for recruitment and activation of NK cells during exercise (Klokker et al. 1995). Although it is not yet clear whether an intact autonomic neural network is required for full mobilization into the circulation of NK cells during exercise, it would appear that at least some of the increase in NK cell number occurs independently of sympathetic neural input.

Exercise and LAK Cytotoxic Activity

As discussed in chapter 2, NKCA may be stimulated by in vitro incubation of NK cells with cytokines such as IL-2 or IFNγ for 72 hours. This increase in activity is attributed to lymphokine-activated killer (LAK) cells, which are considered to be activated NK cells. Several recent studies suggest that, in addition to enhanced NKCA, LAK is also stimulated by acute exercise (Hoffman-Goetz et al. 1994; Nielsen et al. 1996b; Tvede et al. 1993, 1994). For example, LAK activity was significantly enhanced in cells obtained before and during 1 hour of cycling at 25%, 50%, and 75% $\dot{V}O_{2max}$ (Tvede et al. 1993, 1994). Peripheral blood mononuclear cells were isolated from blood obtained before, in the last few minutes of, and 2 hours after exercise and were incubated with IL-2 for 72 hours before the standard NKCA assay. LAK activity increased to a similar extent (two- to threefold) in cells obtained during exercise regardless of exercise intensity. LAK activity returned to preexercise levels by 2 hours postexercise for the two lowest intensities (25% and 50% $\dot{V}O_{2max}$); however, preincubation with IL-2 could not prevent the decline below baseline in LAK activity 2 hours after exercise at 75% $\dot{V}O_{2max}$. The increase in LAK activity during and the decrease 2 hours after exercise appear to be mediated, at least partially, by epinephrine. For example, epinephrine infusion to physiological concentrations observed during exercise elicited a similar pattern of LAK activity during and after exercise (60 minutes of cycle ergometry at 75% $\dot{V}O_{2max}$) (Tvede et al. 1994). However, the elevation of LAK during exercise was greater than during epinephrine infusion, suggesting that other factors in addition to epinephrine may mediate enhancement of LAK activity during exercise.

LAK activity was also reported to increase after maximal exercise in elite rowers (Nielsen et al. 1996b). Ten elite competitive rowers completed 6 minutes all-out exercise on a rowing ergometer; blood was obtained before, during, and 2 hours after exercise. IL-2-stimulated LAK activity increased about 30% during exercise and returned to preexercise values by 2 hours postexercise. This increase in LAK during exercise was smaller than the increase in NKCA (about 80%), mainly due to a greater stimulation of LAK in the preexercise sample compared with NKCA at the same time.

IL-2-stimulated splenic LAK activity was also enhanced in mice allowed access to voluntary exercise training over eight weeks compared with nonexercised mice (Hoffman-Goetz et al. 1994). This increase in LAK occurred despite no change in NKCA, and suggests that regular exercise may enhance the ability of NK cells to respond to stimulation by cytokines. Although observed in a different species, such enhanced responsiveness to IL-2 in mouse NK cells is consistent with the increase in expression of the high-affinity IL-2 receptor noted in exercise-trained humans (Rhind et al. 1994), as discussed above.

Exercise and ADCC

Antibody-dependent cell-mediated cytotoxicity (ADCC) occurs by a different mechanism than does NK cell cytotoxicity, although both activities are mediated by the same cell, the NK cell or LGL. It is expected, then, that exercise-induced changes in ADCC are similar to those observed for NK cells, although there are few reports on ADCC and exercise (Hanson and Flaherty 1981; Hedfors et al. 1978). ADCC increased by 20% to 40% early during exercise (during the first 10 minutes of 25 minutes of moderate cycle ergometry) (Hedfors et al. 1978). The number of cells exhibiting ADCC increases at the same time as does killing activity, although the magnitude of increase in cell number is larger than for killing activity (Hanson and Flaherty 1981), suggesting that inactive cells may be recruited into the circulation during exercise.

Exercise and Monocyte/ Macrophage Cytotoxic Activity

Cytotoxic activity of monocytes/macrophages appears to be enhanced by exercise, although the extent to which cytotoxicity is modified varies with exercise intensity, timing of exercise in relation to measurement of activity, and activation state of the cells (Lotzerich et al. 1990; Woods and Davis 1994; Woods et al. 1993, 1994a, 1994b). In mice, peritoneal macrophages obtained after a single session of exhaustive exercise (45-50 minutes) inhibited in vitro tumor cell proliferation (Lotzerich et al. 1990). However, macrophage binding to and lysis of tumor cells were unaffected by exercise, suggesting that exercise may induce release from macrophages of a soluble factor that inhibits tumor cell growth.

Woods et al. (1993) studied the effects of moderate and intense exercise on the ability of mouse inflammatory peritoneal macrophages to inhibit tumor growth in vitro. Mice were first injected with thioglycate as an inflammatory challenge, since resident macrophages exhibit low cytotoxic activity; primed inflammatory macrophages are representative of the types of cells responding to bacterial, viral, and tumor challenge. Male mice exercised for three consecutive days after injection. Exercise consisted of either 30 minute of moderate treadmill running or running to exhaustion (2-4 hours); control mice did not exercise, but were placed above the treadmill so that they experienced similar conditions (noise, vibration) as the exercised mice. Compared with that in cells from control mice, both exercise conditions stimulated macrophage cytotoxicity, measured as the ability to inhibit tumor cell growth, to a similar extent (mean 50% increase) up to 8 hours postexercise. Postexercise stimulation of cytotoxicity could not be attributed to changes in cell number or adherence, since these did not differ between groups. Moreover, macrophage cytotoxicity was insensitive to corticosterone added to the cultures, indicating no role for glucocorticoids (which increased during exercise). Repeated washings of cells prior to the assay reduced cytotoxicity, suggesting that a soluble mediator was involved. Addition of antibodies to IL-1β to the cultures had no effect, whereas antibodies to TNFα partially abolished exercise-induced stimulation of macrophage cytotoxicity. These data suggest that TNFα and possibly other factors released during exercise may stimulate inflammatory macrophage cytotoxicity; IL-1β is unlikely to be involved.

In subsequent studies using the same exercise protocol as described above, Woods et al. (1994a, 1994b) have shown that the effects of exercise on macrophage cytotoxicity are quite complex and are related to the functional state of macrophages at the time of exercise. Noting that macrophages exist in three general states (resident [resting], inflammatory [primed], and fully activated), Woods et al. (1994a) compared the effects of exercise on cytotoxic activity in inflammatory and activated macrophages. In separate experiments, thioglycate injection was used to elicit inflammatory macrophages (as described above), and injection of inactivated bacteria was used to elicit activated macrophages. When exercise was performed throughout the five to seven days of injections, there was no effect on the number of inflammatory cells; in contrast, exhaustive but not moderate exercise reduced the number of activated cells. Both exercise conditions were equally effective in stimulating cytotoxic activity of activated cells. However, when exercise was introduced after seven days of bacterial injection (i.e., when cells were fully activated), only moderate exercise stimulated cytotoxicity of activated cells. This enhancement of cytotoxicity could not be attributed to release of cytokines such as IL-1β or TNFα. These data suggest that both moderate and exhaustive exercise stimulate macrophage cytotoxicity during the activation process (before cells have become fully activated), but that only moderate exercise enhances cytotoxicity in fully activated cells. The authors noted there is no clear explanation for such differences, but suggested several possibilities, including an exercise-induced increase in the macrophage precursor pool in cells undergoing activation, but not in fully activated cells; effects of prolonged exposure to elevated glucocorticoids on macrophage activation, which may explain the lower number of activated macrophages and lack of stimulation of cytotoxicity by exhaustive but not moderate exercise; or the combined effects of (as yet unidentified) immunoenhancing and immunosuppressing factors that may differentially influence cytotoxicity in primed and activated cells.

In another study using the same exercise protocol as described above, Woods et al. (1994b) questioned whether exercise-induced stimulation of macrophage cytotoxicity influences in vivo tumor growth in a mouse model. Mice were exercised for 3 days before inoculation with a mammary adenocarcinoma, a tumor susceptible to macrophage cytotoxicity; exercise continued for 14 days post-inoculation. Tumor inoculation occurred 3 hours after exercise, a time of maximal enhancement of macrophage cytotoxicity (Woods et al. 1993). Surprisingly, tumor appearance was significantly delayed in control compared with both exercise groups (moderate and exhaustive) during the first 7 days, after which tumor incidence was similar in all groups. Daily tumor size did not differ significantly between the groups. The number of tumor-infiltrating macrophages was higher in moderately-exercised compared with control and exhaustively exercised mice. However, tumor-infiltrating cell number was not significantly correlated with tumor weight. Thus, although moderate exercise appeared to stimulate macrophage infiltration into the tumors, exercise did not alter the progression of tumor growth.

Several possible mechanisms were suggested to explain these results. First, although moderate exercise enhanced the number of cells infiltrating the tumor, cytotoxic activity of these cells may not have been increased. (It was noted that it is very difficult to separate and directly measure cytotoxic activity in tumor-infiltrating macrophages.) Second, although exercise may have stimulated macrophage cytotoxic activity, other aspects of in vivo antitumor defense (e.g., NK cytotoxicity) may have been adversely affected by the exercise protocol. Third, exercise may have altered secretion of soluble factors that influence tumor growth. Finally, it is possible that the time frame of exercise (3 days before and 14 days after inoculation) may have been too short, or the tumor load too high, to show an effect of exercise on in vivo tumor growth. Taken together, these studies suggest that although exercise may stimulate in vitro macrophage cytotoxicity under some conditions, this may not necessarily translate to reduced tumor growth in vivo. Furthermore, the effects of exercise on macrophage cytotoxicity are complex, and likely to be related to exercise intensity and the functional state of cells at the time of exposure to exercise.

Exercise and Cytotoxic T Cells

Although CTL are important effectors in cell-mediated cytotoxicity against tumor and virally infected cells, there do not appear to be any reports on the effects of exercise on CTL cell number or activity.

Summary and Conclusions

The response of NK cells to exercise is among the most extensively studied in the exercise and immunology literature. Exercise induces a rapid and dramatic change in NK cell number and cytotoxic activity, although the pattern of change in cell number and activity is complex and not completely understood. Moreover, exercise-induced changes in circulating NK cell number and cytotoxic activity may occur over different time courses and independently of each other. NK cell number in the circulation increases markedly early in, and remains elevated throughout, exercise. This rise in NK cell number is larger than those observed in other lymphocyte subsets (e.g., T or B cells), such that the proportion of NK cells relative to total lymphocyte number increases from about 10% at rest up to 30% during exercise. The magnitude of increase in NK cell number is related to exercise intensity. NK cell number returns to preexercise values soon after moderate or brief exercise, but may decline below baseline levels for several hours and even days after prolonged intense exercise. Resting NK cell number may also change in response to exercise training; some cross-sectional studies have shown higher NK cell number in endurance athletes compared with nonathletes. However, prolonged periods (e.g., months) of intense exercise training have been associated with declining NK cell number in elite athletes.

NKCA displays a complex response to exercise, varying with exercise mode, duration, and intensity; state of training; and time of blood sample. Total NKCA generally increases during exercise, which appears to be related to both recruitment of cells into the circulation, mediated by catecholamines, and augmentation of cytotoxic activity of each cell, possibly mediated by cytokines or hormones released during exercise. During moderate or brief exercise, the increase in total NKCA is due mainly to an increase in circulating NK cell number, whereas during more intense or prolonged exercise, other factors such as hormones and cytokines may contribute to the increase in NKCA. Total NKCA returns to preexercise values soon after brief or moderate exercise. In contrast, prolonged (>1 hour) or intense exercise (>75% $\dot{V}O_{2max}$) induces a delayed suppression of total NKCA from 1 to 6 hours postexercise, which may persist for up to seven days after intense, prolonged exercise.

Macrophage cytotoxicity against tumor cells appears to be stimulated by exercise, but this enhancement is dependent on exercise intensity, the timing of exercise in relation to assay, and the functional state of the macrophage. In a mouse model, both moderate and exhaustive exercise stimulate cytotoxicity in primed macrophages, but only moderate exercise appears to do so in activated macrophages. Whether enhancement of macrophage cytotoxicity alters in vivo tumor growth has not been clearly demonstrated.

Research Findings

- NK cell number and cytotoxic activity are profoundly affected by acute exercise, more so than for other lymphocyte subsets. NK cell number and activity increase in proportion to exercise intensity during exercise, but may also decline below baseline levels 1 to 6 hours after intense prolonged exercise.
- Exercise-induced changes in some hormones, especially epinephrine, β-endorphin, and prostaglandins, have been associated with changes in NK cell number and activity during and after exercise.
- Although resting NK cell number and cytotoxic activity appear normal in athletes, recent reports suggest decreases during prolonged periods of intense training (10 days to several months). Moderate training may not affect, or may increase, cell number and function.
- LAK cytotoxic activity is also stimulated acutely and chronically by exercise, suggesting further stimulation of already activated cells.
- Monocyte and macrophage cytotoxic activity may be enhanced by moderate exercise, but inconsistent data have been reported and the responses and mechanisms appear to be quite complex.

Possible Applications

- If changes in NKCA are related to changes in immunity, it is possible that regular assessment of NKCA may provide a means of monitoring adaptation to training and susceptibility to illness.
- The observed postexercise suppression of NK cell activity after intense prolonged exercise suggests that those at risk of infection or immunocompromised individuals (e.g., HIV-infected or transplant patients) should avoid this type of exercise.

Yet to Be Explored

- Where do NK cells recruited into the circulation during exercise go after the end of exercise? Do they simply return to marginated pools, or do they migrate to specific tissues (e.g., damaged skeletal muscle)? If the latter, what is their role in tissue damage and repair?
- Why are NK cells affected by exercise to a greater extent than other lymphocyte subsets?
- Do acute and chronic changes in NK cell cytotoxic activity influence immunity in athletes?
- What is the relative contribution of various hormones (e.g., catecholamines, endorphins, prostaglandins) and cytokines to exercise-induced changes in NK cell cytotoxic activity at various times during and after exercise?
- Are the processes of NK cell killing really suppressed 1 to 6 hours postexercise, or does this simply reflect a redistribution of cells in the circulation?
- Are different mechanisms responsible for changes in NK cell cytotoxic activity at different time points during and after exercise?
- Are changes in NK cell and monocyte/macrophage cytotoxic activity related to resistance to infectious illness and/or tumor growth?

Chapter 8

Immune Response to Exercise: Potential Clinical Applications

© CLEO Photography

The recent interest in the immune response to exercise is partially motivated by the potential clinical applications to further understand and possibly prevent, treat, or control certain diseases or conditions. Despite the relative youth of the field as a whole, research in this area has rapidly found a place in applications as diverse as prevention and treatment of cancer, diseases involving the immune system (e.g., HIV/AIDS, autoimmune diseases), spaceflight, chronic fatigue syndrome, and of course, prevention of illness among athletes. Such diversity attests to the potential broad application of knowledge to many issues related to immune function. That physical activity may have positive effects on certain diseases or conditions involving the immune system is implicit in the current clinical interests in exercise immunology research: if regular physical activity exerts positive effects on immune function, then exercise may have a role in stimulating the immune system during certain diseases (e.g., cancer, HIV/AIDS), immune dysfunction (e.g., rheumatoid arthritis, chronic fatigue syndrome), or reduced responsiveness (e.g., aging, spaceflight). It is also possible that a lifetime of regular exercise may help maintain the immune system at its optimum throughout the life span.

However, the possibilities for clinical applications are, at present, not matched by empirical data; that is, although there is great interest in these potential applications, research has only recently begun to quantitatively address these issues. The following discussion on various potential clinical applications is therefore limited by the relatively sparse body of literature that exists at present. This discussion of potential clinical applications focuses specifically on existing data on the interaction between exercise, immune function, and the relevant application, since it is beyond the scope of this book to extensively review the literature on the basic immunology of each particular area (e.g., basic immunology of aging or HIV infection).

Exercise and Cancer

Although interest in this area dates from at least the 1940s and 1950s (Rashkis 1952; Rusch and Kline 1944), there has been renewed interest in recent years. This recent interest focuses on two key research areas—at the basic level, whether and how exercise may reduce risk of cancer, and at the clinical level, the value of appropriately designed exercise programs for cancer patients.

The topic of exercise and cancer in humans has been reviewed several times in recent years (Hoffman-Goetz and Husted 1995; Kramer and Wells 1996; Pedersen 1997; Pinto and Marcus 1994; Sternfeld 1992). Epidemiological evidence indicates an association between regular physical activity and lower incidence of certain cancers, in particular colon cancer in men and breast cancer in women. Animal studies also suggest that exercise training or voluntary activity enhances resistance to experimentally induced tumor growth. In addition, exercise training is often used as adjunct therapy for some cancer patients, primarily to counteract some of the debilitating effects of the disease and treatment. For example, muscle wasting and low functional capacity are two frequent consequences of the disease process itself and subsequent treatment. Regular exercise may help maintain, or prevent the loss of, muscle mass and functional capacity, two important factors to quality of life. Regular exercise may also have beneficial psychological effects in patients facing life-threatening disease or undergoing uncomfortable treatment.

Epidemiological Evidence in Humans

Epidemiological evidence suggests that regular physical activity exerts some level of protection against all-site and some site-specific cancers (reviewed in Pedersen 1997). The association appears to be strongest for occupational compared with recreational activity, although this may be because occupational activity may be more accurately quantified than recreational activity. Of all cancers studied to date, colorectal cancer appears to exhibit the strongest and most consistently observed association with physical activity. A number of studies throughout the world have documented significant elevation of colon cancer risk (by about 50-100%) associated with inactivity especially in those with sedentary jobs; in contrast, physical activity does not appear to influence the risk of rectal cancer (reviewed in Pedersen 1997; Sternfeld 1992). Recent data indicate an inverse relationship between physical activity and estrogen-related cancers (breast, ovarian, endometrial), which are among the leading causes of death in women in developed countries. Studies that have shown such a relationship suggest a reduction in relative risk of about 20-60% in physically active compared with inactive populations (reviewed in Kramer and Wells 1996). As discussed below, it has been suggested that regular physical activity may influence the risk of such cancers either directly or indirectly by altering circulating estrogen levels. Data for other types of cancer are not as clear, possibly because of the lower incidence compared with breast and colon cancer and thus fewer studies and smaller number of cases. There is some evidence that physical activity may lower the risk of lung cancer (Albanes et al. 1989) and prostate cancer in older but not younger men (Lee et al. 1992; Paffenbarger et al. 1992; Vena et al. 1987).

Possible Mechanisms Influencing the Relationship Between Exercise and Cancer

A variety of mechanisms have been proposed to explain the observed association between physical activity and cancer incidence in humans (Bartram and Wynder 1989; Hoffman-Goetz 1994; Hoffman-Goetz and Husted 1995; Kramer and Wells 1996; Sternfeld 1992; Pedersen 1997). The lower risk of colon cancer in physically active populations has been attributed to enhanced gut peristalsis, resulting in faster movement of, and reduced colon exposure to, potential carcinogens in fecal material (Bartram and Wynder 1989; Hoffman-Goetz and Husted 1995; Pedersen 1997).

Regular physical activity may influence cancer risk by reducing body fat and the incidence of obesity. Excess body fat and obesity are risk factors for certain cancers such as endometrial, breast, and colon cancer, possibly via increasing the amount of lipid available for synthesis of sex steroids that may stimulate tumor growth in some individuals (Kramer and Wells 1996). It is well known that moderate exercise mobilizes fatty acids from adipose tissue, and regular activity results in reduced fat mass. Exercise may directly or indirectly influence the level of certain hormones, in particular the sex steroids, for example estradiol, which is considered a causative agent in certain forms of cancer. For example, low circulating levels of estradiol are often found in well-trained female endurance athletes, and long-term reduction in estradiol levels has been implicated as a possible mechanism for the lower incidence of breast cancer among female athletes (Kramer and Wells 1996).

Exercise stimulates release of cytokines such as IL-1, IL-6, and TNFα (discussed in chapter 6) that are capable of stimulating cytotoxic activity of T and NK cells. It has been suggested (Pedersen 1997) that these cytokines may play a role in defense against tumor growth and metastasis via stimulating adhesion of tumor cells to extracellular proteins, possibly enhancing interaction with cytotoxic T and NK cells. In addition, release of IL-2 during exercise may also stimulate lymphokine-activated killer (LAK) activity against tumor cells. Recent evidence suggests the immune system is involved in defense against tumors of viral origin (Herberman 1991; Janeway and Travers 1996). As discussed in chapter 1, exercise appears to influence the immune response to viral infection and it is conceivable, although speculative at present, that there is also a link between regular exercise and defense against malignancy of viral origin.

Individuals who choose to exercise or to work in physically demanding occupations may be more likely to adopt healthier lifestyles overall, thus reducing the risk of cancer associated with factors such as cigarette smoking or high fat diets. Regular physical activity may also reduce stress levels; this may have an indirect positive effect on resistance to cancer or, as discussed in the previous chapter, may augment the immune system's response to spontaneous tumor growth. Finally, some common genetic factors may predispose an individual to both a physically active lifestyle and low risk of cancer.

Animal Studies

Experimental animal models have several advantages over human models for studying the effects of exercise on tumor growth, including shorter duration and control of possible confounding factors such as body mass and the amount of exercise performed. Obviously, direct inoculation with tumor cells for studying the effects of exercise on the growth and spread of these cells is only possible in an experimental animal model. These models may also provide unique information about the mechanisms underlying exercise-induced modulation of tumor resistance.

Several studies have shown an association between physical activity and reduced growth of experimentally induced tumors in animal models (Andrianapoulos et al. 1987; Cohen et al. 1988; Hoffman et al. 1962; MacNeil and Hoffman-Goetz 1993a, 1993b; Rashkis 1952; Rusch and Kline 1944). In studies showing a reduction of tumor growth, animals began exercise days or weeks before introduction of tumors and continued exercise for some time afterward. For example, in mice that had been swim-trained before and then for two weeks after tumor implantation, survival time was increased by 20% compared with that in untrained mice given the same tumor (Rashkis 1952). When training was discontinued 14 days after implantation, tumor growth was greatly accelerated in formerly trained mice, eventually equaling that in the control group. These data suggest that any inhibitory effect of exercise on tumor growth persists only as long as exercise training is maintained. Changes in body composition, body fat levels, or diet can only partially account for the increased resistance to tumor growth in exercise-trained animals (Cohen et al. 1988).

There is some evidence that exercise may influence metastasis of tumor cells, possibly by stimulating cytotoxic activity of some immune cells. As described in chapters 2 and 7, cytotoxic cells (cytotoxic T cells, NK cells, and monocytes/macrophages) are capable of recognizing and killing some tumor cells and are thought to play a role in early surveillance against malignancy primarily via preventing or limiting metastasis (spread) of tumor cells throughout the body. In two related studies using the NK cell-sensitive tumor cell line CIRAS 3, previously trained mice exhibited higher in vitro splenic NKCA, greater in vivo lung clearance of injected radiolabeled tumor cells, and lower incidence of tumors (figure 8.1) (MacNeil and Hoffman-Goetz 1993a, 1993b). This enhanced cytotoxic activity was maintained for three weeks after cessation of exercise training. Woods et al. (1993, 1994a) reported significantly higher number of tumor-infiltrating monocyte/macrophages and cytotoxic activity of these cells against tumor cells in exercise-trained mice compared with untrained mice. However, in these studies, metastatic growth was not inhibited by exercise despite evidence for enhanced cytotoxic activity. In the study by Woods et al.

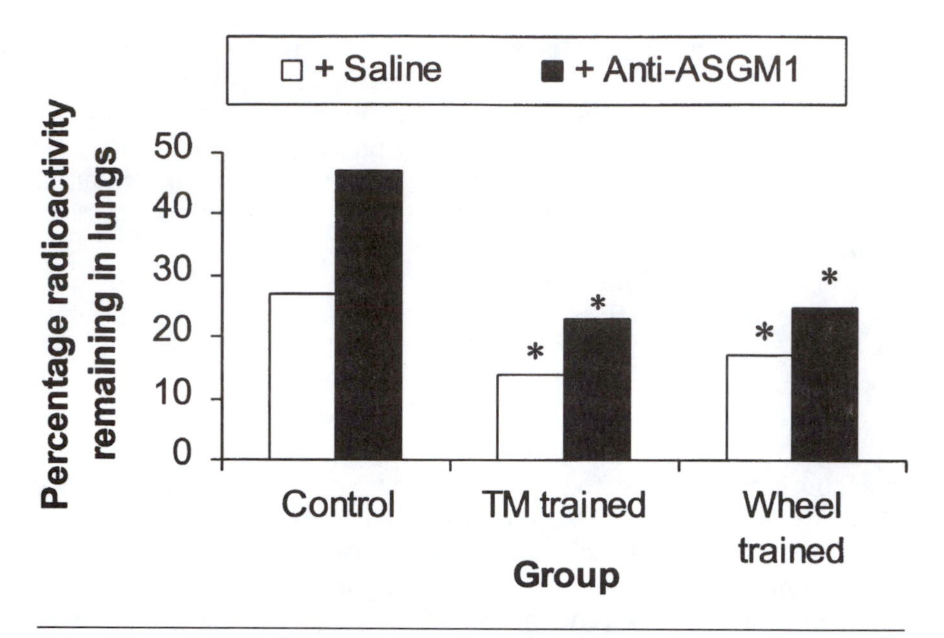

Figure 8.1 Percent radioactivity remaining in lungs after injection of radioactively labeled tumor cells into mice. Mice were grouped according to activity: sedentary controls (Control); treadmill (TM) trained via enforced treadmill running for nine weeks; and voluntary exercisers via access to exercise wheels (Wheel trained) for nine weeks. Anti-$ASGM_1$ = mice were injected with antibody to asialo-GM_1, an antibody to NK cells that blocks their action, 24 hours before sacrifice. Tumor cells were injected 3 hours before sacrifice.

From MacNeil and Hoffman-Goetz 1993a.

(1994a), although exercise training enhanced monocyte/macrophage cytotoxicity, the number of metastases did not differ between control and exercise groups. Similarly, using mice injected with the NK-insensitive MMT66 mammary tumor cell line, Hoffman-Goetz et al. (1994) reported no effect of exercise training on the number of metastases, despite enhanced IL-2-stimulated LAK cell activity in trained compared with untrained mice. It has been noted, however, that the dose and type of tumor cell used in these studies may have been high enough to overwhelm host defense, making any intervention such as exercise ineffective (Pedersen 1997; Woods and Davis 1994). These data suggest that a complex relationship exists between exercise, cytotoxic cells, and malignant growth and that much of the response depends on the type of tumor cell and its sensitivity to effectors of natural immunity (e.g., NK cells and monocytes/macrophages).

Immune Cell Response of Exercise in Cancer Patients

Only a few studies have focused on the effects of exercise on immune cell number and function in cancer patients (Nieman et al. 1995d; Peters et al. 1994). In one study, 24 stage 1 and 2 breast cancer patients, who were at least six months postsurgery, participated in a six-month moderate exercise program (Peters et al. 1994). Exercise consisted of cycle ergometry at 60% heart rate reserve for an average 33 minutes two to three times per week. NK ($CD56^+$) cell percentage and number did not change significantly after training, although NKCA increased significantly after six months of exercise training. Self-reported discomfort and life satisfaction also improved over the six months, and the latter was significantly correlated with training frequency. NK cell number and percentage were within the clinically normal range both before and after training. In contrast, resting NKCA was clinically low in the patients at the start of the study and improved toward the clinically normal range by the end of the study. It was suggested that moderate exercise may enhance NK cell function in breast cancer patients and that such improvement may be related to psychological factors such as discomfort and life satisfaction. However, since no nonexercise patient or healthy groups were included for comparison, the effects of seasonal variability cannot be discounted. Nevertheless, this study raises the possibility that regular moderate exercise may enhance cellular immune function in breast cancer patients.

In contrast, a more recent study did not observe changes in NK cell function after moderate exercise training in breast cancer patients (Nieman et al. 1995d). Twelve breast cancer patients who had undergone surgery and treatment three years before the study were randomly assigned to an exercise or control group. Exercise was performed three times per week for eight weeks and consisted of 30 minutes supervised resistance training, using two sets of 12 repetitions of seven different exercises, and 30 minutes walking at 75% age-predicted maximal heart rate. Total leukocyte, lymphocyte, T ($CD3^+$), and NK ($CD16^+/56^+$) cell numbers and NKCA were clinically normal in patients and did not change significantly over the eight weeks of training in either group. Noting that low NKCA may occur in breast cancer patients early after detection and that low NKCA may be prognostic in breast, Nieman et al. (1995d) suggested that exercise-induced enhancement of NKCA may be of some benefit in breast cancer patients. However, this study suggested that short-term moderate exercise does not influence NKCA in breast cancer patients, although it was acknowledged that the small sample size, relatively long period since diagnosis and treatment, and short exercise training program may have affected the outcome. Since there are so few studies, it is unclear at present whether physical activity alters cellular immune function in cancer patients in any biologically meaningful way.

Taken together, the limited epidemiological and experimental data suggest that regular moderate physical activity affords protection against certain cancers, in particular bowel and breast cancer. However, there are several possible mechanisms to explain this protection, and there is little evidence at present to implicate exercise effects on the immune system as an important mechanism in humans. There is some evidence from animal models suggesting that exercise-induced enhancement of natural immunity may play a role in defense against some types of tumors, but the relationship is complex and not completely understood. While moderate physical activity may enhance functional capacity and muscular strength and help counteract some of the deleterious effects of therapy in cancer patients, there is at present little evidence to suggest that exercise-induced alterations in immune function affect the outcome of disease. Certainly, more work is needed to fully document the immune response to exercise in cancer patients, and possible role of exercise in enhancing the body's natural defenses against tumor growth, and to develop scientifically based guidelines for exercise prescription for cancer patients.

Exercise and Human Immunodeficiency Virus-1 Infection

Human immunodeficiency virus-1 (HIV-1) infection is a chronic infection resulting in a variety of immunological abnormalities, most prominently a gradual decline in CD4 T cell number and function, leading to immune dysfunction and eventually immune deficiency. HIV-1 gains entry into CD4 T cells via the CD4 cell surface receptor, causing incorporation of viral DNA and subsequent cell death. Since CD4 T cells are a major regulatory cell population involved in virtually all aspects of cellular and acquired immunity, depletion of these cells has severe consequence for immune function. Indeed, HIV-1-infected individuals are susceptible to a range of opportunistic infections and autoimmune disorders. At present, HIV infection is a chronic condition that may go undetected for some time and be present for 10 to 15 years before manifestation of immune deficiency (acquired immunodeficiency syndrome or AIDS).

As detailed throughout this book, there is evidence that acute intense exercise causes at least a transitory suppression of a number of immune parameters (e.g., NKCA, CD4:CD8 ratio, mucosal IgA levels, neutrophil function). Since HIV infection is a chronic condition, the acute and long-term responses of immune cells to exercise have important implications for infected individuals. Moreover, anecdotal evidence from HIV-positive individuals and some limited data suggest the possibility that regular moderate exercise may enhance survival and quality of life after infection and that exercise training may also enhance immune function in infected patients (Lawless et al. 1995; Eichner and Calabrese 1994). If, as has been proposed, moderate exercise enhances but vigorous exercise suppresses immune function, it is important to understand the responses to exercise in HIV-1-infected individuals. Current research focuses on describing the immune response to various intensities of exercise, whether such response is similar to or different from that in noninfected individuals, and possible mechanisms responsible for any differences between patients and healthy individuals. One obvious goal is to develop appropriate exercise prescription for various stages of infection that will enhance quality of life while not contributing to, or possibly even counteracting, immune suppression.

Acute Responses to Exercise in HIV-Infected Individuals

In infected individuals, the number and percentage of circulating CD4 T cells are prognostic for predicting the development of immune deficiency (AIDS) and are regularly monitored; there is some debate as to whether the absolute number or percentage of CD4 T cells is the best predictor of further disease development (Ullum et al. 1994b, 1995). As noted in earlier chapters, acute exercise causes marked changes in circulating leukocyte number and subset proportions. Table 8.1 summarizes the acute and training responses to exercise in HIV-1-infected individuals.

In a recent study, the effects of exercise on circulating lymphocyte subset number and percentage were compared in HIV-infected individuals and healthy controls (Ullum et al. 1994b). Eight infected asymptomatic patients and eight matched healthy subjects cycled at 75% $\dot{V}O_{2max}$ for 60 minutes. As expected, CD4 T cell percentage and number were significantly lower at rest and during and after exercise in patients compared with controls. However, in both groups, CD4 T cell percentage did not change significantly during or up to 4 hours

Table 8.1 Exercise Effects in HIV-1-Positive Individuals

Subjects	Exercise	Main results	Reference
50 high-risk asymptomatic males, antibody status unknown at start but not end of study; randomly assigned to exercise or control	10 wk training 3 × wk: 45 min at 80% APMHR	↑ CD4 counts, CD56 cell number maintained in HIV+ exercisers but ↓ in HIV+ controls after notification of antibody status	LaPerriere et al. 1990, 1991
32 seropositive asymptomatic males and matched seronegative controls	Progressive exercise test to volitional fatigue	↓ work capacity, minute ventilation, ventilatory threshold and ↑ slope HR/ $\dot{V}O_2$ curve in HIV+	Johnson et al. 1990
24 seropositive males after acute *pneumocystis carinii* pneumonia randomly assigned to exercise or control	6 wk training 3 × wk 3 sets 10 rep upper and lower body resistance exercise	↑ body mass, lean body mass, ↑ 20-50% muscle strength	Spence et al. 1990
37 seropositive men at various clinical stages	12 wk training 3 × wk: 20 min at 60-80%	↑ 15-20% $\dot{V}O_{2max}$; ↑ 30% strength; no change cell counts,	Rigsby et al. 1992

randomly assigned to exercise or counseling control	HRR + 35 min strength (N = 28 at end 12 week)	CD4:CD8 ratio, or clinical status	
23 seropositive asymptomatic men randomly assigned to exercise or control	10 wk training 3 × wk: 45 min at 80% APMHR	↓ serum antibodies to EBV and HHV-6; no change cell counts or CD4:CD8 ratio	Esterling et al. 1992
25 seropositive men moderately-severely immunocompromised	24 wk training 3 × wk: 24-40 min at 50-60% (low) or 75-85% (high intensity) $\dot{V}O_{2max}$ (N = 6 at end 24 wk)	↑ 24% $\dot{V}O_{2max}$, no change in cell counts or CD4:CD8 ratio in those completing 12 wk	MacArthur et al. 1993
8 seropositive asymptomatic men; 8 seronegative controls	60 min cycling at 75% $\dot{V}O_{2max}$	Less ↑ in NK number and NK and LAK activity after exercise in seropositive vs.-negative	Ullum et al. 1994b
75 seropositive patients of various clinical stages and matched seronegative controls	Progressive exercise test to volitional fatigue	↓ work capacity, ventilatory threshold, $\dot{V}O_{2max}$, O_2 pulse in patients; cardiopulmonary variables correlated with clinical status	Pothoff et al. 1994

*Main results compared with appropriate nonexercise control groups; APMHR = age-predicted maximum heart rate; EBV = Epstein-Barr virus; HHV-6 = human herpes virus-6.

postexercise, and changes in CD4 cell number followed similar patterns in both infected and noninfected individuals (e.g., increase during followed by a decrease 2 hours after exercise). These data suggest that the ability to recruit CD4 T cells into the circulation during exercise is maintained in HIV-positive individuals. In contrast, the ability to recruit neutrophils and NK cells into the circulation and the responsiveness of NK cells appear to be compromised in these patients. Neutrophil number increased far less during and up to 4 hours after exercise in infected compared with noninfected subjects. Moreover, NK ($CD16^+$) cell percentage did not change during exercise, in contrast to the observed doubling of NK cell percentage in healthy subjects. In addition, NKCA increased only marginally during exercise in patients compared with a more than twofold increase in NKCA in noninfected controls. LAK activity in response to IFNα or indomethacin was also significantly lower at rest, and during and after exercise, in infected compared with noninfected individuals. These data suggest an impaired ability of the natural immune system to respond to challenge during and for some time after physical stress. It was suggested that such impairments may be due to alterations in the hormonal responses to stress in patients (e.g., lower catecholamines or growth hormone response to exercise) or to the limited pool of cells that can be recruited into the circulation during physical stress (Pedersen 1997). This impairment of cell recruitment was not related to cytokine release, since pre- and postexercise plasma TNFα and IL-6 concentrations were similar in HIV-positive and -negative individuals.

It has been suggested that alterations in the cardiorespiratory system may limit endurance exercise capacity in untrained HIV-infected individuals (Johnson et al. 1990; Pothoff et al. 1994). In one study, 32 asymptomatic HIV-positive men exercised to volitional fatigue on a cycle ergometer. Subtle abnormal cardiorespiratory responses included lower aerobic power and endurance exercise capacity (measured by exercise time to fatigue), mild tachypnea, lower ventilatory threshold, and increased slope of the heart rate to $\dot{V}O_2$ relationship compared with normal responses for healthy men of the same age (Johnson et al. 1990). It was suggested that these changes reflect an impaired capacity for oxygen delivery to working tissues and that they may be indicative of subclinical cardiac abnormalities associated with HIV-1 infection. If true, this would suggest that the standard exercise prescription based on predictions from healthy populations (e.g., age-predicted maximal heart rate) may not be applicable to HIV-1-infected individuals, and that individual assessment and prescription may be required. It is not clear whether such abnormalities would be minimized or exacerbated by moderate exercise training.

In a recent study, Pothoff et al. (1994) compared the cardiorespiratory responses to a standard progressive cycle ergometry test to exhaustion in three groups of HIV-positive patients and seronegative control subjects. Patients were grouped according to history of respiratory and lung disease and by severity of disease: Group 1 patients (N = 20) had no history of respiratory or lung disease; group 2 (N = 18) had recovered from a former episode of *Pneumocystis carinii* pneumonia (PCP; a common opportunistic infection in HIV) without lung disease; and group 3 (N = 37) had experienced both respiratory and lung disease including PCP. Group 3 patients differed significantly from groups 1 and 2 in terms of spirometry, $\dot{V}O_{2max}$, and exercise capacity. Compared with healthy controls, all groups exhibited reduced exercise capacity (maximum work rate), $\dot{V}O_{2max}$, oxygen pulse ($\dot{V}O_2$/HR), and $\dot{V}O_2$ at anaerobic threshold. In contrast to the study by Johnson et al. (1990) described above, there was no steepening of the $\dot{V}O_2$/heart rate relationship, suggesting no occult heart disease. Importantly, decrements in $\dot{V}O_{2max}$ and exercise capacity were significantly correlated with Walter Reed classification (i.e., stage of disease). Since these data did not indicate the presence of heart disease in these patients, it was suggested that the lower $\dot{V}O_{2max}$ and oxygen pulse reflected impaired oxygen delivery to the working muscles, possibly due to anemia. It was further suggested that the impaired exercise capacity may have resulted from peripheral neuromuscular disease and muscle wasting leading to premature muscular fatigue at higher workloads.

Chronic Response to Exercise Training in HIV-Infected Individuals

Only a few studies have focused on the response of HIV-positive subjects to exercise training (Esterling et al. 1992; LaPerriere et al. 1990a, 1990b, 1991, 1997; Rigsby et al. 1992) (table 8.1). It has been suggested that behavioral interventions such as exercise training or stress management may help restore immunocompetence in asymptomatic HIV-infected patients, especially during the early stages of infection (Antoni et al. 1990; Esterling et al. 1992; LaPerriere et al. 1990a, 1991). These authors presented a model in which stressors (e.g., mood state and social stressors) may alter stress hormone levels, which in turn may adversely influence the immune system (LaPerriere et al. 1994). As part of a large intervention study focusing on social, psychological, and physical dimensions of HIV infection, high-risk but asymptomatic men participated in an aerobic exercise training program consisting of 45-minute cycle ergometry at 80% of age-predicted maximal heart rate, three times per week for 10 weeks (Antoni et al. 1990; LaPerriere et al. 1990a, 1990b, 1991). At the onset of the study, these men were not aware whether they were HIV positive (infected) or negative (not infected), and they were tested for antibodies to HIV-1 infection at the end of the 10-week training program; notification of antibody status is a time of great psychological stress for high-risk individuals. A control group of high-risk asymptomatic men did not exercise during this time, and two groups (exercise and control) of matched low-risk men served as noninfected comparison groups.

In these studies, exercise training improved cardiovascular fitness ($\dot{V}O_{2max}$) in both infected and noninfected men (LaPerriere et al. 1990a, 1990b, 1991). CD4 T cell number increased after training in both exercise groups, although the increase was higher for seronegative compared with -positive (infected) subjects. Interestingly, in exercised HIV-1-positive men, NK cell ($CD56^+$) number was maintained after notification of antibody status; in contrast, NK cell number declined significantly after notification in nonexercised HIV-1-positive men. NK cell number remained constant in both HIV-negative groups (LaPerriere et al. 1990b). Because NK cells are important to defense against viral infection, maintenance of NK cell number by moderate exercise training may enhance defense against HIV-1 as well as opportunistic infections.

In a related study using the same experimental model, antibodies to Epstein-Barr virus (EBV) and human herpes virus type-6 (HHV-6) were measured in seropositive and -negative men before and after a 10-week exercise program similar to the one described above (Esterling et al. 1992). Antibody levels declined and were significantly lower after the 10 weeks in both seropositive and seronegative exercisers compared with serostatus-matched nonexercising controls, who showed no changes in antibody levels over the same period; similar reductions in antibody levels were also seen in a group who participated in a stress management intervention program. Noting that co-infection with EBV or HHV-6 may be a cofactor in development of AIDS in HIV-infected individuals, the authors speculated that these preliminary data suggested better control over latent viral infection possibly via reducing psychological distress, enhancing cellular immune function, or both.

Seropositive exercisers also showed no significant changes in mood state (as assessed by the Profile of Mood States questionnaire) upon notification of infected status, whereas seropositive nonexercisers showed marked increases in tension/anxiety at the same time (Antoni et al. 1990). These limited data suggest that regular moderate endurance exercise may enhance functional capacity (fitness level) as well as immunological and psychological variables during early HIV-1 infection. A recent study comparing the responses of HIV-infected individuals to aerobic, resistance, or stretching exercise training programs suggests that both aerobic and weight training enhance subjective well-being (e.g., physical self-efficacy, mood state, and life satisfaction) (Lox et al. 1995).

Loss of body mass, especially lean body mass due to muscle wasting, is frequent in HIV-infected individuals, and the magnitude of body mass loss is considered prognostic for disease progression. It has been suggested that interventions to maintain lean body mass, such as resistance training, may be effective in limiting some of the physically debilitating consequences of disease and may possibly contribute to slowing disease progression. Two studies indicate that resistance (weight) training increases lean body mass in HIV-1-infected individuals (Rigsby et al. 1992; Spence et al. 1990). To test whether weight training can improve lean body mass, 24 HIV-positive males who had previously recovered from PCP infection were randomly assigned to either a control or resistance training group (Spence et al. 1990). Supervised resistance training was performed using a hydraulic system (i.e., only concentric and no eccentric contractions) three times per week for six weeks. Intensity and volume were progressively increased from one set of 15 repetitions at low resistance at the start to three sets of 10 repetitions at high resistance at the end of the six weeks; exercises included flexion/extension of both lower and upper extremities (e.g., knee, chest-arms, shoulder-arms). Both groups were tested for maximum torque, force, and power of the upper and lower extremities before and after training. Compared with the control group, in which muscular strength and body mass declined somewhat, the trained group exhibited significant increases in 13 of 15 measures of strength, body mass, and limb girth; both upper and lower body strength and dimensions were similarly increased (figure 8.2). No adverse effects were observed in the strength training group. It was concluded that adaptive responses of skeletal muscle protein synthesis and neuromuscular factors are maintained in HIV-infected patients, and that early intervention with progressive resistance exercise training may help counteract muscle atrophy, loss of muscular strength, and functional capacity accompanying later stages of HIV infection.

Improvements in cardiorespiratory fitness may occur in HIV-infected individuals (Macarthur et al. 1993; Rigsby et al. 1992), and the

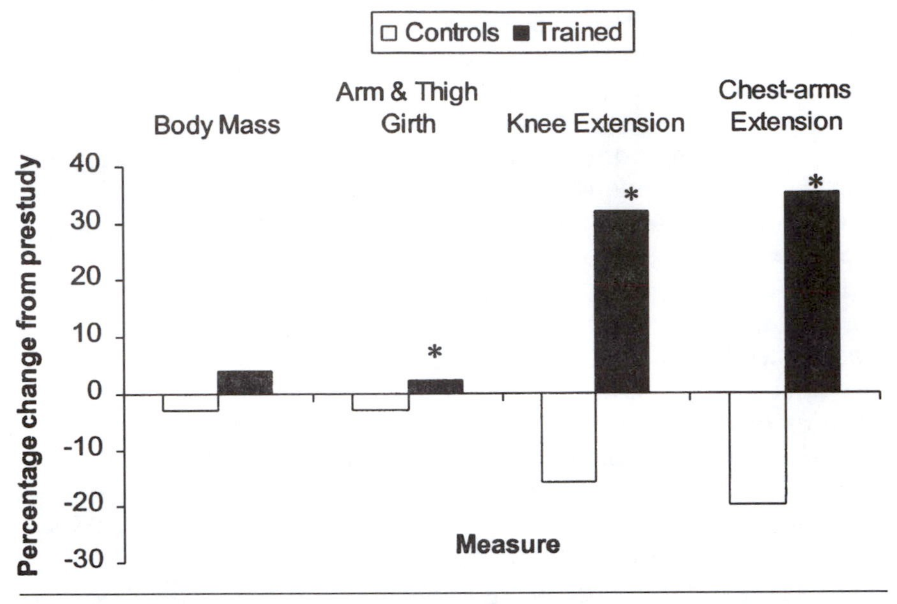

Figure 8.2 Effects of resistance (strength) training on anthropometric and muscular strength measures in HIV-1+ men. Subjects were randomly assigned to a control (no exercise) or trained group. Training consisted of hydraulic resistance (strength) training three times per week for six weeks. Data are expressed as percentage change from prestudy levels. * = Significantly different from controls.

From Spence et al. 1990.

magnitude of improvement appears to be related to exercise intensity (Macarthur et al. 1993). In a 12-week training study on 37 seropositive men with heterogeneous disease classification, exercise training was associated with positive changes in fitness level and strength without changes in immune cell number (Rigsby et al. 1992). Subjects were matched for clinical status and assigned to either an exercise or counseling control group. Exercise consisted of 1-hour sessions three times each week for 12 weeks and included 20 minutes of cycle ergometry at 60-80% heart rate reserve followed by 35 minutes of muscular strength and flexibility training. After training, exercise capacity (time to fatigue), submaximal exercise heart rate, and upper and lower body strength increased significantly in the exercise group but did not change in the counseling group (figure 8.3). No significant changes were observed for leukocyte or lymphocyte subset numbers or the ratio of CD4:CD8 T cells, or for clinical status. These data indicate that in HIV-positive men, moderate exercise training may enhance muscular strength and cardiorespiratory fitness without deleterious effects on immune cell numbers and ratios. It was noted,

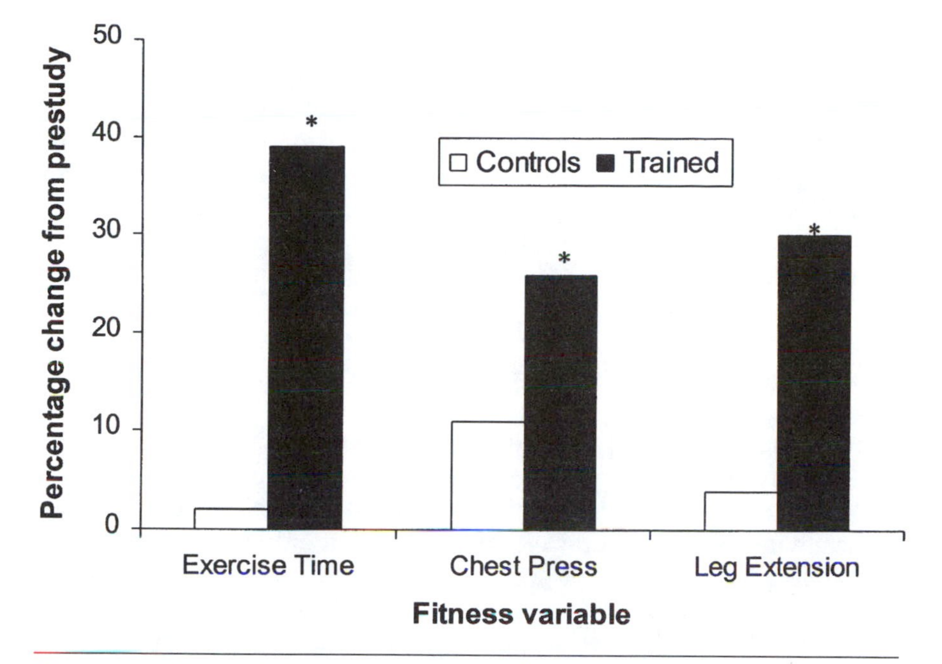

Figure 8.3 Effects of 12-week endurance and resistance training on endurance exercise capacity and muscular strength in HIV-1+ men. Subjects were randomly assigned to a control or trained group. Training was performed three times per week and consisted of 20-minute cycle ergometry at 60-80% heart rate reserve, followed by 35 minutes flexibility and hydraulic resistance strength training. Exercise capacity was assessed by time to fatigue (Exercise Time) in a submaximal cycle ergometry test. Data are expressed as percentage change from prestudy values. * = Significantly different from controls.

From Rigsby et al. 1992.

however, that most subjects were asymptomatic or with AIDS-related complex and that only a few of the subjects had full-blown AIDS, although the few subjects with AIDS exhibited similar training responses to the exercise program as the other subjects.

In another exercise training study on a small group of moderately to severely immunocompromised HIV-infected individuals, subjects who adhered to the program and completed the full 24 weeks exhibited marked improvements in $\dot{V}O_{2max}$, minute ventilation, and oxygen pulse; improvements were larger in a higher- compared with lower-intensity training group (75-85% and 50-60% $\dot{V}O_{2max}$, respectively) (Macarthur et al. 1993). CD4 T cell number and percentage did not change significantly after 12 and 24 weeks of exercise. Interestingly, initial CD4 cell count was higher in compliant subjects who regularly attended and completed the 24-week exercise program compared with those considered noncompliant and somewhat compliant. Although conclusions are limited due to the high dropout rate and heterogeneity of the patients' clinical status, these data suggest that longer-term exercise (i.e., 24 weeks) of even relatively high intensity does not compromise CD4 cell counts in HIV-infected symptomatic individuals.

Although there have been few studies on exercise in HIV-infected individuals, the limited data to date indicate that HIV-1-infected individuals exhibit expected training effects after both endurance and strength training (e.g., increased aerobic power, muscular strength, lean body mass). Anecdotal evidence from long-term survivors of HIV infection associates long-term survival with maintenance of physical fitness, suggesting the possibility that regular moderate exercise that includes both endurance and resistance work may be beneficial. Although some data suggest the possibility that moderate exercise may enhance some aspects of immunity in infected individuals, there are at present too few data to determine whether exercise has any effect on progression of disease. At the very least, moderate exercise appears to be beneficial in maintaining subjective well-being, lean body mass, and functional capacity in the early and possibly later stages of infection, and there is no evidence that regular moderate exercise is harmful to the immune system of HIV-infected individuals.

Aging

In most developed countries, aged individuals compose an increasing relative proportion of the population; average life span in most of these countries is above 70 years for both men and women. One consequence of this aging of the population (and a concomitant relative decrease in the proportion of younger individuals) is an ever increasing cost of medical care, which is borne by the population as a whole. Thus, there is much interest in finding inexpensive nonmedical means to enhance health in older individuals and throughout the life span (and perhaps to prevent many of the diseases that occur later in life).

Aging is associated with decrements in immune function that occur concomitantly with (and are most likely related to) an increased incidence of malignancy, infectious illness, and autoimmune diseases. Age-related changes in immune function include decreases in the T cell proliferative response to mitogenic stimulation, IL-2 production by T cells, number of IL-2-producing T cells, IL-2 receptor affinity, and B cell function secondary to altered T cell stimulation. Monocyte cytokine production, and leukocyte and lymphocyte subset numbers and ratios, may remain unchanged or decline. In contrast, NK cell number and function appear to be maintained during aging (reviewed in Mazzeo and Nasrullah 1992). As with many age-related changes, there is wide individual variation that also increases with age. Moreover, other factors occur concomitantly with aging such as nutritional deficiency, physical inactivity, increased body mass and fat levels, and prevalence of certain diseases, each of which may also independently influence immune function. It is thus difficult to dissect out precisely the extent to which specific immune parameters change during aging, although it is widely accepted that some immunosuppression does occur during aging. It has been suggested that simple interventions such as regular physical activity associated with enhanced immune function in younger populations may help restore immunocompetence in aging individuals (Mazzeo and Nasrullah 1992; Mazzeo 1994; Nieman and Henson 1994).

Immune Responses to Exercise in Aging Individuals

There are few studies on the effects of exercise on immune function in aging humans (Crist et al. 1989; Fiatarone et al. 1989; Nieman et al. 1993a; Rall et al. 1996). Although not studied extensively, it appears that the acute immune cell response to exercise is similar in young and older untrained individuals (Crist et al. 1989; Fiatarone et al. 1989). For example, Fiatarone et al. (1989) compared NKCA and NK cell number in younger and older untrained women (mean ages 30 and 71 years, respectively) before and after a standard graded exercise test to volitional fatigue. As expected, NK cell number and cytotoxic activity increased after exercise, but basal and postexercise NK cell number and cytotoxic activity were similar in both groups. Crist et al. (1989) also measured NKCA before and after a standard progressive exercise test to volitional fatigue in untrained and moderately trained older women (mean age 72 years); the latter had completed a 16-week moderate exercise training program (60 minutes moderate walking and stretching three times per week). In both groups, NKCA increased significantly during exercise, but both basal and postexercise NKCA were higher in the trained compared with untrained women, suggesting the possibility that moderate exercise training may enhance NKCA in older women.

In a 12-week intervention study, 32 healthy elderly women (ages 67-85 years) were randomly assigned to an exercise or calisthenic group; 12 elderly age-matched well-trained women athletes were also included for comparison (Nieman et al. 1993a). Exercise consisted of 30-

40 minutes of brisk walking at 60% heart rate reserve five days per week; the calisthenic group performed only very light flexibility exercise during these times. At the onset of the study, immune cell counts (total leukocyte and differential, lymphocyte subsets) did not differ between untrained and well-trained women. However, basal NK cell cytotoxic activity and T cell proliferative responses to PHA were higher in the trained compared with untrained women. Moreover, NKCA and T cell proliferation were significantly correlated with $\dot{V}O_{2max}$. These differences in cellular activity could not be attributed to differences in cell numbers, suggesting a possible enhancement of cellular activity in the well-trained women. However, despite significant improvements in $\dot{V}O_{2max}$ over the 12 weeks in the exercisers, there were no significant changes in any immune parameter (cell counts, NKCA, T cell proliferation). These data indicate that short-term moderate exercise training does not enhance immune cell function in previously sedentary older women, and suggest that longer-term, more intense exercise (as performed by the well-trained older women) is necessary to influence immune function; the possible influence of other factors such as genetics, lifestyle, or leanness was acknowledged.

Similarly, in a study on older healthy women with rheumatoid arthritis (RA) (ages 65-80 years), 12 weeks progressive resistance (strength) training had no significant effect on a range of immune parameters (Rall et al. 1996). Eight healthy elderly women and eight women with RA performed resistance exercise involving the trunk and the upper and lower body. Sessions were of 45 minutes of duration, performed two to three times per week; for each exercise, subjects performed three sets of eight repetitions at 80%, 1 repetition maximum (considered high-intensity exercise for previously sedentary subjects). A control group of elderly nonexercisers was included to control for possible seasonal variation. Despite significant improvements in strength in the resistance training group after the 12 weeks, no significant changes were observed in the following immune parameters: in vitro production by peripheral blood mononuclear cells of cytokines (IL-1β, TNFα, IL-2, IL-6); circulating lymphocyte subset numbers and percentages; T cell proliferation in response to PHA and ConA stimulation; and delayed type hypersensitivity (a simple noninvasive indicator of in vivo T cell function).

Taken together, these limited data suggest that, despite evidence of immunosuppression during aging, short-term moderate exercise has little if any effect on immune cell function. The effects of longer-term, more intense exercise training have not been extensively studied, but there is some evidence suggesting positive effects on immune function. The acute immune cell response to exercise appears to be similar in older and younger individuals and in older women with RA, suggesting no deleterious effects of exercise on immune function in the elderly. Since exercise does not compromise immune function in the elderly, but is associated with myriad beneficial effects on other systems (e.g., increased muscular strength, bone density, and lean body mass; improved blood lipid profile and glucose tolerance), regular moderate exercise should be encouraged in the elderly.

Spaceflight

Spaceflight presents unique physiological and psychological challenges to humans because of the physical and mental demands, isolation, disruption of normal body rhythms, and effects of microgravity. Although the possible detrimental effects on the immune system were recognized early in the history of human spaceflight, there is renewed urgency to more clearly understand the long-term responses of the human immune system to spaceflight given the recent commitment to extended spaceflight (e.g., to return to the moon, to travel to Mars, and to work in space stations). It is conceivable that compromised immune function may have significant consequences during a mission of extended spaceflight, as well as possible long-term implications for the health of astronauts in the years after return to earth. Exercise has been proposed as a possible intervention (or "countermeasure") to prevent or limit adverse effects of spaceflight on other systems (e.g., loss of muscle and bone mass). Understanding the immune response to spaceflight, and to exercise during spaceflight, is important to ensuring astronaut health and performance during long-term spaceflight. Indeed, based on accumulating data that moderate exercise enhances but strenuous exercise suppresses immune function, it has been suggested that moderate rather than intense exercise training be recommended during long-term spaceflight (Tipton et al. 1996).

Effects of Spaceflight on the Immune Response

The immune response to prolonged intense exercise has been likened to that associated with spaceflight (Gmunder et al. 1988), in particular distribution of leukocytes and circulating cell counts, suppression of lymphocyte proliferative response to mitogenic stimulation (Cogoli 1993; Gmunder et al. 1988; Konstantinova et al. 1993; Meehan et al. 1993), and suppression of NKCA (Fuchs and Medvedev 1993; Meehan et al. 1993; Konstantinova et al. 1993; Taylor 1993). Many of the changes occurring during spaceflight have been attributed to the effects of psychological and physical stress with resultant neuroendocrine changes, fatigue, radiation exposure, and restricted environment and close working conditions of crew members (Greenleaf et al. 1994; Meehan et al. 1993; Schmitt and Schaffar 1993; Sonnenfeld and Miller 1993). Moreover, dramatic plasma and blood volume changes may alter the concentrations of factors such as hormones, cytokines, and immunoglobulins, which may in turn influence immune function. Table 8.2 summarizes the major effects of spaceflight on immune function.

Short-duration spaceflight (i.e., <11 days) has been associated with increased circulating leukocyte number, primarily due to a larger in-

Table 8.2 Major Effects of Spaceflight on the Immune System

Immune parameter	Main results
Leukocyte count	↑ up to 1 wk postflight due to large increase in neutrophil and smaller increases in monocyte number; increase not related to duration of spaceflight
Lymphocyte count	↓ up to 1-2 wk postflight; smaller pattern for T and NK cells, but no changes in B cells; changes not related to duration of spaceflight
Lymphocyte proliferation	↓ after longer flights; little or no change after shorter flights (< 10 d); normal by 1-2 wk postflight
NK activity	Slight, variable ↓ after short flights; larger, more consistently observed ↓ after longer flights; may persist for weeks after return
Cytokines	↓ production of several cytokines including IL-1, IL-2, IL-6, IFNα, and IFNγ; tendency toward normalization after extended flights (wk to mo)
Serum immunoglobulins	Little or no change at any time postflight

Note: Compiled from Cogoli 1993; Konstantinova et al. 1993; Meehan et al. 1993; Sonnenfeld and Miller 1993; Stein and Schluter 1994; Taylor 1993.

crease in neutrophil and a smaller increase in monocyte number (Barger et al. 1995; Cogoli 1993; Meehan et al. 1993; Taylor 1993). Circulating lymphocyte count appears to decrease, due to decreases in numbers of T cells (both CD4 and CD8 cells) and NK cells, without appreciable changes in B cell number. As in the acute response to exercise, the redistribution of leukocyte subsets during spaceflight (i.e., increased neutrophil and decreased lymphocyte counts) has been related to hormonal changes, in particular increasing level of glucocorticoids (Meehan et al. 1993; Stein and Schluter 1994; Taylor 1993).

As with most forms of exercise, neither short- nor long-term spaceflight appears to induce major changes in serum immunoglobulin levels (Fuchs and Medvedev 1993; Konstantinova et al. 1993). A lack of change in Ig levels is consistent with constant B cell number as mentioned above. In contrast, there are reports of significant elevations in serum IgA, IgG, and IgM concentrations after short (2-11 days) and longer (49 days) missions. Increases in these immunoglobulin levels were attributed to secretion of autoantibodies in response to an increase in proteolytic remnants of degraded skeletal muscle (Konstantinova et al. 1993).

Lymphocyte proliferative response to mitogenic stimulation decreases during both short- and long-term spaceflight. Such suppression of lymphocyte responsiveness may be related to several factors including decreased T cell number and cytokine production, alterations in glucocorticoid or catecholamine levels, and hemodynamic changes altering the degree of lymphocyte cell-to-cell contact during spaceflight (Barger et al. 1995; Greenleaf et al. 1994; Meehan et al. 1993; Sonnenfeld and Miller 1993; Stein and Schluter 1994).

NK cell number and cytotoxic activity decrease during and after spaceflight, of both short and long duration. However, the response is more variable (i.e., not shown by all astronauts) after short-term spaceflight (Fuchs and Medvedev 1993; Konstantinova et al. 1993; Meehan et al. 1993; Taylor 1993). Restoration of normal NK cell activity may require several days or weeks after return to earth. There are several possible mechanisms responsible for these changes including alterations in catecholamine or cytokine levels, receptor sensitivity for cytokines or other factors, redistribution of cells between the blood and other tissues, and target cell recognition.

Spaceflight is also associated with decreased secretion of cytokines, including IL-1, IL-2, IL-6, IFNα, and IFNγ (Konstantinova et al. 1993; Sonnenfeld and Miller 1993; Stein and Schluter 1994; Taylor 1993). Since these cytokines are primarily produced by T and NK cells, suppression of secretion is consistent with the observed decreases in cell numbers and functional activities, as described above. However, it also appears that cytokine production may be normalized during extended (i.e., 100-300 days) spaceflight. Sonnenfeld and Miller (1993) noted that the cytokine response to spaceflight is complex; there is much variability in the magnitude, consistency, and time course of changes in cytokines.

Taken together, these data suggest that general immunosuppression may occur during spaceflight and that the magnitude of suppression is related to the duration of spaceflight. Moreover, immune function may not be fully restored for some time after return to earth, indicating potential long-term effects in those involved in extended spaceflight. Similarities in the immune system response to acute exercise and spaceflight suggest that exercise may provide a useful model to further study the effects of spaceflight on immune function. Moreover, exercise training may blunt some aspects of the immune response to exercise, presumably by damping neuroendocrine changes. It is possible that prior exercise training may help limit perturbations in immune function during spaceflight. Certainly, more work is needed to fully understand the interactive effects of exercise and spaceflight on the immune system.

Dietary Intervention to Boost Immunity in Athletes

Athletes are avid consumers of many types of dietary supplements purported to enhance performance, immunity, or other factors. Because dietary supplements are often easily available, relatively inexpensive, and not considered to be detrimental to health or to confer unfair advantage, they are generally not classed as illegal banned substances by sport organizations. A few examples (and their purported benefits) include iron and other mineral supplements (to increase oxidative capacity), antioxidant vitamins (vitamins A, C, and E, to prevent illness and limit muscle damage), creatine (to enhance muscle creatine pool and thus performance in power events), glycerol (to expand plasma volume and performance in endurance events), glutamine and branch chain amino acids (to enhance immunity, prevent overtraining, and delay fatigue during endurance exercise), and carbohydrate (to delay fatigue during endurance exercise). That some of these substances enhance performance is well documented (e.g., carbohydrate supplementation during prolonged exercise, iron supplementation in iron deficiency), although evidence regarding the efficacy of other supplements is equivocal in some instances. The possibility that legal dietary supplements may enhance immunity in athletes has just begun to attract some interest in the exercise immunology community. It is beyond the scope of this chapter to review each of these classes of supplements; only those taken by athletes in the belief that they enhance immunity will be considered here, in particular dietary carbohydrate, glutamine, and vitamins A and C.

Dietary Carbohydrate and Immune Cells

Endurance athletes regularly consume a high carbohydrate diet to enhance muscle glycogen stores. Muscle glycogen depletion is associated with onset of fatigue, and increases in glycogen stores via dietary intervention prolong time to fatigue during endurance exercise (Brooks and Fahey 1995). However, very high carbohydrate diets consumed by endurance athletes often result in low protein intake, particularly from animal sources. To address the question whether such dietary intervention influences immune function in athletes, the immune cell response to two different diets was studied in eight well-trained male athletes (Richter et al. 1991). The two diets were equal in total energy and relative percentages of carbohydrate, protein, and fat but differed in their protein sources: a meat-rich diet with the greatest possible contribution of energy and protein from animal sources, and a lacto-ovo vegetarian diet composed primarily of vegetable protein sources. Athletes consumed one diet for six weeks, followed by a four-week ad libitum diet and then the other diet for an additional six weeks, in balanced order; exercise training was maintained throughout the entire study period. Neither diet influenced any immune parameter measured including the number of circulating leukocytes, lymphocytes, T cell subsets, NK cells, and monocytes. NK and cytokine-stimulated LAK cytotoxic activities were also unchanged by the two diets. These data suggest that immune function is not influenced by simple dietary intervention commonly used by athletes to modulate carbohydrate intake.

In a recent study, Nieman (1997) investigated whether exercise-induced changes in immune cell distribution could be modulated by carbohydrate ingestion during prolonged exercise. As described in chapter 3, recruitment of leukocytes and neutrophils into the circulation during, and removal of lymphocytes from the circulation after, prolonged exercise appear to be mediated at least in part by cortisol. It is well documented that 5-6% carbohydrate (CHO) ingestion in liquid form (e.g., a "sports drink") attenuates the large increase in corticosteroids and catecholamines during intense prolonged exercise. Nieman et al. (1997) reasoned that prevention of such rises in stress hormones may also attenuate large changes in circulating leukocyte and subset numbers. Male distance runners ran on a treadmill for 2.5 hours at 75% $\dot{V}O_{2max}$ and consumed either a 6% CHO sports drink or similar-tasting placebo. In the placebo group, changes in cell number followed the expected pattern, that is, large increases in total leukocyte and neutrophil number during and for several hours after exercise and a decrease in lymphocyte number during recovery. CHO ingestion during and after exercise significantly attenuated both the rise in neutrophil and decline in lymphocyte numbers. CHO ingestion also attenuated the rise in plasma epinephrine and cortisol levels, and changes in these hormones were significantly correlated with changes in cell number. These data support the concept that increases in stress hormones, in particular cortisol, mediate much of the redistribution of leukocyte subsets during and after intense prolonged exercise (discussed further in chapter 3). These data also suggest that these responses may be modulated, at least partially, by dietary carbohydrate supplementation during and after exercise.

Since released stress hormones (e.g., corticosteroids, catecholamines) have been implicated as mediators of exercise-induced alterations in immune parameters, it may be speculated that attenuating release of these hormones may also lessen the immune response to intense exercise. For example, the delayed decreases in peripheral blood lymphocyte number, lymphocyte proliferation, and NK cell cytotoxicity observed 1-4 hours after intense exercise have been attributed, at least partially, to cellular redistribution (i.e., decline of T lymphocyte and NK cell number in the circulation; discussed further in chapters 3 and 7). Increased concentrations of some stress hormones (e.g., catecholamines, growth hormone) have also been implicated as mediators of release of inflammatory mediators such as cytokines and acute phase proteins, and migration of inflammatory cells such as neutrophils to sites of tissue injury (discussed further in chapters 4 and 6). It is possible, although speculative at present, that by attenuating the hormonal response to intense exercise, carbohydrate supplementation may alter some of the suppressive effects on immune cells or may limit the contribution of immune cells to inflammatory processes resulting from intense exercise. Certainly, further work is needed to support or refute this speculation.

Glutamine and Immunity in Athletes

Glutamine, the most prevalent amino acid in the body, is required for normal function by immune cells, and serves a dual role as a nitrogen source for nucleotide synthesis as well as a carbon source for energy production. Although some glutamine is absorbed from protein foods in the gut, skeletal muscle provides the main source of glutamine via degradation of muscle proteins and transamination of amino acids. Glutamine in skeletal muscle comprises more than 50% of the total body amino acid pool and is thus an important source of glutamine for immune cells. Glutamine is required for in vitro maintenance and growth of lymphocytes and also appears to be important to specific immune cell functions such as antibody synthesis, cytokine secretion, and macrophage function. Low plasma glutamine levels have been associated with immunosuppression resulting from various traumas such as burns, surgery, and sepsis (reviewed in Newsholme 1994; Pedersen 1997; Rowbottom et al. 1996).

In vitro experiments with human lymphocytes showed a dose-response relationship between glutamine concentration in the culture medium and lymphocyte proliferative responses to various mitogens including PHA, ConA, purified protein derivative, and IL-2, as well as LAK activity but not NK cell cytotoxic activity (Rohde et al. 1995). In a related study, addition of physiological concentrations of glutamine to immune cell cultures enhanced production of cytokines such as IL-2, IL-6, and IFNγ, suggesting a possible mechanism by which glutamine may enhance lymphocyte function (Rohde et al. 1996b).

Based on the high concentrations measured in skeletal muscle and the high rate of uptake and oxidation by immune cells, it has been proposed that glutamine provides an important metabolic link between skeletal muscle and the immune system (Newsholme 1994). It has been further proposed that exercise-induced fluctuations in plasma glutamine concentration (caused by decreased muscle output or increased lymphocyte uptake or both) are related to immunosuppression and possibly the high rate of infectious illness among well-trained athletes (Newsholme 1994; Rowbottom et al. 1996). Moreover, recent data suggest that plasma glutamine concentration may be reduced in overtrained athletes (i.e., those showing symptoms of poor adaptation to excessive training stress); overtraining syndrome is associated with immunosuppression and increased susceptibility to infectious illness in athletes (Fitzgerald 1991). Consequently, a number of recent studies have focused on the relationship between exercise, illness, and glutamine levels (reviewed in Rowbottom et al. 1996).

Glutamine Response to Acute Exercise

The acute plasma glutamine response depends on exercise intensity, duration, and time of sample. Serum glutamine concentration was unchanged in nine athletes after a 23-hour ultratriathlon (swimming, cycling, running) despite a significant 18% decrease in total amino acid levels (Lehmann et al. 1995). These data suggest that, despite significant protein catabolism during very prolonged exhaustive exercise, plasma glutamine concentration may be maintained. Depletion of muscle glycogen occurs within the first few hours of sustained exercise and is associated with increased mobilization of amino acids to support gluconeogenesis in the liver and oxidative metabolism in skeletal muscle (Brooks and Fahey 1995). During very long duration exercise (e.g., > 3 hours), plasma glutamine concentration is maintained, presumably due to transamination of other amino acids mobilized within skeletal muscle after glycogen depletion.

In contrast, plasma glutamine concentration declined significantly after a 42 km marathon run in athletes given a placebo, whereas glutamine level was unchanged in runners supplemented with branched chain amino acids before the run (Parry-Billings et al. 1992). These data suggest either compromised glutamine release by skeletal muscle or increased lymphocyte uptake during intense prolonged exercise, or some combination of the two. However, although exercise influenced plasma glutamine concentration, the T lymphocyte proliferative response to mitogenic stimulation was unrelated to plasma glutamine level, indicating no direct association between glutamine concentration and lymphocyte function during exercise.

Rohde et al. (1996a) observed a significant nearly 50% decline in plasma glutamine concentration after a triathlon consisting of a 2.5 km swim, 81 km cycle, and 19 km run; glutamine levels reached a nadir 2 hours post-race. NK and IL-2-stimulated LAK activity were also suppressed 2 hours post-race, and the changes in plasma glutamine concentration and LAK activity were significantly correlated ($r = 0.39$), although glutamine level and NKCA were not significantly related. These data suggest a temporal, although not necessarily causal, relationship between the decline in plasma glutamine levels and LAK activity after prolonged exercise.

Glutamine uptake and oxidation by lymphocytes appear to increase after intense interval exercise (Frisina et al. 1994). Seven active males performed 25 1-minute treadmill sprints with 2 minutes rest between each sprint. Lymphocytes were isolated from blood samples obtained before and after the exercise session. Glutamine oxidation by lymphocytes increased by about 20%. The change in glutamine oxidation was significantly correlated with changes in lymphocyte subset numbers—positively correlated with the change in NK cells ($R = 0.78$) and inversely with the change in CD3 T cells ($R = -0.93$). It was concluded that intense interval exercise causes redistribution of lymphocyte subsets (decrease in T and increase in NK cells), bringing into the circulation a higher number of metabolically active NK cells that utilize glutamine at a higher rate than other subsets (T and B cells). Whether such changes are related to compromised immune function is still open to speculation.

Glutamine Response to Training

For the most part, plasma glutamine concentration in athletes appears to be relatively high compared with clinically normal values or those of nonathletes (Keast et al. 1995; Lehmann et al. 1995, 1996; Mackinnon and Hooper 1996; Rowbottom et al. 1995), although levels may decline markedly during periods of very intense training or during overtraining. For example, Rowbottom et al. (1995) reported similar values in healthy, well-trained athletes compared with nonathletes, but 30% lower values in athletes exhibiting overtraining syndrome (discussed further below).

Lehmann et al. (1996) reported relatively constant concentrations of plasma amino acids, including glutamine, during four weeks intensified training in male distance runners. Training was intensified by increasing either volume or intensity, with the increased-volume group showing indications of overtraining syndrome. However, neither protocol altered plasma glutamine levels. In contrast, Keast et al. (1995) observed a progressive decline in plasma glutamine concentration after 10 days intense training in five endurance-trained males. Training consisted of twice-daily repeated maximal treadmill sprints, and subjects showed symptoms of overreaching (Fry et al. 1992a, 1994). Plasma glutamine concentration declined by about 40% after the 10 days and remained low during recovery training (very light exercise), recovering only after 6 days. It has been suggested that although plasma glutamine concentration is generally normal in well-trained athletes, levels may fall below some critical threshold during prolonged periods of intense training and that these very low values may contribute to immunosuppression observed in athletes (Mackinnon and Hooper 1996; Rowbottom et al. 1996).

Glutamine in Overtraining and Chronic Fatigue Syndrome

Overtraining syndrome is a generalized stress response to excessive exercise training observed primarily in high-performance athletes training several hours each day for extended periods (reviewed in Lehmann et al. 1993). It has been suggested for some time that low plasma glutamine concentration may accompany or precede overtraining in athletes (Newsholme 1994; Parry-Billings et al. 1992; Rowbottom et al. 1996). Parry-Billings et al. (1992) noted slightly (9%) but significantly lower plasma glutamine concentration in a variety of elite athletes (runners, cyclists, swimmers, rowers) showing symptoms of overtraining (e.g., poor performance, chronic fatigue) compared with similarly trained athletes who were not diagnosed as overtrained. Mackinnon and Hooper (1996) also observed differences in plasma glutamine concentration between overtrained and well-trained swimmers during four weeks of intensified training. Training was intensified by progressive 10% increases in training volume and resistance training for each of four weeks, resulting in 8 of 24 swimmers exhibiting symptoms of overtraining (e.g., persistent fatigue, poor performance, and sleep disturbances). Plasma glutamine concentration increased by 23% over the four weeks in well-trained swimmers and remained unchanged in overtrained swimmers.

In a series of studies on subjects showing symptoms of chronic fatigue, Rowbottom et al. (1995, 1998a, 1998b) noted that plasma and skeletal muscle glutamine concentrations were the only biochemical or immunological parameters that distinguished athletes with symptoms from those without symptoms or from clinical norms. Compared with values in well-trained athletes and nonathletes, plasma glutamine concentration was more than 30% lower in 10 overtrained athletes from various sports (e.g., distance running, swimming, rowing, tennis) (Rowbottom et al. 1995). Leukocyte, lymphocyte, and subset counts and a variety of other hematological and biochemical measures were clinically normal in these overtrained athletes. However, although plasma glutamine concentration may be low in overtraining and chronic fatigue, it does not appear that glutamine supplementation improves clinical status. For example, in a 26-week randomized double-blind placebo-controlled study, 16 chronic fatigue syndrome patients and 16 matched healthy controls were given either a daily supplement of 2000 mg L-glutamine or placebo (Rowbottom et al. 199a). Although glutamine supplementation significantly increased plasma and skeletal muscle glutamine concentrations to clinically normal levels, there were no improvements in self-reported symptoms in patients (e.g., fatigue, sleep problems, physical complaints). Moreover, glutamine supplementation had no effect on the incidence of URTI during the 26 weeks or on lymphocyte and subset counts. It was concluded that although low plasma and skeletal muscle glutamine concentrations may occur concomitantly with chronic fatigue syndrome, there does not appear to be a direct causal link between symptoms, and increasing glutamine levels does not alleviate symptoms of the syndrome.

Glutamine, Exercise, and Viral Illness

There are few and inconsistent data on glutamine and viral illness. Mackinnon and Hooper (1996) compared plasma glutamine concentration in overtrained and well-trained athletes and in athletes who developed URTI compared with those who remained healthy. Training volume was progressively increased during four weeks intensified training in 24 elite swimmers. As described in chapter 1, an unexpected finding was a higher incidence of URTI in well-trained compared with overtrained swimmers (58% vs. 12.5%, respectively). Comparison between swimmers who developed URTI and those who did not showed no differences in plasma glutamine concentration between the groups at any time during the four weeks. These data suggest that although plasma glutamine concentration may be lower in overtrained athletes, such differences are not related to the appearance of URTI.

In contrast, a recent brief report suggests that glutamine supplementation may prevent URTI during recovery after endurance exercise (Castell et al. 1996). Oral glutamine (5 g L-glutamine in mineral water) was given immediately and 2 hours after a marathon race in 72 runners, with a placebo given at the same times to 79 runners in the same races. Incidence of infectious illness was assessed by questionnaire during the week after the race. The incidence of URTI and influenza was significantly higher in runners consuming the placebo compared with glutamine drinks (81% vs. 49%, respectively), suggesting that glutamine supplementation immediately after strenuous endurance exercise may prevent immunosuppression and illness.

In a brief note, Greig et al. (1995) reported a significant inverse relationship between plasma glutamine concentration and the appearance of viral illness in nonathletes. Fourteen volunteers were monitored over a four-month period with blood samples obtained at least every two weeks. Self-rating of health status with emphasis on symptoms of viral illness was used to develop a viral score (presence of symptoms and duration of illness). Viral score was significantly inversely correlated with plasma glutamine concentration ($R = -0.553$), suggesting that the greater the symptoms and duration of illness, the lower the glutamine concentration. It is unclear, however, whether lower plasma glutamine concentration preceded (and possibly contributed to) or resulted from illness. It was concluded that additional stress (e.g., exercise) during viral illness may further compromise immune function by a combined negative effect on plasma glutamine level. Whether this has implications for athletes who wish to continue training during minor illness has yet to be tested.

In summary, although much recent interest has focused on exercise and glutamine, it is still unclear what role, if any, glutamine plays in mediating changes in immune function associated with exercise. Plasma and skeletal muscle glutamine concentrations may decline acutely during intense exercise and chronically during rigorous training; levels may be clinically low in overtrained athletes and in chronic fatigue syndrome patients. Glutamine is an important substrate required for normal lymphocyte function, and low levels have been associated with immunosuppression during other physically stressful events such as surgery. However, there is at present little evidence to directly link changes in plasma concentration or lymphocyte metabolism of glutamine with immunosuppression associated with intense exercise. One intriguing report suggests the possibility that glutamine supplementation may reduce the incidence of URTI after marathon running, and it is likely this topic will be vigorously pursued in the next few years.

Antioxidant Vitamins: A, C, and E

Vitamin C has long been considered prophylactic against the common cold, although years of research have failed to clearly support such a role for this vitamin. It appears that vitamin C supplementation may lessen the duration and severity, but may not necessarily reduce the incidence, of infection (reviewed in Hemila 1992, 1996). In a recent review, Hemila (1996) suggested that vitamin C supplementation may be beneficial in limiting the severity of infection during periods of physical stress and may therefore have special relevance for competitive athletes. Despite the lack of clear evidence supporting a protective role of vitamin C, many athletes (and nonathletes) regularly consume large amounts of this vitamin in the belief that it will keep them healthy during stressful times such as intense exercise training.

A recent intriguing study on a group of marathon runners supports the idea that vitamin C is prophylactic (Peters et al. 1993). In a double-blind procedure, 84 ultramarathon runners and 73 nonathletes consumed either 600 mg per day vitamin C supplements or placebo for three weeks before a 90 km race. The incidence and symptoms of URTI were assessed by questionnaire during the two weeks after the race. The number of subjects reporting symptoms was greatly reduced in the vitamin C-supplemented runners but not controls. For example, 68% of runners taking placebo exhibited URTI compared with 33% of supplemented runners; vitamin C supplementation had no effect on the incidence of URTI among nonrunners (45% and 53% for placebo and supplemented controls, respectively), although the incidence of infection appears rather high for nonathletes. The authors concluded that intense exercise training may deplete the body's stores of vitamin C, increasing susceptibility to URTI. Nonathletes, presumably, would not suffer such depletion and thus would not benefit from supplementation. The data from this intriguing study await further verification in other studies on other types of athletes. A related study from the same laboratory showed that vitamin A supplementation had no effect on the incidence of URTI in distance runners (Peters et al. 1992).

There is some suggestion that vitamin E supplementation may influence circulating leukocyte distribution during and after eccentric exercise (Cannon et al. 1990, 1991, 1995). In one double-blind study, young sedentary men took either a placebo or vitamin E supplement (800 IU per day) for 48 days, then ran downhill for 45 minutes at –10% grade at 75% $\dot{V}O_{2max}$ (Cannon et al. 1991). Cytokine release was measured in lymphocytes obtained before, after, and then every 2 days up to 12 days after exercise; cells were stimulated in vitro with endotoxin. The postexercise rise in IL-1β was blunted in the vitamin E- compared with placebo-supplemented group, and this difference persisted for 12 days postexercise (figure 8.4). Basal and postexercise IL-6 production was also lower in supplemented compared with placebo subjects. TNFα secretion was unaffected by vitamin E supplementation. It was suggested that vitamin E may influence production of mediators of inflammation such as prostaglandins, leukotrienes, and oxygen radicals, which

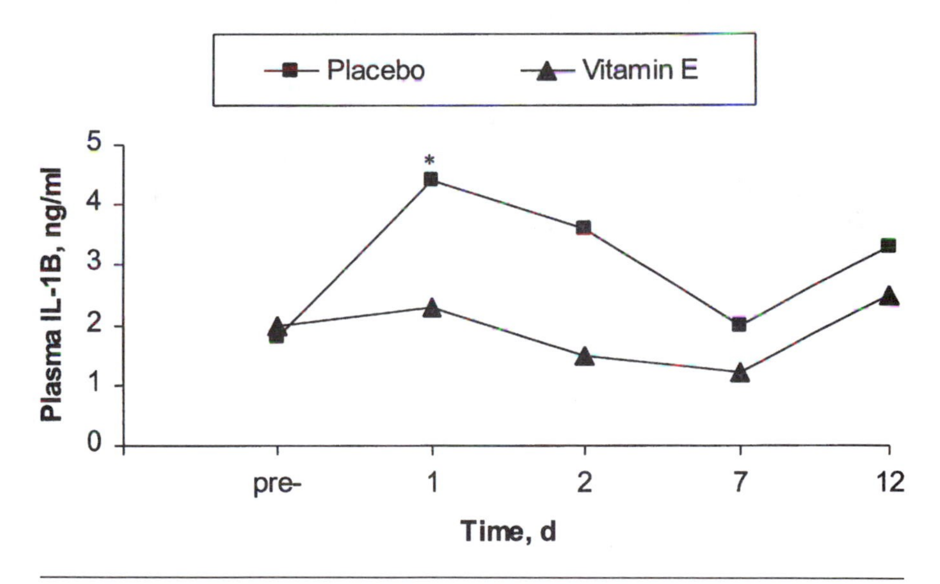

Figure 8.4 Effects of vitamin E on the IL-1β response to exercise. IL-1β secretion by LPS-stimulated mononuclear cells isolated from subjects before (pre-) and 1, 2, 7, and 12 hours postexercise. Exercise consisted of 45 minutes downhill treadmill running. For 48 days prior to exercise, subjects consumed either a vitamin E supplement (800 IU/day) or placebo.

From Cannon et al. 1991.

enhance cytokine release. In a related study comparing the responses to downhill running in young and older adults, vitamin E supplementation blunted the postexercise increase in neutrophil number in older but not younger subjects (61-72 vs. 20-32 years, respectively) (Cannon et al. 1990). These data suggest that prior vitamin E supplementation may help limit postexercise increases in inflammatory mediators and cytokines and thus reduce infiltration of inflammatory cells into skeletal muscle after eccentric exercise.

Summary and Conclusions

There are wide-ranging potential clinical applications of research on exercise and immunology. Regular physical activity appears to reduce the risk of certain cancers, although it is unclear whether the immune system is involved. Moderate exercise training may be a beneficial adjunct therapy in treatment of HIV-1-positive individuals, by increasing muscular strength and lean body mass and possibly by reducing psychological stress and enhancing immune function. Regular exercise has been suggested as a possible intervention to limit or prevent immunosuppression during aging and spaceflight, although there are, at present, few data to support this notion. High-performance athletes regularly consume a variety of dietary supplements purported to enhance performance and to reduce the risk of illness or injury (e.g., carbohydrate supplements, vitamins C and E, glutamine and branch chain amino acids). Although far from conclusive, recent evidence suggests the possibility of some beneficial effects of these supplements on immune function in athletes, in particular vitamin C and glutamine. None of these applications has been extensively studied to date, but future work is likely to focus on these exciting possibilities.

Research Findings

- Human epidemiological and experimental animal studies show that regular moderate physical activity reduces the risk of certain types of cancer (bowel cancer in particular), although it is unclear whether the immune system is involved.
- There is limited evidence suggesting that moderate exercise may be beneficial in HIV-infected patients, helping to maintain lean body mass and perhaps to enhance immune cell number in the circulation.
- The limited evidence to date suggest that short-term moderate exercise training has no beneficial effects on immunity in older individuals, despite improvements in physical capacity and muscular strength.

The longer-term (i.e., years) effects of physical activity on immune function in the elderly are not known.

- Spaceflight is associated with suppression of a number of immune parameters, and the immune response to spaceflight is similar to that observed after prolonged exercise. It has been speculated that prior exercise training may provide a "countermeasure" against immune suppression during spaceflight, although there is no supporting evidence at present.
- Recent data suggest that some dietary supplements (e.g., carbohydrate drinks, vitamin C, glutamine) may be beneficial in altering the suppressive effects of intense exercise on immune function.

Practical Applications

- If physical activity can be shown to reduce the risk of cancer, there is one more valid reason (beyond those for prevention of other diseases) for the health profession to advocate regular moderate exercise as part of a healthy lifestyle.
- Information from research on the immune system response to exercise in cancer, HIV-infected, and other immunocompromised patients should help provide a scientific basis for exercise prescription in these individuals.
- If exercise is shown to be effective in preventing some of the adverse immune system changes during spaceflight, then training before and during spaceflight may be recommended, especially for long flights.
- Intense exercise, which has been shown to suppress some aspects of immune function, is a necessary part of training for high-performance athletes. If dietary supplements are shown to be effective in attenuating some of the negative effects of intense exercise on immune function, this may provide an easily administered means to avoid illness during periods of intense training.

Yet to Be Explored

- Is the apparent association between physical activity and reduced cancer risk due to exercise-induced changes in immune function? If so, what specific aspects of immune function are involved, and what are the minimal and optimal amounts of exercise to reduce risk of cancer?
- Are there long-term beneficial effects of resistance exercise training (e.g., maintenance of lean body mass) in treatment of HIV-infected individuals?

- Does regular moderate aerobic exercise training enhance immune function in HIV-infected individuals, and does this result in long-term benefits? If so, what are the minimal and optimal amounts of exercise for eliciting a positive response? Is vigorous exercise detrimental to the immune system in HIV-infected patients?
- To what extent are changes in immune function related to an increasingly sedentary lifestyle among the elderly? Can a lifetime of regular moderate physical activity help prevent these age-related changes in immune function?
- Do exercise training prior to, and continued training during, spaceflight provide a countermeasure against adverse immune responses during spaceflight? If so, what are the minimal and optimal types and amount of exercise needed to elicit such protection? If so, are the mechanisms related to direct stimulation of immune function or to indirect effects such as via hormonal responses or maintenance of muscle mass?
- Can dietary supplements such as carbohydrate, vitamin C, or glutamine prevent deleterious effects of intense exercise training on immune function and thus help prevent URTI in athletes?

Chapter 9

Exercise and Immunology: Current Knowledge and Future Directions

© Human Kinetics

Having reviewed the current literature describing the effects of exercise on various aspects of immune function, and having touched on possible mechanisms underlying such changes and some of the complexities of the immune system, I now wish to summarize the current state of knowledge and then look to the future. Although it is difficult to predict directions with any certainty in such a rapidly expanding and changing field, I think it is instructive to look at some of the most recent and novel work and then to speculate where this might lead.

Current Knowledge

Tables 9.1 and 9.2 summarize our current state of knowledge on the effects of acute exercise and exercise training on immune function.

Table 9.1 Potentially Positive, Negative, and Neutral Effects of Acute Exercise on Immune Parameters

Potentially positive	Potentially negative	Neutral (no effect)
	Resistance to illness ↑ incidence of URTI[d] ↑ paralysis polio infection[d]	*Resistance to illness* Incidence of URTI[e]
Leukocyte distribution Recruitment into circulation	*Leukocyte distribution* Neutrophil migration to skeletal muscle[d]	
Neutrophils ↑ priming/activation[a,b*] ↑ bactericidal activity[a,b]		
Lymphocytes ↑ number in circulation	*Lymphocytes* ↓ number[c] ↓ CD4:CD8 T cell ratio[a,c] ↓ proliferation response to mitogen[a]	*Lymphocytes* Proliferation[b] Proliferation/T cell
Natural killer cells ↑ number in circulation[a,b] ↑ cytotoxic activity (NKCA)[a,c]	*Natural killer cells* ↓ number in circulation[c] ↓ NKCA[c]	*Natural killer cells* NKCA/cell[b]
Cytokines ↑ release of IL-1, IL-6, TNFα, IFN[c,d]		
Other soluble factors ↑ specific antibody response[b,e] ↑ acute phase proteins[d]	*Soluble factors* ↓ specific antibody reponse[a,c] ↓ secretory IgA response[a,c]	*Soluble factors* Secretory IgA response[b]

a = Immediately after intense or prolonged exercise; b = immediately after moderate exercise; c = delayed response 1-6 h after intense exercise; d = delayed response for days after intense exercise; e = delayed response for days after moderate exercise; no symbol = general response after all types of exercise.

Table 9.2 Potentially Positive, Negative, and Neutral Effects of Exercise Training on Immune Parameters

Potentially positive	Potentially negative	Neutral (no effect)
Resistance to illness ↑ survival viral infection[b] ↑ survival bacterial infection[b] ↓ incidence of cancer[b] ↓ incidence of URTI[b]	*Resistance to illness* ↑ incidence of URTI related to training volume ↑ paralysis polio infection[a]	*Resistance to illness* Incidence of URTI[b]
	Leukocyte distribution ↓ resting number in circulation[d]	*Leukocyte distribution* Resting number[a,b,c]
	Neutrophils ↓ priming/activation at rest and after exercise[a,c,d]	*Neutrophils* Resting number
		Lymphocytes Resting number proliferation
Natural killer cells ↑ NKCA at rest[c]	*Natural killer cells* ↓ Cell number[d]	*Natural killer cells* Resting number[a,b,c] NKCA[a,b,c]
	Cytokines ↑ resting IL-1 plasma level[c]	*Cytokines* Resting concentrations
Other soluble factors ↑ specific antibody response[b]	*Other Soluble factors* ↓ serum Ig levels[c,d] ↓ secretory IgA levels[c,d] ↓ acute phase protein release after exercise	*Other soluble factors* Serum and secretory Ig levels[b]

a = With intense training; b = with moderate training; c = in athletes compared with nonathletes; d = in athletes after periods of intensified training; no symbol = general response after all types of exercise training.

These effects have been categorized by their potential to influence immune function, that is, potentially positive, negative, or neutral (no effect).

From general trends in the literature, four major points are apparent:

1. Exercise alters many aspects of immune function, potentially in both positive and negative directions; some immune parameters are unaffected by exercise. It is not yet clear whether some of the observed changes are beneficial or detrimental (e.g., elevated cytokines during recovery from exercise, downregulation of neutrophil priming).
2. Exercise influences the immune response at the level of the intact organism (e.g., survival during infection, incidence of cancer), as well as individual immune parameters (e.g., lymphocyte proliferation).
3. There is often a dose-response relationship between the amount of exercise and a specific immune response; there appears to be significant interaction of exercise duration, intensity, and, to a certain extent, fitness level in determining the response of a particular immune parameter to exercise.
4. There are inconsistencies in the immune response to exercise that cannot yet be fully explained (e.g., some studies report increases in cytokine production, whereas others do not).

Future Directions

Although considerable progress has been made over the past few years in describing and understanding the mechanisms underlying the immune response to exercise, much work is still needed to generate an integrated model. The following discussion focuses on what I consider to be some key questions that I believe point to new directions for future work in this field.

Are Athletes Immunocompromised?

As described in chapter 1, recent epidemiological data are consistent in suggesting that male endurance athletes (mainly distance runners) experience elevated rates of URTI in the few weeks after major competition. However, for the most part these studies have not been extended to other types of athletes (female endurance athletes; strength and power athletes of both sexes). We still do not know if this high rate of URTI relates specifically to the type of activity or to some more general phenomenon associated with training and competing at the elite level (e.g., psychological stress). Certainly, larger longer-term prospective studies are needed to more completely document the risk of URTI among all types of athletes. Such data are likely to also shed light on which specific aspects of training and/or competition are related to the increased incidence of illness. Recent data from Peters and coworkers (1992, 1996), showing the lowest post-race incidence of URTI in runners who train moderately compared with those training too little or excessively are consistent with the proposed J-curve discussed in Chapter 1 (Nieman and Nehlsen-Cannarella 1992). Obviously, there are practical implications for athletes and coaches who must balance the need to train intensely over many months with the desire to maintain athletes' health. However, such data would have more far-reaching implications for understanding the mechanisms by which physical and psychological stress may influence immune function.

It does not appear that athletes are immunocompromised by a strict clinical definition; that is, athletes do not present with serious illnesses normally associated with immunodeficiency. Indeed, the only illness to which athletes appear to be more susceptible is URTI. However, in the past few years, evidence has begun to suggest suppression of several immune parameters during long periods of intense training. As described throughout this book and summarized in table 9.2, intense training has been associated with low resting leukocyte and lymphocyte counts compared with clinical norms, decreasing NK cell number, low serum and secretory Ig concentrations, and suppression of neutrophil antimicrobial activity. To date, the only exercise-induced change in an immune parameter that has been directly associated with the appearance of URTI in athletes is secretory IgA (Mackinnon et al. 1993b); however, few studies have attempted to correlate changes in specific immune parameters with the incidence of URTI (Mackinnon et al. 1993b; Mackinnon and Hooper 1996; Pyne et al. 1995). Although many of these alterations are relatively small, it is possible that immune function is compromised by the additive and interactive effects of small changes in several parameters important to host defense.

According to Pedersen and Ullum's (1994) "open window" hypothesis, some aspects of immune function (e.g., NK cytotoxic activity, neutrophil priming) appear to be stimulated during, but suppressed for some time after, intense exercise. It was proposed that athletes are susceptible to infection during the extended period of immune suppression after exercise (the "open window"). High-performance athletes train intensely at least once, and often twice, each day for many months without appreciable breaks. Figure 9.1 presents a theoretical

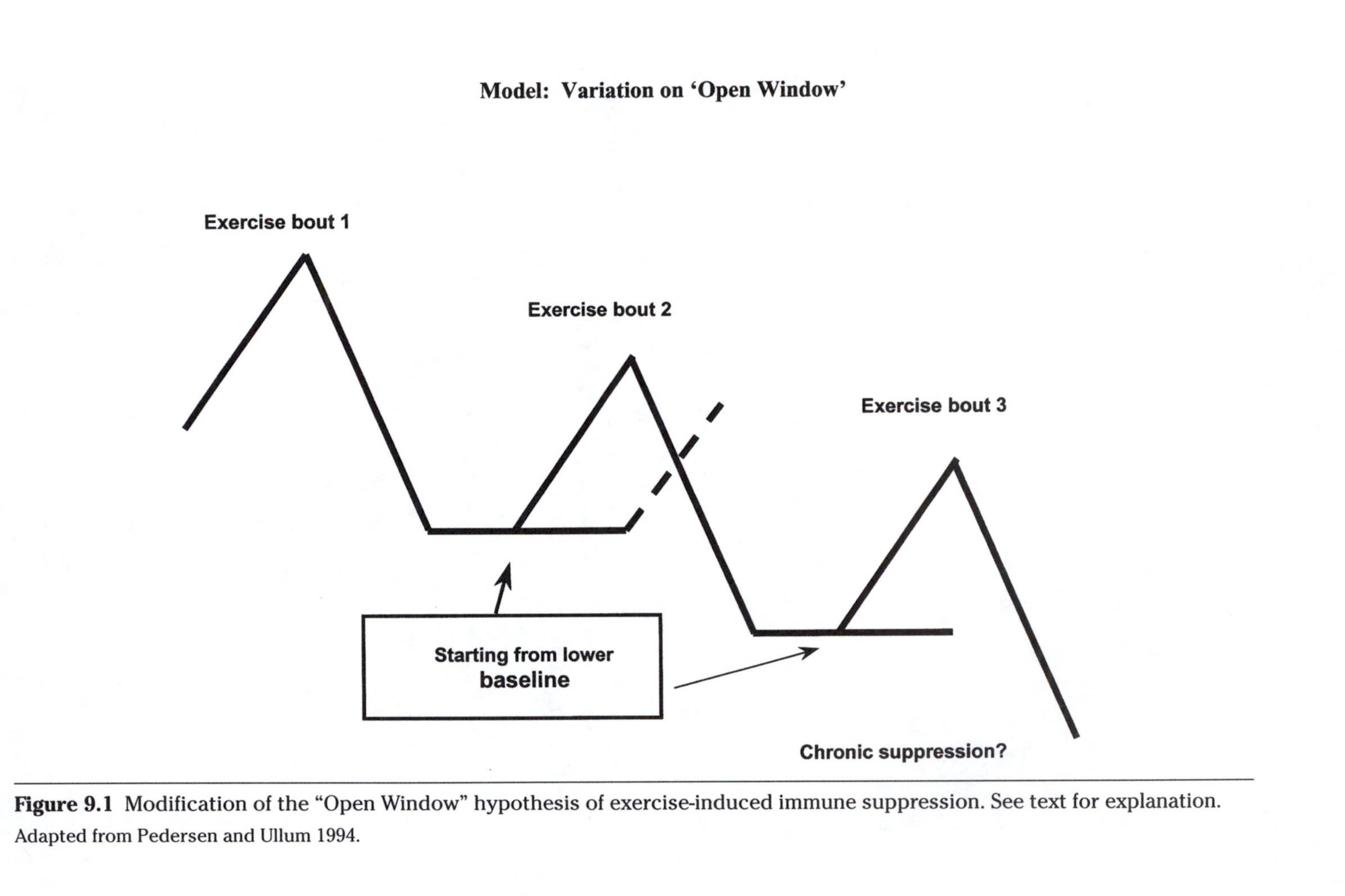

Figure 9.1 Modification of the "Open Window" hypothesis of exercise-induced immune suppression. See text for explanation.
Adapted from Pedersen and Ullum 1994.

model modified from Pedersen and Ullum's open window hypothesis. In this modified model, it is possible that an athlete may begin a training session before full recovery from the previous session (i.e., during the open window of immunosuppression). Repeating this pattern over time may eventually lead to chronic suppression, if intense training sessions are spaced such that there is incomplete recovery of immune function between sessions. Recent research suggests that there is a cumulative suppression of at least some aspects of immune function, such as mucosal IgA levels (Gleeson et al. 1995; Mackinnon and Hooper 1994), leukocyte number (Lehmann et al. 1996), and NK cell number (Fry et al. 1992) due to daily intense exercise. Future work focusing on the cumulative effects of daily or more frequent intense exercise would help identify whether such training patterns increase risk of infection among athletes.

Is Downregulation of Nonspecific Immunity in Athletes Beneficial or Harmful?

As noted in chapter 4, certain aspects of nonspecific immunity, in particular neutrophil activation/priming, appear to be downregulated in athletes undergoing intense training compared with nonathletes or with athletes performing more moderate training (see Hack et al. 1992, 1994; Pyne et al. 1995; Smith et al. 1990 for examples). Neutrophils are important effector cells in the body's early defense against a variety of pathogens, and it is possible that such downregulation adversely influences resistance to illness in athletes. On the other hand, neutrophils have also been implicated as mediators of tissue damage during inflammation via their release of reactive oxygen species and other toxic factors (Pedersen 1997; Smith 1994). As discussed in chapters 3, 4, and 6, intense exercise stimulates neutrophil migration (Espersen et al. 1991) and infiltration into a variety of tissues including skeletal muscle (Belcastro et al. 1996; Cannon et al. 1990, 1991; Fielding et al. 1993), cardiac muscle and liver (Belcastro et al. 1996), and the nasal mucosa (Muns et al. 1996). Migration to tissues may be harmful because activated neutrophils release damaging reactive oxygen species. Muns et al. (1996) suggested that migration of activated neutrophils to the nasal mucosa may elicit local inflammation, which may render the mucosa more susceptible to infection. Thus, downregulation of neutrophil activity may serve a protective function by limiting chronic inflammation caused by intense daily exercise as performed by athletes. It is possible that, in athletes, mild immunosuppression reflects a compromise between the body's attempts to limit chronic inflammation and the maintenance of immune responsiveness. It can be expected that future work will continue to focus on the relationship between mediators of inflammation and immunity and their responses to exercise.

What Mediates Communication Between Events in Skeletal Muscle and the Immune System?

The immune response to exercise is, in many ways, dose dependent; that is, the magnitude of change in many immune parameters is often a function of exercise intensity and duration. As discussed throughout this book, released stress hormones such as cortisol and epinephrine have been implicated as mediators of exercise-induced changes in immune parameters, such as leukocytosis and increased NK cell cytotoxicity. There are some interesting recent observations that imply close communication between immune function and metabolic and/or structural events in skeletal muscle. For example, as discussed in chapter 8, Nieman (1997) noted that the influx of leukocytes and neutrophils into the circulation during prolonged exercise was attenuated by carbohydrate supplementation and that this effect appeared to be reflected in alterations in cortisol levels. Presumably, cortisol was released in response to hypoglycemia as a result of glycogen depletion in, and uptake of blood glucose by, working skeletal muscle. As discussed in chapter 3, Pizza et al. (1995) reported greater recruitment of neutrophils and lymphocytes into the circulation during exercise with a large eccentric bias (downhill running) compared with level running, despite similar metabolic cost of the two exercises. In a series of studies, Cannon, Fielding, and coworkers (Cannon et al. 1990, 1991; Fielding et al. 1993) showed marked infiltration of neutrophils into skeletal muscle damaged during downhill running (discussed in chapters 3 and 6). Neutrophil infiltration was correlated with the appearance of IL-1β in skeletal muscle (Fielding et al. 1993). Although stress hormones such as cortisol and cytokines such as IL-1β have been implicated as mediators of such effects, the mechanisms by which these factors influence recruitment of immune cells during and after exercise remain to be determined. Other studies have noted increased expression of some adhesion molecules (e.g., ICAM, LFA-1) in response to exercise (Baum et al. 1994; Gabriel et al. 1994a). Since expression of adhesion molecules may be influenced by hormones and cytokines, one exciting avenue for future work is to further understand the role of adhesion molecule expression in mediating exercise-induced changes in immune function.

Can Exercise Training Provide a "Countermeasure" Against Immunosuppressive Events?

As discussed in chapters 1 and 8, there is far-reaching clinical significance to understanding how exercise influences immune function. Examples of such applications include working with elite athletes to optimize training without compromising resistance to URTI, exercise prescription for immunocompromised patients (e.g., HIV-1 infection), and the potential of moderate exercise training in prevention of compromised immune function (e.g., spaceflight, aging). Although it is commonly believed that moderate exercise stimulates immune function (e.g., the proposed J-curve relationship between exercise dose and URTI risk; Nieman and Nehlsen-Cannarella 1992), there is at present very limited supporting empirical evidence, and most long-term studies suggest that moderate exercise training exerts little effect on immune function in healthy populations. It is possible, however, that moderate exercise training beneficially influences immune function in other groups. For example, while moderate exercise may be advocated as adjunct therapy for certain cancer patients to help maintain functional capacity and muscular strength, there is little empirical evidence on which to scientifically base an exercise prescription. There are at present too few studies to provide clear guidance as to the level of exercise tolerated by cancer patients undergoing radiation and chemical therapies that may adversely affect immune cell number and function. Whether moderate physical activity may help to stimulate, or at least maintain, cell number and function, or may enhance natural immunity in these cancer patients, has yet to be fully addressed in the research literature.

Similarly, as discussed in chapter 8, recent evidence suggests positive effects on immune cell number and function in HIV-1 infection (LaPerriere et al. 1990a, 1997). If these data are confirmed by further study, there may be sound scientific evidence for incorporating physical activity in the routine treatment of these patients. It is important, however, to clearly identify optimal, minimal, and possible adverse "doses" of exercise (i.e., duration, frequency, intensity) to maximize positive while minimizing negative effects on immune function. For example, the question of whether an HIV-infected individual should participate in (noncontact) competitive or rigorous sport requiring maximal effort has yet to be fully addressed. Common sense would dictate avoidance of maximal exercise, but this conservative approach is based on assumptions and extrapolations from the literature on noninfected individuals. More complete understanding of the immune response to various exercise intensities in HIV-infected individuals would provide a more sound rationale for any decision faced by an athlete and physician.

Also described in chapter 8, spaceflight is associated with general immunosuppression that has been compared with the effect observed after intense prolonged exercise (Gmunder et al. 1988). Exercise may thus serve as a useful model for further understanding long-term responses to physical stressors such as spaceflight. Moreover, moderate exercise training has been proposed as a potential "countermeasure" against immunosuppression during spaceflight. Work has already begun on studying whether and how exercise training before and during spaceflight may attenuate immunosuppression during spaceflight.

Aging, with its associated declines in certain immune functions, is another yet-to-be-explored area rich with potential relevance given the increase in the relative proportion of older individuals in most developed countries. A lifetime of regular moderate physical activity is now advocated in the prevention of a host of diseases such as heart disease, hypertension, obesity, adult onset diabetes, and osteoporosis, diseases generally of middle-aged and older individuals. It is naturally of interest to these individuals (and society in general, which bears the ever increasing medical costs) whether physical activity will enhance resistance to infectious illness or other diseases involving immune function. Since older individuals generally exhibit lower functional capacity, it may be reasoned that even very moderate activity (e.g., 40-60% $\dot{V}O_{2max}$) may provide some enhancement of immune function. As with many other clinically relevant applications, however, such statements are at present speculative, requiring careful experimentation to clarify the relationship between exercise dose and immune function in older individuals.

Do Changes in Immune Function Underlie the Protective Effect of Regular Physical Activity Against Cancer?

As discussed in chapter 8, epidemiological and experimental data increasingly show a relationship between moderate exercise training and reduced incidence of certain forms of cancer. Also discussed in chapter 8 are the many possible mechanisms by which exercise may influence resistance to cancer; few of these relate to training-induced changes in immune function. However, exercise has been shown to exert a profound influence on several factors important to defense against neoplastic growth, such as monocyte and NK cell cytotoxic activity and cytokine release. Two recent studies (MacNeil and Hoffman-Goetz 1993a, 1993b) showed a relationship between lung clearance of tumor cells and increased splenic NK cytotoxic activity in trained mice. Work from Woods and Davis (Woods et al. 1993, 1994a) also shows stimulation of monocyte tumor cell cytotoxicity by exercise training, although this was not directly related to inhibition of tumor growth. Since neoplastic growth is influenced by many factors, in particular the origin and location of tumor cells, there is unlikely to be a direct and generalized relationship between exercise-induced effects on tumor growth and immune cell cytotoxicity. One area for future work might focus on identifying which types of tumors are sensitive to the effects of exercise, and then determining whether exercise effects on immune cell cytotoxicity are consistent across different models.

To What Extent Are Exercise-Induced Changes in Immune Function Mediated at the Molecular Level?

Although much recent attention has focused on the mechanisms responsible for exercise-induced changes in immune function at the whole body, systemic, and cellular levels, to date there has been only limited study at the molecular level. It may then be speculated that future study will apply molecular biology techniques to further identify mechanisms by which exercise influences immune function. As discussed in various sections throughout this book, some research groups have begun to study the effects of exercise on gene expression of certain regulatory factors (e.g., cytokines), and it is likely that future work will also include other mediators of immune function. For example, much recent attention has focused on measuring cytokine levels in blood, urine, or, less frequently, tissues in response to exercise. Many regulatory factors, including cytokines, are released in very small quantities, act locally, and are rapidly metabolized once released. Thus, simply measuring concentrations in large pools such as blood or urine may not yield a comprehensive view of important cellular and subcellular events. Studying gene expression of certain key factors gives more complete information about the sequence, location, and mechanisms underlying immune system responses to exercise.

Closing Comments

In the preface to this book I commented that the field of exercise immunology has, over the past decade, rapidly progressed from a newly emerging to a mature area of scientific inquiry. Along with this process of maturation has come extension of initially descriptive studies to research aimed at broader clinical applications and more complete understanding of regulatory mechanisms. There has been a concomitant broadening of interest, from the initial group of exercise scientists interested in athletes' health, to include scientists and clinicians from diverse fields such as immunology, medicine, physiology, endocrinology, psychiatry, psychology, and gerontology. It is easy to predict, then, that future work will encompass multidisciplinary and integrated approaches to more completely document the immune response to exercise at all levels, from whole body to molecular, which will allow us not only to extend our knowledge to clinical applications but also to enhance our overall understanding of regulation of immune function.

Glossary

acquired immunity—"Adaptive" immunity involving antibody and immune cell responses specific to the infectious agent. Acquired immunity results in "memory," preventing later disease by the same agent.

ADCC (antibody-dependent cell-mediated cytotoxicity)—Cytotoxic activity by large granular lymphocytes that recognize certain antibodies on the target cell.

adherent cell—Primarily monocytes isolated from PBMC by their ability to adhere to a plastic culture dish/vessel; nonadherent cells are decanted and the adherent cells removed from the plastic.

adhesion molecules—Membrane or transmembrane proteins that mediate binding between cells or between cells and extracellular matrix and that are important to leukocyte trafficking and migration between the circulation and tissues.

aerobic exercise—Endurance-type exercise, lasting up to hours (e.g., distance running or cycling), that relies on oxidative metabolism as the major source of energy production.

antibody—An immunoglobulin molecule that can bind specifically to a particular antigen.

antigen—A protein that induces an antibody response.

antigen-presenting cells—Diverse cell types able to present antigen on their cell surface and thus stimulate immune cells.

APP/APR (acute phase proteins or reactants)—A heterogeneous class of serum glycoproteins that increase in concentration during inflammation and infection.

B cell—A type of lymphocyte capable of producing antibody. B cells differentiate into plasma cells.

bactericidal—Able to kill bacteria.

CD (cluster designation)—Cell surface proteins, identified with monoclonal antibodies, that are used to classify different types of leukocytes.

CD4 T cell (helper/inflammatory T lymphocyte)—A subset of T lymphocyte capable of recognizing antigen and producing several cytokines that activate other immune cells.

CD8 (suppressor T lymphocyte)—A subset of T lymphocyte capable of suppressing activity of other immune cells or processes.

complement—A group of 20 serum proteins involved in inflammation and humoral immunity.

ConA (concanavalin A)—A substance that stimulates T cell proliferation.

CRP (C-reactive protein)—An acute phase protein found in serum during inflammation and infection.

CSF (colony-stimulating factors)—A group of cytokines that stimulate hematopoietic stem cell proliferation and differentiation.

CTL, TC (cytotoxic T lymphocyte)—A subset of T cell capable of killing certain tumor and virally infected cells.

cytokine—A soluble factor produced by myriad cells involved in communication between immune cells. Many cytokines are growth factors.

cytostatic—Able to inhibit cellular growth.

cytotoxic—Able to kill cells.

eccentric exercise—Movement in which a muscle generates tension while lengthening. Eccentric exercise occurs mainly in stabilizing the body against gravity and is associated with muscle fiber damage and delayed muscle soreness.

granulocyte—A heterogeneous class of leukocytes characterized by a multilobed nucleus and intracellular granules. Granulocytes are composed of neutrophils, eosinophils, basophils, and mast cells.

granulocytosis—Increase in circulating granulocyte number.

HIV-1 (human immunodeficiency virus-1)—Retrovirus first identified in 1983 as the cause of acquired immune deficiency syndrome (AIDS), which gains entry into cells via the CD4 receptor.

humoral immunity—Immune function via soluble factors found in blood and other body fluids.

IFN (interferon)—A class of unrelated cytokines with antiviral activity and the capacity to stimulate certain immune cells.

Ig (immunoglobulin)—Glycoprotein found in blood and other body fluids that may exert antibody activity. All antibodies are Ig molecules, but not all Ig molecules exhibit antibody activity.

IL (interleukin)—A class of unrelated cytokines involved in communication between immune cells. At least 12 interleukin molecules have been identified.

IL-2R (interleukin-2 receptor)—Receptor for IL-2 found on activated lymphocytes.

interval exercise—Exercise performed in alternating work and rest intervals, commonly used by athletes during training.

isometric—Static muscle contraction in which the muscle generates tension but does not change length.

leukocyte—Heterogeneous cells found in the circulation and various tissues with diverse functions related to the immune response.

leukocytosis—Increase in circulating leukocyte number.

LPS (lipopolysaccharide)—A substance that stimulates B cell proliferation.

lymphocyte—Mononuclear immune cell.

lymphocytosis—Increase in circulating lymphocyte number.

macrophage—A phagocytic cell residing in tissues and derived from the monocyte.

maximum heart rate (age-predicted)—Highest possible heart rate, usually achieved during maximal exercise. Maximum heart rate decreases with age and can be estimated as 220 minus age.

MHC (major histocompatibility complex)—Membrane proteins, existing in two forms in humans (MHCI and MHCII), which are important to recognition of self or foreign antigens/cells.

mitogen—A substance that stimulates cell division (mitosis) in lymphocytes.

monocyte—A circulating phagocytic leukocyte, which can differentiate into a macrophage upon migration into tissue.

mucosal immunity—Immune function related to the body's external surfaces of the gut and the oronasal, respiratory, and genitourinary tracts.

neutrophil—A phagocytic leukocyte characterized by a multilobed nucleus and many intracellular granules.

NK (natural killer) cell—A large granular lymphocyte capable of killing certain tumor and virally infected cells.

nonadherent cell—Primarily lymphocytes and NK cells isolated from PBMC by short-term (1 hour) culture in a plastic dish/vessel during which time the monocytes adhere to the plastic.

overtraining—A neuroendocrine disorder related to a stress response to excessive training in athletes, characterized by persistent fatigue, depletion of catecholamines, poor performance in sport, and mood state changes.

peripheral blood mononuclear cell (PBMC)—A mixture of leukocytes including monocytes, lymphocytes, and NK cells obtained by isodensity centrifugation of whole blood. Adherent cells (primarily monocytes) and nonadherent cells (primarily lymphocytes and NK cells) can be further isolated from PBMC fractions.

PHA (phytohemagglutinin)—A substance that stimulates T cell proliferation.

phagocytosis—A process by which a leukocyte (monocyte, neutrophil) engulfs, ingests, and degrades a foreign particle or organism.

plasma cell—Mature antibody-secreting cell derived from the B cell.

PWM (pokeweed mitogen)—A substance that stimulates T cell-dependent B cell proliferation.

repetition maximum (RM)—In resistance training, the amount of weight that can be lifted a specified number of times before fatigue; for example, 5 RM is the weight that can be lifted only five times.

resistance training—Training against some type of resistance, usually accomplished by lifting or moving a weight or the athlete's body mass, to improve muscular strength, endurance, and size.

T cell (lymphocyte)—A heterogeneous population of lymphocytes comprising helper/inflammatory T cells and cytotoxic/suppressor T cells.

TNF (tumor necrosis factor)—A cytokine with many actions, such as antiviral and antitumor activity, increasing body temperature, and activating certain immune cells.

URTI (upper respiratory tract infection)—Infectious illness involving the oral and nasal regions (e.g., common cold, sore throat).

$\dot{V}O_{2max}$—Maximum oxygen consumption, usually expressed as a volume of oxygen consumed per minute. $\dot{V}O_{2max}$ is used as an indicator of maximal exercise power and to standardize exercise rate between individuals (e.g., exercising at 60% of one's $\dot{V}O_{2max}$).

References

Ahlborg, B., and G. Ahlborg. 1970. Exercise leukocytosis with and without beta-adrenergic blockade. *Acta Medica Scandinavica* 187: 241-246.

Albanes, D., A. Blair, and P.R. Taylor. 1989. Physical activity and risk of cancer in the NHANES I population. *American Journal of Public Health* 79: 744-750.

Anderson, T.D. 1992. Structure and function of interleukin-2. In *Cytokines in Health and Disease,* ed. S.L. Kunkel and D.G. Remick, 27-60. New York: Dekker.

Andrianopoulus, G., R.L. Nelson, C.T. Bombeck, and G. Souza. 1987. The influence of physical activity in 1,2 dimethylhydrazine induced colon carcinogenesis in the rat. *Anticancer Research* 7: 849-852.

Antoni, M.H., N. Schneiderman, M.A. Fletcher, D.A. Goldstein, G. Ironson, and A. LaPerriere. 1990. Psychoneuroimmunology and HIV-1. *Journal of Consulting and Clinical Psychology* 58: 38-49.

Arnold, D.L., P.J. Bore, G.K. Radda, P. Styles, and D.J. Taylor. 1984. Excessive intracellular acidosis of skeletal muscle on exercise in a patient with a post-viral exhaustion/fatigue syndrome: A ^{31}P nuclear magnetic resonance study. *Lancet* 2: 1367-1369.

Bagby, G.J., D.E. Sawaya, L.D. Crouch, and R.E. Shepherd. 1994. Prior exercise suppresses the plasma tumor necrosis factor response to bacterial lipopolysaccharide. *Journal of Applied Physiology* 77: 1542-1547.

Baggiolini, M. 1993. Novel aspects of inflammation: Interleukin-8 and related chemotactic cytokines. *Clinical Investigator* 71: 812-814.

Bailey, G.H. 1925. The effect of fatigue upon the susceptibility of rabbits to intratracheal injections of type I pneumococcus. *American Journal of Hygiene* 5: 175-295.

Baj, Z., J. Kantorski, E. Majewska, K. Zeman, L. Pokoca, E. Fornalczyk, H. Tchorzeski, A. Sulowska, and R. Lewicki. 1994. Immunological status of competitive cyclists before and after the training season. *International Journal of Sports Medicine* 15: 319-324.

Ballard-Barbash, R., A. Schatzkin, D. Albanes, M.H. Schiffman, B.E. Kreger, W.B. Kannel, K.M. Anderson, and W.E. Helsel. 1990. Physical activity and risk of large bowel cancer in the Framingham study. *Cancer Research* 50: 3610-3613.

Barger, L.K., J.E. Greenleaf, F. Baldini, and D. Huff. 1995. Effects of space missions on the human immune system: A meta-analysis. *Sports Medicine, Training and Rehabilitation* 5: 293-310.

Bartram, H.P., and E.L. Wynder. 1989. Physical activity and colon cancer risk? Physiological considerations. *The American Journal of Gastroenterology* 84: 109-112.

Baum, M., H. Liesen, and J. Enneper. 1994. Leucocytes, lymphocytes, activation parameters and cell adhesion molecules in middle-distance runners under different training conditions. *International Journal of Sports Medicine* 15: S122-S126.

Baumann, H., and J. Gauldie. 1994. The acute phase response. *Immunology Today* 15: 74-80.

Beagley, K.W., J.H. Eldridge, F. Lee, H. Kiyono, M.P. Everson, W.J. Koopman, T. Hirano, T. Kishimoto, and J.R. McGhee. 1989. Interleukins and IgA synthesis. *Journal of Experimental Medicine* 169: 2133-48.

Beisel, W.R., B.B. Morgan, P.J. Bertelloni, G.D. Coates, F.R. DeRubertis, and E.A. Allusi. 1974. Symptomatic therapy in viral illness: A controlled study of effects on work performance. *JAMA* 228: 581-584.

Belcastro, A.N., G.D. Arthur, R.A. Albisser, and D.A. Raj. 1996. Heart, liver, and skeletal muscle myeloperoxidase activity during exercise. *Journal of Applied Physiology* 80: 1331-1335.

Benoni, G., P. Bellavite, A. Adami, S. Chirumbolo, G. Lippi, G. Brocco, G.M. Guilini, and L. Cuzzolin. 1995a. Changes in several neutrophil functions in basketball players before, during and after the sports season. *International Journal of Sports Medicine* 16: 34-37.

Benoni, G., P. Bellavite, A. Adami, S. Chirumbolo, G. Lippi, G. Brocco, G.M. Guilini, and L. Cuzzolin. 1995b. Effect of acute exercise on some haematological parameters and neutrophil functions in active and inactive subjects. *European Journal of Applied Physiology* 70: 187-191.

Benschop, R.J., F.G. Oostveen, C.J. Heijen, and R.E. Ballieux. 1993. β_2-adrenergic stimulation causes detachment of natural killer cells from cultured endothelium. *European Journal of Immunology* 23: 3242-3247.

Berglund, B., and P. Hemmingsson. 1990. Infectious disease in elite cross-country skiers: A one-year incidence study. *Clinical Sports Medicine* 2: 19-23.

Berk, L.S., D.C. Nieman, W.S. Youngberg, K. Arabatzis, M. Simpson-Westerberg, J.W. Lee, S.A. Tan, and W.C. Eby. 1990. The effect of long endurance running on natural killer cells in marathoners. *Medicine and Science in Sports and Exercise* 22: 207-212.

Berne, R.M., and M.N. Levy. 1988. *Physiology.* 2nd ed. St. Louis: Mosby.

Bieger, W.P., M. Weiss, G. Michel, and H. Weicker. 1980. Exercise-induced monocytosis and modulation of monocyte function. *International Journal of Sports Medicine* 1: 30-36.

Blair, S.N., H.W. Kohl, R.S. Paffenbarger, D.G. Clark, K.H. Cooper, and L.W. Gibbons. 1989. Physical fitness and all-cause mortality: A prospective study of healthy men and women. *JAMA* 262: 2395-2401.

Blannin, A.K., L.J. Chatwin, R. Cave, and M. Gleeson. 1996. Effects of submaximal cycling and long-term endurance training on neutrophil phagocytic activity in middle aged men. *British Journal of Sports Medicine* 30: 125-129.

Blecha, R., and H.C. Minocha. 1983. Suppressed lymphocyte blastogenic responses and enhanced in vitro growth of infectious bovine rhinotracheitis virus in stressed feeder calves. *American Journal of Veterinary Research* 44: 2145-2148.

Bosenberg, A.T., J.G. Brock-Utne, S.L. Gaffin, M.T.B. Wells, and G.T.W. Blake. 1988. Strenuous exercise causes systemic endotoxemia. *Journal of Applied Physiology* 65: 106-108.

Brahmi, Z., J.E. Thomas, M. Park, M. Park, and I.R.G. Dowdeswell. 1985. The effect of acute exercise on natural killer-cell activity of trained and sedentary human subjects. *Journal of Clinical Immunology* 5: 321-328.

Brooks, G.A., T.D. Fahey, and T.P. White. 1996. *Exercise Physiology: Human Bioenergetics and Its Applications.* 2nd ed. New York: Macmillan.

Bury, T.B., and F. Pirnay. 1995. Effect of prolonged exercise on neutrophil myeloperoxidase secretion. *International Journal of Sports Medicine* 6: 410-412.

Busse, W.W., C.L. Anderson, P.G. Hanson, and J.D. Folts. 1980. The effect of exercise on the granulocyte response to isoproterenol in the trained athlete and unconditioned individual. *Journal of Allergy and Clinical Immunology* 65: 358-364.

Cabinian, A.E., R.J. Kiel, F. Smith, K.L. Ho, R. Khatib, and M.P. Reyes. 1990. Modification of exercise-aggravated coxsackievirus B3 murine myocarditis by T lymphocyte suppression in an inbred model. *Journal of Laboratory and Clinical Medicine* 115: 454-462.

Calabrese, J.R., M.A. Kling, and P.W. Gold. 1987. Alterations in immunocompetence during stress, bereavement, and depression: Focus on neuroendocrine regulation. *American Journal of Psychiatry* 144: 1123-34.

Calabrese, L.H., S.M. Kleiner, B.P. Barna, C.I. Skibinski, D.T. Kirkendall, R.G. Lahita, and J.A. Lombardo. 1989. The effects of anabolic steroids and strength training on the human immune response. *Medicine and Science in Sports and Exercise* 21: 386-392.

Camus, G., J. Duchateau, G. Deby-Dupont, J. Pincemail, C. Deby, A. Juchmes-Ferir, F. Feron, and M. Lamy. 1994. Anaphylatoxin C5a production during short-term submaximal dynamic exercise in man. *International Journal of Sports Medicine* 15: 32-35.

Camus, G., J. Pincemail, M. Ledent, A. Juchmes-Ferir, M. Lamy, and G. Deby-Dupont. 1992. Plasma levels of polymorphonuclear elastase and myeloperoxidase after uphill walking and downhill running at similar energy cost. *International Journal of Sports Medicine* 13: 443-446.

Cannon, J.G., W.J. Evans, V.A. Hughes, C.N. Meredith, and C.A. Dinarello. 1986. Physiological mechanisms contributing to increased interleukin-1 secretion. *Journal of Applied Physiology* 61: 1869-1874.

Cannon, J.G., M.A. Fiatarone, R.A. Fielding, and W.J. Evans. 1994. Aging and stress-induced changes in complement activation and neutrophil mobilization. *Journal of Applied Physiology* 76: 2616-2620.

Cannon, J.G., M.A. Fiatarone, M. Meydani, J. Gong, L. Scott, J.B. Blumberg, and W.J. Evans. 1995. Aging and dietary modulation of elastase and interleukin-1β secretion. *American Journal of Physiology* 268 (*Regulatory Integrative Comparative Physiology* 37): R20-R213.

Cannon, J.G., R.A. Fielding, M.A. Fiatarone, S.F. Orencole, C.A. Dinarello, and W.J. Evans. 1989. Increased interleukin 1β in human skeletal muscle. *American Journal of Physiology* 257 (*Regulatory Integrative Comparative Physiology* 26): R451-R455.

Cannon, J.G., and M.J. Kluger. 1983. Endogenous pyrogen activity in human plasma after exercise. *Science* 220: 617-619.

Cannon, J.G., and M.J. Kluger. 1984. Exercise enhances survival rate in mice infected with Salmonella typhimurium. *Proceedings of the Society for Experimental Biology and Medicine* 175: 518-521.

Cannon, J.G., S.N. Meydani, R.A. Fielding, M.A. Fiatarone, M. Meydani, N. Farhangmehr, S.F. Orencole, J.B. Blumberg, and W.J. Evans. 1991. Acute phase response in exercise. II. Association between vitamin E, cytokines and muscle proteolysis. *American Journal of Physiology* 260 (*Regulatory Integrative Comparative Physiology* 29): R1235-R1240.

Cannon, J.G., S.F. Orencole, R.A. Fielding, M. Meydani, S.N. Meydani, M.A. Fiatarone, J.B. Blumberg, and W.J. Evans. 1990. Acute phase response in exercise: Interaction of age and vitamin E on neutrophils and muscle enzyme release. *American Journal of Physiology* 259 (*Regulatory Integrative Comparative Physiology* 28): R1214-R1219.

Castell, L.M., E.A. Newsholme, and J.R. Poortmans. 1996. Does glutamine have a role in reducing infections in athletes? *European Journal of Applied Physiology* 73: 488-490.

Castell, L.M., J.R. Poortmans, R. Leclercq, M. Brasseur, J. Duchateau, and E.A. Newsholme. 1997. Some aspects of the acute phase response after a marathon race, and the effects of glutamine supplementation. *International Journal of Sports Medicine* 75: 47-53.

Cate, T.R., R.D. Rosen, R.G. Douglas, W.T. Butler, and R.B. Couch. 1966. The role of nasal secretion and serum antibody in the rhinovirus common cold. *American Journal of Epidemiology* 84: 352-63.

Chao, C.C., F. Strgar, M. Tsang, and P.K. Peterson. 1992. Effects of swimming exercise on the pathogenesis of acute murine *Toxoplasma gondii* Me49 infection. *Clinical Immunology and Immunopathology* 62: 220-226.

Christensen, R.D., and H.H. Hill. 1987. Exercise-induced changes in the blood concentration of leukocyte populations in teenage athletes. *The American Journal of Pediatric Hematology/Oncology* 9: 140-142.

Cogoli, A. 1993. The effect of hypogravity and hypergravity on cells of the immune system. *Journal of Leukocyte Biology* 54: 259-268.

Cohen, L.A., K. Choi, and C.-X Wang. 1988. Influence of dietary fat, caloric restriction, and voluntary exercise on N-nitrosomethylurea-induced mammary tumorigenesis in rats. *Cancer Research* 48: 4276-4283.

Coleman, K.J., and D.R. Rager. 1993. Effects of voluntary exercise on immune function in rats. *Physiology and Behavior* 54: 771-774.

Conn, C.A., W.E. Kozak, P.C.J. Tooten, E. Gruys, K.T. Borer, and M.J. Kluger. 1995a. Effect of voluntary exercise and food restriction in response to lipopolysaccharide in hamsters. *Journal of Applied Physiology* 78: 466-477.

Conn, C.A., W.E. Kozak, P.C.J. Tooten, T.A. Niewold, K.T. Borer, and M.J. Kluger. 1995b. Effect of exercise and food restriction on selected markers of the acute phase response in hamsters. *Journal of Applied Physiology* 78: 458-465.

Cowles, W.N. 1918. Fatigue as a contributory cause of pneumonias. *Boston Medical and Surgery Journal* 179: 555.

Cox, G., and J. Gauldie. 1992. Structure and function of interleukin-6. In *Cytokines in Health and Disease,* ed. S.L. Kunkel and D.G. Remick, 97-120. New York: Dekker.

Crist, D.M., and J.C. Kraner. 1990. Supplemental growth hormone increases the tumor cytotoxic activity of natural killer cells in healthy adults with normal growth hormone secretion. *Metabolism* 39: 1320-1324.

Crist, D.M., L.T. Mackinnon, R.F. Thompson, H.A. Atterbom, and P.A. Egan. 1989. Physical exercise increases natural cellular-mediated tumor cytotoxicity in elderly women. *Gerontology* 35: 66-71.

Crist, D.M., G.T. Peake, L.T. Mackinnon, W.L. Sibbit, and J.C. Kraner. 1987. Exogenous growth hormone treatment alters body composition and increases natural killer activity in women with impaired growth hormone secretion. *Metabolism* 36: 1115-17.

Dacie, J.V., and S.M. Lewis. 1984. *Practical Haematology.* 6th ed., 8-10. Edinburgh: Churchill Livingstone.

Daniels, W.L., D.S. Sharp, J.E. Wright, J.A. Vogel, G. Friman, W.R. Beisel, and J.J. Knapik. 1985. Effects of virus infection on physical performance in man. *Military Medicine* 150: 1-8.

Davidson, R.J.L., J.D. Robertson, G. Galea, and R.J. Maughan. 1987. Hematological changes associated with marathon running. *International Journal of Sports Medicine* 8: 19-25.

De la Fuente, M., I. Martin, and E. Ortega. 1993. Effect of physical exercise on the phagocytic function of peritoneal macrophages from Swiss mice. *Comparative Immunology, Microbiology and Infectious Disease* 16: 29-37.

Deuster, P.A., A.M. Curiale, M.L. Cowan, and F.D. Finkelman. 1988. Exercise-induced changes in populations of peripheral blood mononuclear cells. *Medicine and Science in Sports and Exercise* 20: 276-280.

Douglas, D.J., and P.G. Hanson. 1978. Upper respiratory infections in the conditioned athlete. *Medicine and Science in Sport* 10: 55.

Douglass, J.H. 1974. The effects of physical tracing *(sic)* on the immunological response in mice. *Journal of Sports Medicine* 14: 48-54.

Drenth, J.P.H., S.H.M. van Uum, M. van Dueren, G.J. Pesman, J. van der ven-Jondekrigh, and J.W.M. van der Meer. 1995. Endurance run increases circulating IL-6 and Il-1ra but downregulates ex vivo TNFα and IL-1β production. *Journal of Applied Physiology* 79: 1497-1503.

Dufaux, B., K. Hoffken, and W. Hollmann. 1983. Acute phase proteins and immune complexes during several days of severe physical exercise. In *Biochemistry of Exercise,* vol. 13, ed. H.G. Knuttgen, J.A. Vogel, and J. Poortmans, 356-362. Champaign, IL: Human Kinetics.

Dufaux, B., and U. Order. 1989a. Complement activation after prolonged exercise. *Clinica Chimica Acta* 179: 45-50.

Dufaux, B., and U. Order. 1989b. Plasma elastase-α1-antitrypsin, neopterin, tumor necrosis factor, and soluble interleukin-2 receptor after prolonged exercise. *International Journal of Sports Medicine* 10: 434-438.

Dufaux, B., U. Order, H. Geyer, and W. Hollmann. 1984. C-reactive protein serum concentration in well-trained athletes. *International Journal of Sports Medicine* 5: 102-106.

Dufaux, B., U. Order, and H. Liesen. 1991. Effect of a short maximal physical exercise on coagulation, fibrinolysis, and complement system. *International Journal of Sports Medicine* 12: S38-S42.

Eberhardt, A. 1971. Influence of motor activity on some serologic mechanisms of nonspecific immunity of the organism. *Acta Physiologica Polonica* 22: 185-194.

Edwards, A.J., T.H. Bacon, C.A. Elms, R. Verardi, M. Felder, and S.C. Knight. 1984. Changes in the populations of lymphoid cells in human peripheral blood following physical exercise. *Clinical Experimental Immunology* 58: 420-427.

Eichner, E.R. 1987. Infectious mononucleosis: Recognition and management in athletes. *The Physician and Sportsmedicine* 15: 61-70.

Eichner, E.R., and L.H. Calabrese. 1994. Immunology and exercise: Physiology, pathophysiology and implications for HIV infection. *Medical Clinics of North America* 78: 377-387.

Eskola, J., O. Ruuskanen, E. Soppi, M.K. Viljanen, M. Jarvinen, H. Toivonen, and K. Kouvalainen. 1978. Effect of sport stress on lymphocyte transformation and antibody formation. *Clinical Experimental Immunology* 32: 339-345.

Espersen, G.T., A. Elbaek, E. Ernst, E. Toft, S. Kaalund, C. Jersild, and N. Grunnet. 1990. Effect of physical exercise on cytokines and lymphocyte subpopulations in human peripheral blood. *APMIS* 98: 395-400.

Espersen, G.T., E. Toft, E. Ernst, S. Kaalund, and N. Grunnet. 1991. Changes of polymorphonuclear granulocyte migration and lymphocyte proliferative responses in elite runners undergoing intense exercise. *Scandinavian Journal of Medicine and Science in Sports* 1: 158-162.

Esterling, B.A., M.H. Antoni, N. Schneiderman, P.S. Carver, A. LaPerriere, G. Ironson, N.G. Klimas, and M.A. Fletcher. 1992. Psychosocial modulation of antibody to

Epstein-Barr viral capsid antigen and Human Herpesvirus Type-6 in HIV-1-infected and at-risk gay men. *Psychosomatic Medicine* 54: 354-371.

Evans, W.J., C.N. Meredith, J.G. Cannon, C.A. Dinarello, W.R. Frontera, V.A. Hughes, B.H. Jones, and H.G. Knuttgen. 1986. Metabolic changes following eccentric exercise in trained and untrained men. *Journal of Applied Physiology* 61: 1864-1868.

Fairbarn, M.S., S.P. Blackie, R.L. Pardy, and J.C. Hogg. 1993. Comparison of effects of exercise and hyperventilation on leukocyte kinetics in humans. *Journal of Applied Physiology* 75: 2425-2428.

Fehr, H.-G., H. Lotzerich, and H. Michna. 1988. The influence of physical exercise on peritoneal macrophage functions: Histochemical and phagocytic studies. *International Journal of Sports Medicine* 9: 77-81.

Fehr, H.-G., H. Lotzerich, and H. Michna. 1989. Human macrophage function and physical exercise: Phagocytic and histochemical studies. *European Journal of Applied Physiology* 58: 613-617.

Ferry, A., F. Picard, A. Duvallet, B. Weill, and M. Rieu. 1990. Changes in blood leucocyte populations induced by acute maximal and chronic submaximal exercise. *European Journal of Applied Physiology* 59: 435-442.

Fiatarone, M.A., J.E. Morley, E.T. Bloom, D. Benton, T. Makinodan, and G.F. Solomon. 1988. Endogenous opioids and the exercise-induced augmentation of natural killer cell activity. *Journal of Clinical Laboratory Medicine* 112: 544-552.

Fiatarone, M.A., J.E. Morley, E.T. Bloom, D. Benton, G.F. Solomon, and T. Makinodan. 1989. The effect of exercise on natural killer cell activity in young and old subjects. *Journal of Gerontology* 44: M37-45.

Field, C.J., R. Gougeon, and E.B. Marliss. 1991. Circulating mononuclear cell numbers and function during intense exercise and recovery. *Journal of Applied Physiology* 71: 1089-1097.

Fielding, R.A., T.J. Manfredi, W. Ding, M.A. Fiatarone, W.J. Evans, and J.G. Cannon. 1993. Acute phase response in exercise III. Neutrophil and IL-1β accumulation in skeletal muscle. *American Journal of Physiology (Regulatory Integrative Comparative Physiology* 34): R166-R172.

Fitzgerald, L. 1991. Overtraining increases the susceptibility to infection. *International Journal of Sports Medicine* 12: S5-S8.

Forner, M.A., M.E. Collazos, C. Barriga, M. De la Fuente, A.B. Rodriguez, and E. Ortega. 1994. Effect of age on adherence and chemotaxis capacities of peritoneal macrophages. Influence of physical activity stress. *Mechanisms of Aging and Development* 75: 179-189.

Foster, C., M. Pollock, P. Farrell, M. Maksud, J. Anholm, and J. Hare. 1982. Training responses of speed skaters during a competitive season. *Research Quarterly for Exercise and Sport* 53: 243-246.

Foster, N.K., J.B. Martyn, R.E. Rangno, J.C. Hogg, and R.L. Pardy. 1986. Leukocytosis of exercise: Role of cardiac output and catecholamines. *Journal of Applied Physiology* 61: 2218-2223.

Frey, M.J., D. Mancini, D. Fishberg, J.R. Wilson, and P.B. Molinoff. 1989. Effect of exercise duration on density and coupling of β-adrenergic receptors on human mononuclear cells. *Journal of Applied Physiology* 66: 1494-1500.

Friman, G. 1977. Effect of acute infectious disease on isometric muscle strength. *Scandinavian Journal of Clinical Laboratory Investigation* 37: 303-308.

Friman, G., and N.-G. Ilback. 1992. Exercise and infection—interaction, risks and benefits. *Scandinavian Journal of Medicine and Science in Sports* 2: 177-189.

Friman, G., N.-G. Ilback, W.R. Beisel, and D.J. Crawford. 1982. The effects of strenuous exercise on infection with *Francisella tularensis* in rats. *The Journal of Infectious Diseases* 145: 706-714.

Friman, G., N.-G. Ilback, D.J. Crawford, and H.A. Neufeld. 1991. Metabolic responses to swimming exercise in *Streptococcus pneumoniae* infected rats. *Medicine and Science in Sports and Exercise* 23: 415-421.

Friman, G., H.H. Schiller, and M.S. Schwartz. 1977. Disturbed neuromuscular transmission in viral infections. *Scandinavian Journal of Infectious Disease* 9: 99-103.

Friman, G., J.E. Wright, N.-G. Ilback, W.R. Beisel, J.D. White, D.S. Sharp, E.L. Stephen, W.L. Daniels, and J.A. Vogel. 1985. Does fever of myalgia indicate reduced physical performance capacity in viral infections? *Acta Medica Scandinavica* 217: 353-361.

Frisina, J.P., S. Gaudieri, T. Cable, D. Keast, and T.N. Palmer. 1994. Effect of acute exercise on lymphocyte subsets and metabolic activity. *International Journal of Sports Medicine* 15: 36-41.

Frohlich, J., G. Simon, A. Schmidt, T. Hitschhold, and M. Bierther. 1987. (Abstract) Disposition to infections of athletes during treatment with immunoglobulins. *International Journal of Sports Medicine* 8: 119.

Fry, R.W., J.R. Grove, A.R. Morton, P.M. Zeroni, S. Gaudieri, and D. Keast. 1994. Psychological and immunological correlates of acute overtraining. *British Journal of Sports Medicine* 28: 241-246.

Fry, R.W., A.R. Morton, P. Garcia-Webb, G.P.M. Crawford, and D. Keast. 1992a. Biological responses to overload training in endurance sports. *European Journal of Applied Physiology* 64: 335-344.

Fry, R.W., A.R. Morton, and D. Keast. 1991. Overtraining in athletes: An update. *Sports Medicine* 12: 32-65.

Fry, R.W., A.R. Morton, and D. Keast. 1992b. Acute intensive interval training and T-lymphocyte function. *Medicine and Science in Sports and Exercise* 24: 339-345.

Fuchs, B.B., and A.E. Medvedev. 1993. Countermeasures for ameliorating in-flight immune dysfunction. *Journal of Leukocyte Biology* 54: 245-252.

Fukatsu, A., N. Sato, and H. Shimizu. 1996. 50-mile walking race suppresses neutrophil bactericidal function by inducing increases in cortisol and ketone bodies. *Life Sciences* 25: 2237-2343.

Gabriel, H., L. Brechtel, A. Urhausen, and W. Kindermann. 1994a. Recruitment and recirculation of leukocytes after an ultramarathon run: Preferential homing of cells expressing high levels of the adhesion molecule LFA-1. *International Journal of Sports Medicine* 15: S148-S153.

Gabriel, H., and W. Kindermann. 1995. Flow cytometry: Principles and application in exercise immunology. *Sports Medicine* 20: 1-15.

Gabriel, H., H.J. Miller, A. Urhausen, and W. Kindermann. 1994b. Suppressed PMA-induced oxidative burst and unimpaired phagocytosis of circulating granulocytes one week after a long endurance exercise. *International Journal of Sports Medicine* 15: 441-445.

Gabriel, H., B. Schmitt, and W. Kindermann. 1991a. Age-related increase of $CD45RO^+$ lymphocytes in physically active adults. *European Journal of Immunology* 23: 2704-2706.

Gabriel, H., B. Schmitt, A. Urhausen, and W. Kindermann. 1993. Increased $CD45RA^+CD45RO^+$ cells indicate activated T cells after endurance exercise. *Medicine and Science in Sports and Exercise* 25: 1352-1357.

Gabriel, H., L. Schwarz, P. Born, and W. Kindermann. 1992a. Differential mobilization of leucocyte and lymphocyte subpopulations into the circulation during endurance exercise. *European Journal of Applied Physiology* 65: 529-534.

Gabriel, H., L. Schwarz, G. Steffens, and W. Kindermann. 1992b. Immunoregulatory hormones, circulating leucocyte and lymphocyte subpopulations before and after endurance exercise of different intensities. *International Journal of Sports Medicine* 13: 359-366.

Gabriel, H., A. Urhausen, and W. Kindermann. 1991b. Circulating leucocyte and lymphocyte subpopulations before and after intensive endurance exercise to exhaustion. *European Journal of Applied Physiology* 63: 449-457.

Gabriel, H., A. Urhausen, and W. Kindermann. 1992c. Mobilization of circulating leucocyte and lymphocyte subpopulations during and after short, anaerobic exercise. *European Journal of Applied Physiology* 65: 164-170.

Galun, E., R. Burstein, E. Assia, I. Tur-Kaspa, J. Rosenblum, and Y. Epstein. 1987. Changes of white blood cell count during prolonged exercise. *International Journal of Sports Medicine* 8: 253-255.

Garabrant, D.H., J.M. Peters, T.M. Mack, and L. Bernstein. 1984. Job activity and colon cancer risk. *American Journal of Epidemiology* 119: 1105-1114.

Garagiola, U., M. Buzzetti, E. Cardella, F. Confalonieri, E. Giani, V. Polini, P. Ferrante, R. Mancuso, M. Montanari, E. Grossi, and A. Pecori. 1995. Immunological patterns during regular intensive training in athletes: Quantification and evaluation of a preventive pharmacological approach. *Journal of International Medical Research* 23: 85-95.

Gatmaitan, B.G., J.L. Chason, and A.M. Lerner. 1970. Augmentation of the virulence of murine coxsackie-virus B-3 myocardiopathy by exercise. *Journal of Experimental Medicine* 131: 1121-1136.

Gimenez, M., T. Mohan-Kumar, J.C. Humbert, N. de Talance, and J. Buisine. 1986. Leukocyte, lymphocyte and platelet response to dynamic exercise: Duration or intensity effect? *European Journal of Applied Physiology* 55: 465-470.

Gimenez, M., T. Mohan-Kumar, J.C. Humbert, N. de Talance, M. Teboul, and F.J.A. Belenguer. 1987. Training and leucocyte, lymphocyte and platelet response to dynamic exercise. *Journal of Sports Medicine* 27: 172-177.

Gleeson, M., W.A. McDonald, A.W. Cripps, D.B. Pyne, R.L. Clancy, and P.A. Fricker. 1995. The effect on immunity of long term intensive training in elite swimmers. *Clinical and Experimental Immunology* 102: 210-216.

Gleeson, M., D.B. Pyne, W.A. McDonald, R.L. Clancy, A.W. Cripps, P.L. Horn, and P.A. Fricker. 1996. Pneumococcal antibody response in elite swimmers. *Clinical and Experimental Immunology* 105: 238-244.

Gmunder, F.K., G. Lorenzi, B. Bechler, P. Joller, J. Muller, W.H. Ziegler, and A. Cogoli. 1988. Effect of long-term physical exercise on lymphocyte reactivity: Similarity to space-flight reactions. *Aviation, Space, and Environmental Medicine* 59: 146-151.

Gmunder, F.K., P.W. Joller, H.I. Joller-Jemelka, B. Bechler, M. Cogoli, W.H. Ziegler, J. Muller, R.E. Aeppli, and A. Cogoli. 1990. Effect of a herbal yeast food supplements and long-distance running on immunological parameters. *British Journal of Sports Medicine* 24: 103-112.

Good, R.A., and G. Fernandes. 1981. Enhancement of immunologic function and resistance to tumor growth in Balb/c mice by exercise. *Federation Proceedings* 40: 1040.

Graham, N.M.H., R.M. Douglas, and P. Ryan. 1986. Stress and acute respiratory infection. *American Journal of Epidemiology* 124: 389-401.

Gray, A.B., R.D. Telford, M. Collins, M.S. Baker, and M.J. Weidemann. 1993a. Granulocyte activation induced by intense interval running. *Journal of Leukocyte Biology* 53: 591-597.

Gray, A.B., R.D. Telford, M. Collins, and M.J. Weidemann. 1993b. The response of leukocyte subsets and plasma hormones to interval exercise. *Medicine and Science in Sports and Exercise* 25: 1252-1258.

Grazzi, L., A. Salmaggi, A. Dufour, C. Ariano, A.M. Colangelo, E. Parati, M. Lazzaroni, A. Nespolo, G. Bordin, and C. Castellazzi. 1993. Physical effort-induced changes in immune parameters. *International Journal of Neuroscience* 68: 133-140.

Green, R.L., S.S. Kaplan, B.S. Rabin, C.L. Stanitski, and U. Zdziarski. 1981. Immune function in marathon runners. *Annals of Allergy* 47: 73-75.

Greenleaf, J.E., C.G.R. Jackson, and D. Lawless. 1994. Immune responses and function: Exercise conditioning versus bed-rest and spaceflight deconditioning. *Sports Medicine, Training and Rehabilitation* 5: 223-241.

Greenleaf, J.E., C.G.R. Jackson, and D. Lawless. 1995. $CD4^+/CD8^+$ T-lymphocyte ratio: Effects of rehydration before exercise in dehydrated men. *Medicine and Science in Sports and Exercise* 27: 194-199.

Greig, J.E., D.G. Rowbottom, and D. Keast. 1995. The effect of a natural (viral) stress on plasma glutamine concentration. *Medical Journal of Australia* 163: 385-388.

Haahr, P.M., B.K. Pedersen, A. Fomsgaard, N. Tvede, M. Diamant, K. Klarlund, J. Halkjaer-Kristensen, and K. Bendtzen. 1991. Effect of physical exercise on in vitro production of interleukin 1, interleukin 6, tumour necrosis factor-α, interleukin 2, and interferon-α. *International Journal of Sports Medicine* 12: 223-227.

Hack, B., G. Strobel, J.-P. Rau, and H. Weicker. 1992. The effect of maximal exercise on the activity of neutrophil granulocytes in highly trained athletes in a moderate training period. *European Journal of Applied Physiology* 65: 520-524.

Hack, B., G. Strobel, M. Weiss, and H. Weicker. 1994. PMN cell counts and phagocytic activity of highly trained athletes depend on training period. *Journal of Applied Physiology* 77: 1731-1735.

Hanley, D.F. 1976. Medical care of the US Olympic Team. *JAMA* 236: 147-148.

Hansen, J.-B., L. Wilsgard, and B. Osterud. 1991. Biphasic changes in leukocytes induced by strenuous exercise. *European Journal of Applied Physiology* 62: 157-161.

Hanson, P.G., and D.K. Flaherty. 1981. Immunological responses to training in conditioned runners. *Clinical Science* 60: 225-228.

Haq, A., K. Al-Hussein, J. Lee, and S. Al-Sedairy. 1993. Changes in peripheral blood lymphocyte subsets associated with marathon running. *Medicine and Science in Sports and Exercise* 25: 186-190.

Haralambie, G., and J. Keul. 1970. Serum glycoprotein levels in athletes in training. *Experientia* 26: 959-960.

Heath, G.W., E.S. Ford, T.E. Craven, C.A. Macera, K.L. Jackson, and R.R. Pate. 1991. Exercise and the incidence of upper respiratory tract infections. *Medicine and Science in Sports and Exercise* 23: 152-157.

Heath, G.W., C.A. Macera, and D.C. Nieman. 1992. Exercise and upper respiratory tract infections: Is there a relationship? *Sports Medicine* 14: 353-365.

Hebert, J.R., J. Barone, M.M. Reddy, and J.-Y.C Backlund. 1990. Natural killer cell activity in a longitudinal dietary fat intervention trial. *Clinical Immunology and Immunopathology* 54: 103-116.

Hedfors, E., P. Biberfeld, and J. Wahren. 1978. Mobilization to the blood of human non-T and K lymphocytes during physical exercise. *Journal of Clinical Laboratory Immunology* 1: 159-162.

Hedfors, E., G. Holm, M. Ivansen, and J. Wahren. 1983. Physiological variation of blood lymphocyte reactivity: T-cell subsets, immunoglobulin production, and mixed-lymphocyte reactivity. *Clinical Immunology and Immunopathology* 27: 9-14.

Hedfors, E., G. Holm, and B. Ohnell. 1976. Variations of blood lymphocytes during work studied by cell surface markers, DNA synthesis and cytotoxicity. *Clinical Experimental Immunology* 24: 328-335.

Hedin, G., and G. Friman. 1982. Orthostatic reactions and blood volumes after moderate physical activation during acute febrile infections. *International Rehabilitation Medicine* 4: 107-109.

Hemila, H. 1992. Vitamin C and the common cold. *British Journal of Nutrition* 67: 3-16.

Hemila, H. 1996. Vitamin C and common cold incidence: A review of studies with subjects under heavy physical stress. *International Journal of Sports Medicine* 17: 379-383.

Herberman, R.B. 1991. Principles of tumor immunology. In *Textbook of Clinical Oncology,* ed. A.I. Holleb, D.J. Fink, and G.P. Murphy. Atlanta: American Cancer Society.

Hinton, J.R., D.G. Rowbottom, D. Keast, and A.R. Morton. 1997. Acute intensive interval training and in vitro T-lymphocyte function. *International Journal of Sports Medicine* 18: 132-137.

Hoffman, S.A., K.E. Paschkis, D.A. DeBias, A. Cantarow, and T.L. Williams. 1962. The influence of exercise on the growth of transplanted rat tumors. *Cancer Research* 22: 597-599.

Hoffman-Goetz, L. 1994. Exercise, natural immunity, and tumor metastasis. *Medicine and Science in Sports and Exercise* 26: 157-163.

Hoffman-Goetz, L. 1995. Serine esterase (BLT-esterase) activity in murine splenocytes is increased with exercise but not training. *International Journal of Sports Medicine* 16: 94-98.

Hoffman-Goetz, L., and J. Husted. 1995. Exercise and cancer: Do the biology and epidemiology correspond? *Exercise Immunology Review* 1: 81-96.

Hoffman-Goetz, L., R. Keir, M.E. Thorne, and C. Houston. 1986. Chronic exercise in mice depresses splenic T lymphocyte mitogenesis in vitro. *Clinical Experimental Immunology* 66: 551-557.

Hoffman-Goetz, L., K.M. May, and Y. Arumugam. 1994. Exercise training and mouse mammary tumor metastasis. *Anticancer Research* 14: 2627-2632.

Hoffman-Goetz, L., R.J. Thorne, and M.E. Houston. 1988. Splenic immune responses following treadmill exercise in mice. *Canadian Journal of Physiology and Pharmacology* 66: 1415-1419.

Hoffman-Goetz, L., R. Thorne, J.A. Randall-Simpson, and Y. Arumugam. 1989. Exercise stress alters murine lymphocyte subset distribution in spleen, lymph nodes and thymus. *Clinical Experimental Immunology* 76: 307-310.

Hooper, S., L.T. Mackinnon, A. Howard, R.D. Gordon, and A.W. Bachmann. 1995. Markers for monitoring overtraining and recovery in elite swimmers. *Medicine and Science in Sports and Exercise* 27: 106-112.

Horstmann, D.M. 1950. Acute poliomyelitis: Relation of physical activity at the time of onset to the course of the disease. *JAMA* 142: 236-241.

Housh, T.J., G.O. Johnson, D.J. Housh, S.L. Evans, and G.D. Tharp. 1991. The effect of exercise at various temperatures on salivary levels of immunoglobulin A. *International Journal of Sports Medicine* 12: 498-500.

Hubinger, L.M., L.T. Mackinnon, L. Barber, J. McCosker, A. Howard, and F. Lepre. 1997. The acute effects of treadmill running on lipoprotein(a) levels in males and females. *Medicine and Science in Sports and Exercise* 29: 436-442.

Huupponen, M.R.H., L.H. Makinen, P.M. Hyvonen, C.K. Sen, T. Rankinen, S. Vaisanen, and R. Rauramaa. 1995. The effect of N-acetylcysteine on exercise-induced priming of human neutrophils. *International Journal of Sports Medicine* 16: 399-403.

Ilback, N.-G., J. Fohlman, and G. Friman. 1989. Exercise in coxsackie B3 myocarditis: Effects on heart lymphocyte subpopulations and the inflammatory reaction. *American Heart Journal* 117: 1298-1302.

Ilback, N.-G., G. Friman, W.R. Beisel, A.J. Johnson, and R.F. Berendt. 1984. Modifying effects of exercise on clinical course and biochemical response of the myocardium in influenza and tularemia in mice. *Infection and Immunity* 45: 498-504.

Ilback, N.-G., G. Friman, D.J. Crawford, and H.A. Neufeld. 1991. Effects of training on metabolic responses and performance capacity in Streptococcus pneumoniae infected rats. *Medicine and Science in Sports and Exercise* 23: 422-427.

Iversen, P.O., B.L. Arvesen, and H.B. Benestad. 1994. No mandatory role for the spleen in the exercise-induced leucocytosis in man. *Clinical Science* 86: 505-510.

Janeway, C.A., and P. Travers. 1996. *Immunobiology: The Immune System in Health and Disease.* 2nd ed. London: Current Biology Ltd.

Janssen, G.M.E., J.W.J. van Wersch, V. Kaiser, and R.J.M.M Does. 1989. White cell system changes associated with a training period of 18-20 months: A transverse and longitudinal approach. *International Journal of Sports Medicine* 10: S176-S180.

Jemmott, J.B., M. Borysenko, R. Chapman, J.Z. Borysenko, D.C. McClelland, D. Meyer, and H. Benson. 1983. Academic stress, power motivation, and decrease in secretion rate of salivary secretory immunoglobulin A. *Lancet* 1: 1400-1402.

Johnson, J.E., G.T. Anders, H.M. Blanton, C.E. Hawkes, B.A. Bush, C.K. McAllister, and J.I. Matthews. 1990. Exercise dysfunction in patients seropositive for the human immunodeficiency virus. *American Review of Respiratory Diseases* 141: 618-622.

Jordan, S.C. 1992. Cytokines and lymphocytes. In *Cytokines in Health and Disease,* ed. S.L. Kunkel and D.G. Remick, 309-325. New York: Dekker.

Kajiura, J.S., J.D. MacDougall, P.B. Ernst, and E.V. Younglai. 1995. Immune response to changes in training intensity and volume in runners. *Medicine and Science in Sports and Exercise* 27: 1111-1117.

Kappel, M., M. Diamant, M.B. Hansen, M. Klokker, and B.K. Pedersen. 1991a. Effects of in vitro hyperthermia on the proliferative response of blood mononuclear cell subsets, and detection of interleukins 1 and 6, tumour necrosis factor-alpha and interferon-gamma. *Immunology* 73: 304-308.

Kappel, M., M.B. Hansen, M. Diamant, J.O.L. Jergensen, A. Gyhrs, and B.K. Pedersen. 1993. Effects of an acute bolus growth hormone infusion on the human immune system. *Hormone and Metabolism Research* 25: 579-585.

Kappel, M., C. Stadeager, N. Tvede, H. Galbo, and B.K. Pedersen. 1991b. Effects of in vivo hyperthermia on natural killer cell activity, in vitro proliferative responses and blood mononuclear cell subpopulations. *Clinical and Experimental Immunology* 84: 175-180.

Kappel, M., N. Tvede, H. Galbo, P.M. Haahr, M. Kjaer, M. Linstouw, K. Klarlund, and B.K. Pedersen. 1991c. Evidence that the effect of physical exercise on natural killer cell activity is mediated by epinephrine. *Journal of Applied Physiology* 70: 2350-2354.

Kaufman, J.C., T.J. Harris, J. Higgins, and A.S. Maisel. 1994 Exercise-induced enhancement of immune function in the rat. *Circulation* 90: 525-532.

Keast, D., D. Arstein, W. Harper, R.W. Fry, and A.R. Morton. 1995. Depression of plasma glutamine concentration after exercise stress and its possible influence on the immune system. *Medical Journal of Australia* 162: 15-18.

Keen, P., D.A. McCarthy, L. Passfield, H.A.A. Shaker, and A.J. Wade. 1995. Leucocyte and erythrocyte counts during a multi-stage cycling race ('The Milk Race'). *British Journal of Sports Medicine* 29: 61-65.

Kendall, A., L. Hoffman-Goetz, M. Houston, B. MacNeil, and Y. Arumugam. 1990. Exercise and blood lymphocyte subset responses: Intensity, duration and subject fitness level. *Journal of Applied Physiology* 69: 251-260.

Kiel, R.J., F.F. Smith, J. Chason, R. Khatib, and M.P. Reyes. 1989. Coxsackievirus B3 myocarditis in C3H/HeJ mice: Description of an inbred model and the effect of exercise on virulence. *European Journal of Epidemiology* 5: 348-350.

Kirwan, J.P., D.L. Costill, M.G. Flynn, J.B. Mitchell, W.J. Fink, P.D. Neuffer, and J.A. Houmard. 1988. Physiological responses to successive days of intense training in competitive swimmers. *Medicine and Science in Sports and Exercise* 20: 255-259.

Klokker, M., M. Kjaer, N.H. Secher, B. Hanel, L. Worm, M. Kappel, and B.K. Pedersen. 1995. Natural killer cell response to exercise in humans: Effect of hypoxia and epidural anesthesia. *Journal of Applied Physiology* 78: 709-716.

Konstantinova, I.V., M.P. Rykova, A.T. Lesnyak, and E.A. Antropova. 1993. Immune changes during long-duration missions. *Journal of Leukocyte Biology* 54: 189-201.

Kotani, T., Y. Aratake, R. Ishiguro, I. Yamamoto, Y. Uemura, K. Tamura, and S. Ohtaki. 1987. Influence of physical exercise on large granular lymphocytes, Leu-7 bearing mononuclear cells and natural killer activity in peripheral blood—NK-cell and NK-activity after physical exercise. Acta Haematologica *Journal of Applied Physiology* 50: 1210-1216.

Kramer, M.M., and C.L. Wells. 1996. Does physical activity reduce the risk of estrogen-dependent cancer in women? *Medicine and Science in Sports and Exercise* 28: 322-334.

Kreider, R., A.C. Fry, and M. O'Toole (eds.). 1997. *Overtraining and Overreaching in Sport.* Champaign, IL: Human Kinetics (in press).

Kunkel, S.L., and D.G. Remick (eds.). 1992. *Cytokines in Health and Disease.* New York: Dekker.

Kunkel, S.L., T.J. Standiford, S.W. Chensue, R.M. Strieter, and J. Westwick. 1992. Interleukin-8 and the inflammatory response. In *Cytokines in Health and Disease,* ed. S.L. Kunkel and D.G. Remick, 121-130. New York: Dekker.

Kusaka, Y., H. Kondou, and K. Morimoto. 1992. Healthy lifestyles are associated with higher natural killer cell activity. *Preventive Medicine* 21: 602-615.

Kvernmo, H., J.O. Olsen, and B. Osterud. 1992. Changes in blood cell response following strenuous physical exercise. *European Journal of Applied Physiology* 64: 318-322.

Lamont, A.G., and L. Adorini. 1996. IL-12: A key cytokine in immune regulation. *Immunology Today* 17: 214-217.

Landmann, R., M. Portenier, M. Staehelin, M. Wesp, and R. Box. 1988. Changes in β-adrenoceptors and leukocyte subpopulations after physical exercise in normal subjects. *Naunyn-Schmiedeberg's Archives of Pharmacology* 337: 261-266.

LaPerriere, A., M.A. Fletcher, M.H. Antoni, N.G. Klimas, G. Ironson, and N. Schneiderman. 1991. Aerobic exercise training in an AIDS risk group. *International Journal of Sports Medicine* 12: S53-S57.

LaPerriere, A., G. Ironson, M.H. Antoni, N. Schneiderman, N. Klimas, and M.A. Fletcher. 1994. Exercise and psychoneuroimmunology. *Medicine and Science in Sports and Exercise* 26: 182-190.

LaPerriere, A., N. Klimas, M.A. Fletcher, A. Perry, G. Ironson, F. Perna, and N. Schneiderman. 1997. Changes in CD4+ cell enumeration following aerobic exercise training in HIV-1 disease: Possible mechanisms and practical applications. *International Journal of Sports Medicine* 18: S56-S61.

LaPerriere, A., H. Schneiderman, M.H. Antoni, and M.A. Fletcher. 1990a. Aerobic exercise training and psychoneuroimmunology in AIDS research. In *Psychological Aspects of AIDS,* ed. A. Baum and L. Temoshok, 259-286. Hillsdale, NJ: Erlbaum.

LaPerriere, A.R., M.H. Antoni, N. Schneiderman, G. Ironson, N. Klimas, P. Caralis, and M.A. Fletcher. 1990b. Exercise intervention attenuates emotional distress and natural killer cell decrements following notification of positive serologic status for HIV-1. *Biofeedback and Self-Regulation* 15: 229-242.

Lawless, D., C.G.R. Jackson, and J.E. Greenleaf. 1995. Exercise and human immunodeficiency virus (HIV-1) infection. *Sports Medicine* 19: 235-239.

Lee, D.J., R.T. Meehan, C. Robinson, T.R. Mabry, and M.L. Smith. 1992. Immune responsiveness and risk of illness in U.S. Air Force Academy cadets during basic cadet training. *Aviation, Space, and Environmental Medicine* 63: 517-523.

Lee, I.M., R.S. Paffenbarger, and C.C. Hsieh. 1992. Physical activity and the risk of prostatic cancer among college alumni. *American Journal of Epidemiology* 135: 169-179.

Lehmann, M., P. Baumgartl, C. Wiesenack, A. Seidel, J. Baumann, S. Fischer, U. Spori, G. Genmdrisch, R. Kaminski, and J. Keul. 1992. Training-overtraining: Influence of a defined increase in training volume vs training intensity on performance, catecholamines and some metabolic parameters in experienced middle- and long-distance runners. *European Journal of Applied Physiology* 64: 169-170.

Lehmann, M., C. Foster, and J. Keul. 1993. Overtraining in endurance athletes; A brief review. *Medicine and Science in Sports and Exercise* 25: 854-862.

Lehmann, M., M. Huonker, F. Dimeo, N. Heinz, U. Gastmann, N. Treis, J.M. Steinacker, J. Keul, R. Kajewski, and D. Haussinger. 1995. Serum amino acid concentrations in nine athletes before and after the 1993 Colmar Ultra Triathlon. *International Journal of Sports Medicine* 16: 155-159.

Lehmann, M., H. Mann, U. Gastmann, J. Keul, D. Vetter, J.M. Steinacker, and D. Haussinger. 1996. Unaccustomed high-mileage vs intensity training-related changes in performance and serum amino acid levels. *International Journal of Sports Medicine* 17: 187-192.

Levando, V.A., R.S. Suzdal'nitskii, B.B. Pershin, and M.P. Zykov. 1988. Study of secretory and antiviral immunity in sportsmen. *Sports Training, Medicine and Rehabilitation* 1: 49-52.

Levinson, S.O., A. Milzer, and P. Lewin. 1945. Effect of fatigue, chilling and mechanical trauma on resistance to experimental poliomyelitis. *American Journal of Hygiene* 42: 204-213.

Lewicki, R., H. Tchorzewski, A. Denys, M. Kowalska, and A. Golinska. 1987. Effect of physical exercise on some parameters of immunity in conditioned sportsmen. *International Journal of Sports Medicine* 8: 309-314.

Lewicki, R., H. Tchorzewski, E. Majewska, Z. Nowak, and Z. Baj. 1988. Effect of maximal physical exercise on T-lymphocyte subpopulations and on interleukin 1 IL 1 and interleukin 2 IL 2 production in vitro. *International Journal of Sports Medicine* 9: 114-117.

Liesen, H., B. Dufaux, and W. Hollmann. 1977. Modifications of serum glycoproteins on the days following a prolonged physical exercise and the influence of physical training. *European Journal of Applied Physiology* 37: 243-254.

Liew, F.Y., S.M. Russell, G. Appleyard, G.M. Brand, and J. Beale. 1984. Cross-protection in mice infected with influenza A virus by the respiratory route is correlated with local IgA antibody rather than serum antibody or cytotoxic T cell reactivity. *European Journal of Immunology* 14: 350-356.

Liles, W.C., and W.C. Van Voorhis. 1995. Review: Nomenclature and biologic significance of cytokines involved in inflammation and the host immune response. *Journal of Infectious Diseases* 172: 1573-1580.

Linde, F. 1987. Running and upper respiratory tract infections. *Scandinavian Journal of Sports Science* 9: 21-23.

Lindena, J., W. Kupper, and I. Trautschold. 1984. Enzyme activities in thoracic duct lymph and plasma of anaesthetized, conscious resting and exercising dogs. *European Journal of Applied Physiology* 52: 188-195.

Liu, Y.G., and S.Y. Wang. 1986/87. The enhancing effect of exercise on the production of antibody to Salmonella typhi in mice. *Immunology Letters* 14: 117-120.

Lotzerich, H., H.-G. Fehr, and H.-J. Appell. 1990. Potentiation of cytostatic but not cytolytic activity of murine macrophages after running stress. *International Journal of Sports Medicine* 11: 61-65.

Lox, C.L., E. McAuley, and R.S. Tucker. 1995. Exercise as an intervention for enhancing subjective well-being in an HIV-1 population. *Journal of Sport and Exercise Psychology* 17: 345-362.

Macarthur, R.D., S.D. Levine, and T.J. Birk. 1993. Supervised exercise training improves cardiopulmonary fitness in HIV-infected persons. *Medicine and Science in Sports and Exercise* 25: 684-688.

Macha, M., M. Shlafer, and M.J. Kluger. 1990. Human neutrophil hydrogen peroxide generation following physical exercise. *The Journal of Sports Medicine and Physical Fitness* 30: 412-419.

Mackinnon, L.T. 1989. Exercise and natural killer cells: What is the relationship? *Sports Medicine* 7: 141-149.

Mackinnon, L.T. 1994. Current challenges and future expectations in exercise immunology: Back to the future. *Medicine and Science in Sports and Exercise* 26: 191-194.

Mackinnon, L.T. 1996. Exercise, immunoglobulin and antibody. *Exercise Immunology Review* 2: 1-32.

Mackinnon, L.T., T.W. Chick, A. van As, and T.B. Tomasi. 1988. Effects of prolonged intense exercise on natural killer cell number and function. In *Exercise Physiology: Current Selected Research,* ed. C.O. Dotson and J.H. Humphrey, vol. 3, 77-89. New York: AMS Press.

Mackinnon, L.T., T.W. Chick, A. van As, and T.B. Tomasi. 1989. Decreased secretory immunoglobulins following intense endurance exercise. *Sports Training, Medicine and Rehabilitation* 1: 209-218.

Mackinnon, L.T., E. Ginn, and G. Seymour. 1991. Effects of exercise during sports training and competition on salivary IgA levels. In *Behaviour and Immunity,* ed. A.J. Husband, 169-177. Boca Raton, FL: CRC Press.

Mackinnon, L.T., E. Ginn, and G.J. Seymour. 1993a. Decreased salivary immunoglobulin A secretion rate after intense interval training in elite kayakers. *European Journal of Applied Physiology* 67: 180-184.

Mackinnon, L.T., E. Ginn, and G.J. Seymour. 1993b. Temporal relationship between exercise-induced decreases in salivary IgA and subsequent appearance of upper respiratory tract infection in elite athletes. *Australian Journal of Science and Medicine in Sport* 25: 94-99.

Mackinnon, L.T., and S.L. Hooper. 1994. Mucosal (secretory) immune system responses to exercise of varying intensity and during overtraining. *International Journal of Sports Medicine* 15: S179-S183.

Mackinnon, L.T, and S.L. Hooper. 1996. Plasma glutamine concentration and upper respiratory tract infection during overtraining in elite swimmers. *Medicine and Science in Sports and Exercise* 28: 285-290.

Mackinnon, L.T., S.L. Hooper, S. Jones, A.W. Bachmann, and R.D. Gordon. 1997. Hormonal, immunological and hematological responses to intensified training in elite swimmers. *Medicine and Science in Sports and Exercise* 29: 1637-1645.

Mackinnon, L.T., and D.G. Jenkins. 1993. Decreased salivary IgA after intense interval exercise before and after training. *Medicine and Science in Sports and Exercise* 25: 678-683.

MacNeil, B., and L. Hoffman-Goetz. 1993a. Chronic exercise enhances in vivo and in vitro cytotoxic mechanisms of natural immunity in mice. *Journal of Applied Physiology* 74: 388-395.

MacNeil, B., and L. Hoffman-Goetz. 1993b. Exercise training and tumor metastasis in mice: Influence of time of exercise onset. *Anticancer Research* 13(6A): 2085-2088.

MacNeil, B., L. Hoffman-Goetz, A. Kendall, M. Houston, and Y. Arumugam. 1991. Lymphocyte proliferation responses after exercise in men: Fitness, intensity, and duration effects. *Journal of Applied Physiology* 70: 179-185.

MacVicar, M.G., and M.L. Winningham. 1986. Promoting the functional capacity of cancer patients. *The Cancer Bulletin* 38: 235-239.

MacVicar, M.G., M.L. Winningham, and J.L. Nickel. 1989. Effects of aerobic interval training on cancer patients' functional capacity. *Nursing Research* 38: 348-351.

Maffulli, N., V. Testa, and G. Capasso. 1993. Post-viral fatigue syndrome. A longitudinal assessment in varsity athletes. *The Journal of Sports Medicine and Physical Fitness* 33: 392-399.

Mahan, M.P., and M.R. Young. 1989. Immune parameters of untrained or exercise-trained rats after exhaustive exercise. *Journal of Applied Physiology* 66: 282-287.

Maisel, A.S., T. Harris, C.A. Rearden, and M.C. Michel. 1990. β-adrenergic receptors in lymphocyte subsets after exercise. *Circulation* 82: 2003-2010.

Marlin, S.D., D.E. Staunton, T.A. Springer, C. Stratowa, W. Sommergruber, and V.J. Merluzzi. 1990. A soluble form of intercellular adhesion molecule-1 inhibits rhinovirus infection. *Nature* 344: 70-72.

Mazzeo, R.S. 1994. The influence of exercise and aging on immune function. *Medicine and Science in Sports and Exercise* 26: 586-592.

Mazzeo, R.S., and I. Nasrullah. 1992. Exercise and age-related decline in immune functions. In *Exercise and Disease,* ed. R.R. Watson and M. Eisinger, 159-178. Boca Raton, FL: CRC Press.

McCarthy, D.A., and Dale, M.M. 1988. The leucocytosis of exercise: A review and model. *Sports Medicine* 6: 333-363.

McCarthy, D.A., I.A. Macdonald, H.A. Shaker, P. Hart, S. Geogiannos, J. Deeks, and A.J. Wade. 1992. Changes in the leucocyte count during and after brief intense exercise. *European Journal of Applied Physiology* 64: 518-522.

McCully, K.K., S.A. Sisto, and B.H. Natelson. 1996. Use of exercise for treatment of chronic fatigue syndrome. *Sports Medicine* 21: 35-48.

McDowell, S.L., K. Chalos, T.J. Housh, G.D. Tharp, and G.O. Johnson. 1991. The effect of exercise intensity and duration on salivary immunoglobulin A. *European Journal of Applied Physiology* 63: 108-111.

McDowell, S.L., R.A. Hughes, R.J. Hughes, D.J. Housh, T.J. Housh, and G.O. Johnson. 1992a. The effect of exhaustive exercise on salivary immunoglobulin A. *The Journal of Sports Medicine and Physical Fitness* 32: 412-415.

McDowell, S.L., R.A. Hughes, R.J. Hughes, T.J. Housh, and G.O. Johnson. 1992b. The effect of exercise training on salivary immunoglobulin A and cortisol responses to maximal exercise. *International Journal of Sports Medicine* 13: 577-580.

McDowell, S.L., J.P. Weir, J.M. Eckerson, L.L. Wagner, T.J. Housh, and G.O. Johnson. 1992c. A preliminary investigation of the effect of weight training on salivary immunoglobulin A. *Research Quarterly for Exercise Science* 64: 348-51.

McLelland, D.C., E. Floor, R.J. Davidson, and C. Saron. 1980. Stressed power motivation, sympathetic activation, immune function and illness. *Journal of Human Stress* 6: 11-19.

Meehan, R., P. Whitson, and C. Sams. 1993. The role of psychoneuro-endocrine factors on spaceflight-induced immunological alterations. *Journal of Leukocyte Biology* 54: 236-244.

Mitchell, J.B., A.J. Paquet, F.X. Pizza, R.D. Starling, R.W. Holtz, and P.W. Grandjean. 1996. The effect of moderate aerobic training on lymphocyte proliferation. *International Journal of Sports Medicine* 17: 384-389.

Moldawer, L.L. 1992. The beneficial role of cytokines, particularly interleukin-1, in the host response to injury, infection, and inflammation. In *Cytokines in Health and Disease,* ed. S.L. Kunkel and D.G. Remick, 217-234. New York: Dekker.

Moorthy, A.V., and S.W. Zimmerman. 1978. Human leukocyte response to an endurance race. *European Journal of Applied Physiology* 38: 271-276.

Muir, A.L., A. Cruz, B.A. Martin, H. Thommasen, A. Belzberg, and J.C. Hogg. 1984. Leukocyte kinetics in the human lung: Role of exercise and catecholamines. *Journal of Applied Physiology* 57: 711-719.

Mujika, I., J.-C. Chatard, and A. Geyssant. 1996. Effects of training and taper on blood leucocyte populations in competitive swimmers: Relationships with cortisol and performance. *International Journal of Sports Medicine* 17: 213-217.

Muns, G. 1993. Effect of long-distance running on polymorphonuclear neutrophil phagocytic function of the upper airways. *International Journal of Sports Medicine* 15: 96-99.

Muns, G., H. Liesen, H. Riedel, and K.-Ch Bergmann. 1989. Influence of long-distance running of IgA in nasal secretion and saliva. *Deutsche Zeitschrift fur Sportmedizin* 40: 63-65.

Muns, G., I. Rubinstein, and P. Singer. 1996. Neutrophil chemotactic activity is increased in nasal secretions of long-distance runners. *International Journal of Sports Medicine* 17: 56-59.

Muns, G., P. Singer, F. Wolf, and I. Rubinstein. 1995. Impaired nasal mucociliary clearance in long-distance runners. *International Journal of Sports Medicine* 16: 209-213.

Murphy, B.R., D.L. Nelson, P.F. Wright, E.L. Tierney, M.A. Phelan, and R.M. Chanock. 1982. Secretory and systemic immunological response in children infected with live attenuated influenza A virus vaccines. *Infection and Immunology* 36: 1102-1108.

Nash, M.S. 1994. Immune responses to nervous system decentralization and exercise in quadriplegia. *Medicine and Science in Sports and Exercise* 26: 164-171.

Natelson, B.H., X. Zhour, J.E. Ottenweller, M.T. Bergen, S.A. Sisto, S. Drastal, W.N. Tapp, and W.L. Gause. 1996. Effect of acute exhausting exercise on cytokine gene expression in men. *International Journal of Sports Medicine* 17: 299-302.

Ndon, J.A., A.C. Snyder, C. Foster, and W.B. Wehrenberg. 1992. Effects of chronic intensive exercise training on the leukocyte response to acute exercise. *International Journal of Sports Medicine* 13: 176-182.

Nehlsen-Cannarella, S.L., D.C. Nieman, A.J. Balk-Lamberton, P.A. Markoff, D.B.W. Chritton, G. Gusewitch, and J.W. Lee. 1991a. The effect of moderate exercise training on immune response. *Medicine and Science in Sports and Exercise* 23: 64-70.

Nehlsen-Cannarella, S.L., D.C. Nieman, J. Jessen, L. Chang, G. Gusewitch, G.G. Blix, and E. Ashley. 1991b. The effects of acute moderate exercise on lymphocyte function and serum immunoglobulin levels. *International Journal of Sports Medicine* 12: 391-398.

Newsholme, E.A. 1994. Biochemical mechanisms to explain immunosuppression in well-trained and overtrained athletes. *International Journal of Sports Medicine* 15: S142-S147.

Nicholls, E.E., and R.A. Spaeth. 1922. The relation between fatigue and the susceptibility of guinea pigs to infections of type I pneumococcus. *American Journal of Hygiene* 2: 527-535.

Nielsen, H.B., N.H. Secher, N.J. Christensen, and B.K. Pedersen. 1996a. Lymphocytes and NK cell activity during repeated bouts of maximal exercise. *American Journal of Physiology* 271 (*Regulatory Integrative Comparative Physiology* 40): R222-R227.

Nielsen, H.B., N.H. Secher, M. Kappel, B. Hanel, and B.K. Pedersen. 1996b. Lymphocyte, NK and LAK cell responses to maximal exercise. *International Journal of Sports Medicine* 17: 60-65.

Nieman, D.C. 1994. Exercise, upper respiratory tract infection, and the immune system. *Medicine and Science in Sports and Exercise* 26: 128-139.

Nieman, D.C. 1997. Immune responses to heavy exertion. *Journal of Applied Physiology* 82: 1385-1394.

Nieman, D.C., J.C. Ahle, D.A. Henson, B.J. Warren, J. Suttles, J.M. Davis, K.S. Buckley, S. Simandle, D.E. Butterworth, O.R. Fagoaga, and S.L. Nehlsen-Cannarella. 1995a. Indomethacin does not alter natural killer cell response to 2.5 h of running. *Journal of Applied Physiology* 79: 748-755.

Nieman, D.C., L.S. Berk, M. Simpson-Westerberg, K. Arabatzis, S. Youngberg, S.A. Tan, J.W. Lee, and W.C. Eby. 1989a. Effects of long-endurance running on immune system parameters and lymphocyte function in experienced marathoners. *International Journal of Sports Medicine* 10: 317-323.

Nieman, D.C., D. Brendle, D.A. Henson, J. Suttles, V.D. Cook, B.J. Warren, D.E. Butterworth, O.R. Fagoaga, and S.L. Nehlsen-Cannarella. 1995b. Immune function in athletes versus nonathletes. *International Journal of Sports Medicine* 16: 329-333.

Nieman, D.C., K.S. Buckley, D.A. Henson, B.J. Warren, J. Suttles, J.C. Ahle, S. Simandle, O.R. Fagoaga, and S.L. Nehlsen-Cannarella. 1995c. Immune function in marathon runners versus sedentary controls. *Medicine and Science in Sports and Exercise* 27: 986-992.

Nieman, D.C., V.D. Cook, D.A. Henson, J. Suttles, W.J. Rejeski, P.M. Ribisl, O.R. Fagoaga, and S.L. Nehlsen-Cannarella. 1995d. Moderate exercise training and natural killer cell cytotoxic activity in breast cancer patients. *International Journal of Sports Medicine* 16: 334-337.

Nieman, D.C., and D.A. Henson. 1994. Role of endurance exercise in immune senescence. *Medicine and Science in Sports and Exercise* 26: 172-181.

Nieman, D.C., D.A. Henson, G. Gusewitch, B.J. Warren, R.C. Dotson, D.E. Butterworth, and S.L. Nehlsen-Cannarella. 1993a. Physical activity and immune function in elderly women. *Medicine and Science in Sports and Exercise* 25: 823-831.

Nieman, D.C., D.A. Henson, R. Johnson, L. Lebeck, J.M. Davis, and S.L. Nehlsen-Cannarella. 1992. Effects of brief, heavy exertion on circulating lymphocyte subpopulations and proliferative response. *Medicine and Science in Sports and Exercise* 24: 1339-1345.

Nieman, D.C., D.A. Henson, C.S. Sampson, J.L. Herring, J. Stulles, M. Conley, M.H. Stone, D.E. Butterworth, and J.M. Davis. 1995e. The acute immune response to exhaustive resistance exercise. *International Journal of Sports Medicine* 16: 322-328.

Nieman, D.C., L.M. Johanssen, and J.W. Lee. 1989b. Infectious episodes in runners before and after a roadrace. *The Journal of Sports Medicine and Physical Fitness* 29: 289-296.

Nieman, D.C., L.M. Johanssen, J.W. Lee, and K. Arabatzis. 1990a. Infectious episodes in runners before and after the Los Angeles Marathon. *The Journal of Sports Medicine and Physical Fitness* 30: 316-328.

Nieman, D.C., A.R. Miller, D.A. Henson, B.J. Warren, G. Gusewitch, R.L. Johnson, J.M. Davis, D.E. Butterworth, J.L. Herring, and S.L. Nehlsen-Cannarella. 1994. Effect of high- versus moderate-intensity exercise on lymphocyte subpopulations and proliferative response. *International Journal of Sports Medicine* 15: 199-206.

Nieman, D.C., A.R. Miller, D.A. Henson, B.J. Warren, G. Gusewitch, R.L. Johnson, J.M. Davis, D.E. Butterworth, and S.L. Nehlsen-Cannarella. 1993b. Effects of high- vs moderate-intensity exercise on natural killer activity. *Medicine and Science in Sports and Exercise* 25: 1126-1134.

Nieman, D.C., and S.L. Nehlsen-Canarella. 1991. The effects of acute and chronic exercise on immunoglobulins. *Sports Medicine* 11: 183-201.

Nieman, D.C., and S.L. Nehlsen-Cannarella. 1992. Exercise and infection. In *Exercise and Disease,* ed. R.R. Watson and M. Eisinger, 122-148. Boca Raton, FL: CRC Press.

Nieman, D.C., S.L. Nehlsen-Cannarella, P.A. Markoff, A.J. Balk-Lamberton, H. Yang, D.B.W. Chritton, J.W. Lee, and K. Arabatzis. 1990b. The effects of moderate exercise training on natural killer cells and acute upper respiratory tract infections. *International Journal of Sports Medicine* 11: 467-473.

Nieman, D.C., S. Simandle, D.A. Henson, B.J. Warren, J. Suttles, J.M. Davis, K.S. Buckley, J.C. Ahle, D.E. Butterworth, O.R. Fagoaga, and S.L. Nehlsen-Cannarella. 1995f. Lymphocyte proliferative response to 2.5 hours of running. *International Journal of Sports Medicine* 16: 404-408.

Nieman, D.C., S.A. Tan, J.W. Lee, and L.S. Berk. 1989c. Complement and immunoglobulin levels in athletes and sedentary controls. *International Journal of Sports Medicine* 10: 124-128.

Northoff, H., C. Weinstock, and A. Berg. 1994. The cytokine response to strenuous exercise. *International Journal of Sports Medicine* 15: S167-S171.

Nosaka, K., and P.M. Clarkson. 1996. Changes in indicators of inflammation after eccentric exercise of the elbow flexors. *Medicine and Science in Sports and Exercise* 28: 953-961.

Oppenheimer, E.H., and R.A. Spaeth. 1922. The relation between fatigue and the susceptibility of rats towards a toxin and an infection. *American Journal of Hygiene* 2: 51-66.

Ortega, E. 1994. Physiology and biochemistry: Influence of exercise on phagocytosis. *International Journal of Sports Medicine* 15: S172-S178.

Ortega, E., M.E. Collazos, C. Barriga, and M. De la Fuente. 1992. Stimulation of the phagocytic function in guinea pig peritoneal macrophages by physical activity stress. *European Journal of Applied Physiology* 64: 323-327.

Oshida, Y., K. Yamanouchi, S. Hayamizu, and Y. Sato. 1988. Effect of acute physical exercise on lymphocyte subpopulations in trained and untrained subjects. *International Journal of Sports Medicine* 9: 137-140.

Osterback, L., and Y. Qvarnberg. 1987. A prospective study of respiratory infections in 12-year-old children actively engaged in sports. *Acta Paediatrica Scandinavica* 76: 944-949.

Osterud, B., J.O. Olsen, and L. Wilsgard. 1989. Effect of strenuous exercise on blood monocytes and their relation to coagulation. *Medicine and Science in Sports and Exercise* 21: 374-378.

Paffenbarger, R.S., R.T. Hyde, and A.L. Wing. 1987. Physical activity and incidence of cancer in diverse populations: A preliminary report. *American Journal of Clinical Nutrition* 45: 312-317.

Paffenbarger, R.S., I.M. Lee, and A.L. Wing. 1992. The influence of physical activity on the incidence of site-specific cancers in college alumni. *Advances in Experimental Medicine and Biology* 322: 7-15.

Palmo, J., A. Asp, J.R. Daugaard, E.A. Richter, M. Klokker, and B.K. Pedersen. 1995. Effect of eccentric exercise on natural killer cell activity. *Journal of Applied Physiology* 78: 1442-1446.

Parker, S., P.D. Brukner, and M. Rosier. 1996. Chronic fatigue syndrome and the athlete. *Sports Medicine, Training and Rehabilitation* 6: 269-278.

Parry-Billings, M., R. Budgett, Y. Koutedakis, E. Blomstrand, S. Brooks, C. Williams, P.C. Calder, S. Pilling, R. Baigrie, and E.A. Newsholme. 1992. Plasma amino acid concentrations in the overtraining syndrome: Possible effects on the immune system. *Medicine and Science in Sports and Exercise* 24: 1353-1358.

Pedersen, B.K. 1997. *Exercise Immunology*. Heidelberg: Springer.

Pedersen, B.K., N. Tvede, L.D. Christensen, K. Klarlund, S. Kragbak, and J. Halkjaer-Kristensen. 1989. Natural killer cell activity in peripheral blood of highly trained and untrained persons. *International Journal of Sports Medicine* 10: 129-131.

Pedersen, B.K., N. Tvede, F.R. Hansen, V. Andersen, T. Bendix, G. Bendixen, K. Bendtzen, H. Galbo, P.M. Haahr, K. Klarlund, J. Sylvest, B.S. Thomsen, and J. Halkjaer-Kristensen. 1988. Modulation of natural killer cell activity in peripheral blood by physical exercise. *Scandinavian Journal of Immunology* 27: 673-678.

Pedersen, B.K., N. Tvede, K. Klarlund, L.D. Christensen, F.R. Hansen, H. Galbo, and A. Kharazmi. 1990. Indomethacin in vitro and in vivo abolishes post-exercise suppression of natural killer cell activity in peripheral blood. *International Journal of Sports Medicine* 11: 127-131.

Pedersen, B.K., and H. Ullum. 1994. NK cell response to physical activity: Possible mechanisms of action. *Medicine and Science in Sports and Exercise* 26: 140-146.

Perkins, J.C., D.N. Tucker, H.L.S. Knopf, R.P. Wenzel, A.Z. Kapikian, and R.M. Chanock. 1969. Comparison of protective effect of neutralizing antibody in serum and nasal secretions in experimental rhinovirus type 13 illness. *American Journal of Epidemiology* 90: 519-526.

Pershin, B.B., S.N. Kuz'min, R.S. Suzdal'nitskii, and V.A. Levando. 1988. Reserve potential of immunity. *Sports Training, Medicine and Rehabilitation* 1: 53-60.

Peters, C., H. Lotzerich, B. Niemeier, K. Schule, and G. Uhlenbruck. 1994. Influence of moderate exercise training on natural killer cytotoxicity and personality traits in cancer patients. *Anticancer Research* 14: 1033-1036.

Peters, E.M., and E.D. Bateman. 1983. Ultramarathon running and upper respiratory tract infections. *South African Medical Journal* 64: 582-584.

Peters, E.M., A. Campbell, and L. Pawley. 1992. Vitamin A supplementation fails to increase resistance to upper respiratory tract infection in athletes. *South African Journal of Sports Medicine* 7: 3-7.

Peters, E.M., J.M. Goetzsche, B. Grobbelaar, and T.D. Noakes. 1993. Vitamin C supplementation reduces the incidence of postrace symptoms of upper respiratory tract infection in ultramarathon runners. *American Journal of Clinical Nutrition* 57: 170-174.

Peters, E.M., J.M. Goetzsche, L.E. Joseph, and T.D. Noakes. 1996. Vitamin C as effective as combinations of anti-oxidant nutrients in reducing symptoms of upper respiratory tract infection in ultramarathon runners. *South African Sportsmedicine* 3: 23-27.

Peters, R.K., D.H. Garabrant, M.C. Yu, and T.M. Mack. 1989. A case-control study of occupational and dietary factors in colorectal cancer in young men by subsite. *Cancer Research* 49: 5459-5468.

Pinto, B.M., and B.H. Marcus. 1994. Physical activity, exercise and cancer in women. *Medicine, Exercise, Nutrition and Health* 3: 102-111.

Pizza, F.X., J.B. Mitchell, B.H. Davis, R.D. Starling, R.W. Holtz, and N. Bigelow. 1995. Exercise-induced muscle damage: Effect on circulating leukocyte and lymphocyte subsets. *Medicine and Science in Sports and Exercise* 27: 363-370.

Poortmans, J.R. 1971. Serum protein determination during short exhaustive physical activity. *Journal of Applied Physiology* 30: 190-92.

Poortmans, J.R., and G. Haralambie. 1979. Biochemical changes in a 100 km run: Proteins in serum and urine. *European Journal of Applied Physiology* 40: 245-254.

Pothoff, G., K. Wasserman, and H. Ostmann. 1994. Impairment of exercise capacity in various groups of HIV-infected patients. *Respiration* 61: 80-85.

Priest, J.B., T.O. Oei, and W.R. Moorehead. 1982. Exercise-induced changes in common laboratory tests. *American Journal of Clinical Pathology* 77: 285-289.

Pyne, D.B. 1994. Regulation of neutrophil function during exercise. *Sports Medicine* 17: 245-258.

Pyne, D.B., M.S. Baker, P.A. Fricker, W.A. McDonald, R.D. Telford, and M.J. Weidemann. 1995. Effects of an intensive 12-wk training program by elite swimmers on neutrophil oxidative activity. *Medicine and Science in Sports and Exercise* 27: 536-542.

Rall, L.C., R. Roubenoff, J.G. Cannon, L.W. Abad, C.A. Dinarello, and S.N. Meydani. 1996. Effects of progressive resistance training on immune response in aging and chronic inflammation. *Medicine and Science in Sports and Exercise* 28: 1356-1365.

Rashkis, H.A. 1952. Systemic stress as an inhibitor of experimental tumors in Swiss mice. *Science* 116: 169-171.

Repsher, L.H., and R.K. Freebern. 1969. Effects of early and vigorous exercise on recovery from infectious hepatitis. *New England Journal of Medicine* 281: 1393-1396.

Reyes, M.P., and A.M. Lerner. 1976. Interferon and neutralizing antibody in sera of exercised mice with coxsackievirus B-3 myocarditis. *Proceedings of the Society for Experimental Biology and Medicine* 151: 333-338.

Rhind, S.G., P.N. Shek, S. Shinkai, and R.J. Shephard. 1994. Differential expression of interleukin-2 receptor alpha and beta chains in relation to natural killer cell subsets and aerobic fitness. *International Journal of Sports Medicine* 15: 911-918.

Richter, E.A., B. Kiens, A. Raben, N. Tvede, and B.K. Pedersen. 1991. Immune parameters in male athletes after a lacto-ovo vegetarian diet and a mixed Western diet. *Medicine and Science in Sports and Exercise* 23: 517-521.

Ricken, K.-H., T. Rieder, G. Hauck, and W. Kindermann. 1990. Changes in lymphocyte subpopulations after prolonged exercise. *International Journal of Sports Medicine* 11: 132-135.

Rigsby, L., R.K. Dishman, Q.W. Jackson, G.S. Maclean, and P.B. Raven. 1992. Effects of exercise training on men seropositive for the human immunodeficiency virus-1. *Medicine and Science in Sports and Exercise* 24: 6-12.

Rivier, A., J. Pene, P. Chanez, F. Anselme, C. Caillaud, C. Prefaut, Ph. Godard, and J. Bousquet. 1994. Release of cytokines by blood monocytes during strenuous exercise. *International Journal of Sports Medicine* 15: 192-198.

Roberts, J.A. 1985. Loss of form in young athletes due to viral infection. *British Medical Journal* 290: 357-358.

Roberts, J.A. 1986. Viral illnesses and sports performance. *Sports Medicine* 3: 296-303.

Robertson, A.J., K.C.R.B. Ramesar, R.C. Potts, J.H. Gibbs, M.C.K. Browning, R.A. Brown, P.C. Hayes, and J.S. Beck. 1981. The effect of strenuous physical exercise on circulating blood lymphocytes and serum cortisol levels. *Journal of Clinical Laboratory Immunology* 5: 53-57.

Rocker, L., K.A. Krisch, and H. Stoboy. 1976. Plasma volume, albumin and globulin concentrations and their intravascular masses. *European Journal of Applied Physiology* 36: 57-64.

Rohde, T., D.A. Maclean, A. Hartkopp, and B.K. Pedersen. 1996a. The immune system and serum glutamine during a triathlon. *European Journal of Applied Physiology* 74: 428-434.

Rohde, T., D.A. MacLean, and B.K. Pedersen. 1996b. Glutamine, lymphocyte proliferation and cytokine production. *Scandinavian Journal of Immunology* 44: 648-650.

Rohde, T., H. Ullum, J. Palmo, J. Halkjaer Kristensen, E.A. Newsholme, and B.K. Pedersen. 1995. Effects of glutamine on the immune system-influence of muscular exercise and HIV infection. *Journal of Applied Physiology* 79: 146-150.

Roitt, I., J. Brostoff, and D. Male. 1993. *Immunology*. 3rd ed. St Louis: Mosby.

Rose, R.J., and M.S. Bloomberg. 1989. Response to sprint exercise in the greyhound: Effects on haematology, serum biochemistry and muscle metabolites. *Research in Veterinary Science* 47: 212-218.

Rosenbaum, H.E., and C.G. Harford. 1953. Effects of fatigue on susceptibility of mice to poliomyelitis. *Proceedings of the Society for Experimental Biology and Medicine* 83: 678-681.

Round, J.M., D.A. Jones, and G. Cambridge. 1987. Cellular infiltrates in human skeletal muscle: Exercise induced damage as a model for inflammatory muscle disease? *Journal of the Neurological Sciences* 82: 1-11.

Rowbottom, D.G., D. Keast, C. Goodman, and A.R. Morton. 1995. The haematological, biochemical and immunological profile of athletes suffering from the overtraining syndrome. *European Journal of Applied Physiology* 70: 502-509.

Rowbottom, D.G., D. Keast, and A.R. Morton. 1996. The emerging role of glutamine as an indicator of exercise stress and overtraining. *Sports Medicine* 21: 80-97.

Rowbottom, D.G., D. Keast, Z. Pervan, C. Goodman, D. Bhagat, B. Kakulas, and A.R. Morton (1988a). The role of glutamine in the aetiology of the chronic fatigue syndrome: A prospective study. *Journal of Chronic Fatigue Syndrome* 4: 3-22.

Rowbottom, D.G., D. Keast, Z. Pervan, and A.R. Morton. (1988b) The physiological response to exercise of Chronic Fatigue Syndrome sufferers. *Journal of Chronic Fatigue Syndrome* 4: 3-22.

Rusch, H.P., and B.E. Kline. 1944. The effect of exercise on the growth of a mouse tumor. *Cancer Research* 4: 33-49.

Russell, W.R. 1947. Poliomyletis: The preparalytic stage, and the effect of physical activity on the severity of paralysis [preliminary report]. *British Medical Journal* December 27, 1947: 1023-1028.

Russell, W.R. 1949. Paralytic poliomyelitis: The early symptoms and the effect of physical activity on the course of the disease. *British Medical Journal* March 19, 1949: 465-471.

Sanders, V.M., and A.E. Munson. 1985. Norepinephrine and the antibody response. *Pharmacological Reviews* 37: 229-48.

Sanders, V.M., and F.E. Powell-Oliver. 1992. β_2-adrenoceptor stimulation increases the number of antigen-specific precursor B lymphocytes that differentiate into IgM-secreting cells without affecting burst size. *Journal of Immunology* 148: 1822-28.

Sawka, M.N., A.J. Young, R.C. Dennis, R.R. Gonzalez, K.B. Pandolf, and C.R. Valeri. 1989. Human intravascular immunoglobulin response to exercise-heat and hypohydration. *Aviation, Space, and Environmental Medicine* 60: 634-638.

Scales, W.E. 1992. Structure and function of interleukin-1. In *Cytokines in Health and Disease,* ed. S.L. Kunkel and D.G. Remick, 15-26. New York: Dekker.

Schaefer, R.M., K. Kokot, A. Heidland, and R. Plass. 1987. Joggers' leukocytes. *New England Journal of Medicine* 316: 223-224.

Schmitt, D.A., and L. Schaffar. 1993. Isolation and confinement as a model for spaceflight immune changes. *Journal of Leukocyte Biology* 54: 209-213.

Schouten, W.J., R. Verschuur, and H.C.G. Kemper. 1988. Physical activity and upper respiratory tract infections in a normal population of young men and women: The

Amsterdam growth and health study. *International Journal of Sports Medicine* 9: 451-455.

Seversson, R.K., A.M.Y. Nomura, J.S. Grove, and G.N. Stemmermann. 1989. A prospective analysis of physical activity and cancer. *American Journal of Epidemiology* 130: 522-529.

Shek, P.N., B.H. Sabiston, A. Buguet, and M.W. Radomski. 1995. Strenuous exercise and immunological changes: A multiple-time-point analysis of leukocyte subsets, CD4/CD8 ratio, immunoglobulin production and NK cell response. *International Journal of Sports Medicine* 16: 466-474.

Shephard, R.J., S. Rhind, and P.N. Shek. 1994a. Exercise and the immune system: Natural killer cells, interleukins, and related responses. *Sports Medicine* 18: 340-369.

Shephard, R.J., S. Rhind, and P.N. Shek. 1994b. Exercise and training: Influences on cytotoxicity, interleukin-1, interleukin-2 and receptor structures. *International Journal of Sports Medicine* 15: S154-S166.

Shinkai, S., S. Shore, P.N. Shek, and R.J. Shephard. 1992. Acute exercise and immune function: Relationship between lymphocyte activity and changes in subset counts. *International Journal of Sports Medicine* 13: 452-461.

Simon, H.B. 1987. Exercise and infection. *The Physician and Sportsmedicine* 15: 135-141.

Simpson, J.A.R., and L. Hoffman-Goetz. 1991. Exercise, serum zinc, and interleukin-1 concentrations in man: Some methodological considerations. *Nutrition Research* 11: 309-323.

Simpson, J.A.R., L. Hoffman-Goetz, R. Thorne, and Y. Arumugam. 1989. Exercise stress alters the percentage of splenic lymphocyte subsets in response to mitogen but not in response to interleukin-1. *Brain, Behavior, and Immunology* 3: 119-128.

Singh, A., E.B. Zelazowska, J.S. Petrides, R.B. Raybourne, E.M. Sternberg, P.W. Gold, and P.A. Deuster. 1996. Lymphocyte subset responses to exercise and glucocorticoid suppression in healthy men. *Medicine and Science in Sports and Exercise* 28: 822-828.

Smith, C.B., R.H. Purcell, J.A.A. Bellanti, and R.M. Chanock. 1966. Protective effect of antibody to parainfluenza type I virus. *New England Journal of Medicine* 275: 1145-52.

Smith, J., D. Chi, S. Salazar, G. Krish, S. Berk, S. Reynolds, and G. Cambron. 1993. Effect of moderate exercise on proliferative responses of peripheral blood mononuclear cells. *Journal of Sports Medicine and Physical Fitness* 33: 152-158.

Smith, J.A. 1994. Neutrophils, host defense, and inflammation: A double-edged sword. *Journal of Leukocyte Biology* 56: 672-686.

Smith, J.A. 1995. Guidelines, standards and perspectives in exercise immunology. *Medicine and Science in Sports and Exercise* 27: 497-506.

Smith, J.A., A.B. Gray, D.B. Pyne, M.S. Baker, R.D. Telford, and M.J. Weidemann. 1996. Moderate exercise triggers both priming and activation of neutrophil subpopulations. *American Journal of Physiology* 270 (*Regulatory Integrative Comparative Physiology* 39): R838-R845.

Smith, J.A., R.D. Telford, M.S. Baker, A.J. Hapel, and M.J. Weidemann. 1992. Cytokine immunoreactivity in plasma does not change after moderate endurance-exercise. *Journal of Applied Physiology* 71: 1396-1401.

Smith, J.A., R.D. Telford, I.B. Mason, and M.J. Weidemann. 1990. Exercise, training and neutrophil microbicidal activity. *International Journal of Sports Medicine* 11: 179-187.

Smith, J.A., and M.J. Weidemann. 1990. The exercise and immunity paradox: A neuroendocrine/cytokine hypothesis. *Medical Science Research* 18: 749-753.

Smith, L.L. 1991. Acute inflammation: The underlying mechanism in delayed onset muscle soreness? *Medicine and Science in Sports and Exercise* 23: 542-551.

Smith, L.L., M. McCammon, S. Smith, M. Chamness, R.G. Israel, and K.F. O'Brien. 1989. White blood cell response to uphill walking and downhill jogging at similar metabolic loads. *European Journal of Applied Physiology* 58: 833-837.

Solari, R., and J.-P. Kraehenbuhl. 1985. The biosynthesis of secretory component and its role in the transepithelial transport of IgA dimer. *Immunology Today* 6: 17-20.

Soman, V.R., V.A. Koivisto, D. Diebert, P. Felig, and R.A. DeFronzo. 1979. Increased insulin sensitivity and insulin binding to monocytes after physical training. *New England Journal of Medicine* 301: 1200-1204.

Sonnenfeld, G., and E.S. Miller. 1993. The role of cytokines in immune changes induced by spaceflight. *Journal of Leukocyte Biology* 54: 253-258.

Soppi, E., P. Varjo, J. Eskola, and L.A. Laitinen. 1982. Effect of strenuous physical stress on circulating lymphocyte number and function before and after training. *Journal of Clinical Laboratory Immunology* 8: 43-46.

Spence, D.W., M.L.A. Galatino, K.A. Mossberg, and S.O. Zimmerman. 1990. Progressive resistance exercise: Effect on muscle function and anthropometry of a select AIDS population. *Archives of Physical Medicine and Rehabilitation* 71: 644-648.

Sprenger, H., C. Jacobs, M. Nain, A.M. Gressner, H. Prinz, W. Wesemann, and D. Gemsa. 1992. Enhanced release of cytokines, interleukin-2 receptors, and neopterin after long-distance running. *Clinical Immunology and Immunopathology* 53: 188-195.

Steel, C.M., J. Evans, and M.A. Smith. 1974. Physiological variation in circulating B cell:T cell ratio in man. *Nature* 247: 387-389.

Steel, D.M., and A.S. Whitehead. 1994. The major acute phase reactants: C-reactive protein, serum amyloid P component and serum amyloid A protein. *Immunology Today* 15: 81-88.

Stein, T.P., and M.D. Schluter. 1994. Excretion of IL-6 by astronauts during spaceflight. *American Journal of Physiology 266 (Endocrinology and Metabolism)*: E448-E452.

Sternfeld, B. 1992. Cancer and the protective effect of physical activity: The epidemiological evidence. *Medicine and Science in Sports and Exercise* 24: 1995-1209.

Strahan, A.R., T.D. Noakes, G. Kotzenberg, A.E. Nel, and F.C. de Beer. 1984. C reactive protein concentrations during long distance running. *British Medical Journal* 289: 1249-1251.

Suzuki, K., S. Naganuma, M. Totsuka, K.-J. Suzuki, M. Mochizuki, M. Shiraishi, S. Nakaji, and K. Sugawara. 1996. Effects of exhaustive endurance exercise and its one-week daily repetition on neutrophil count and functional status in untrained men. *International Journal of Sports Medicine* 17: 205-212.

Targan, S., L. Britvan, and F. Dorey. 1981. Activation of human NKCC by moderate exercise: Increased frequency of NK cells with enhanced capability of effector-target lytic interactions. *Clinical Experimental Immunology* 45: 352-360.

Taylor, C., G. Rogers, C. Goodman, R.D. Baynes, T.H. Rothwell, W.R. Bezwoda, F. Kramer, and J. Hattingh. 1987. Hematologic, iron-related, and acute-phase protein responses to sustained strenuous exercise. *Journal of Applied Physiology* 62: 464-469.

Taylor, G.R. 1993. Immune changes during short-duration mission. *Journal of Leukocyte Biology* 54: 202-208.

Tharp, G.D. 1991. Basketball exercise and secretory immunoglobulin A. *European Journal of Applied Physiology* 63: 312-314.

Tharp, G.D., and M.W. Barnes. 1990. Reduction of saliva immunoglobulin levels by swim training. *European Journal of Applied Physiology* 60: 61-64.

Thompson, H.J., A.M. Ronan, K.A. Ritacco, and A.R. Tagliafero. 1989. Effect of type and amount of dietary fat on the enhancement of rat mammary tumorigenesis by exercise. *Cancer Research* 49: 1904-1908.

Tidball, J.G. 1995. Inflammatory cell response to acute muscle injury. *Medicine and Science in Sports and Exercise* 27: 1022-1032.

Tipton, C.M., J.E. Greenleaf, and C.G.R. Jackson. 1996. Neuroendocrine and immune system responses with spaceflight. *Medicine and Science in Sports and Exercise* 28: 988-998.

Tomasi, T.B., and A.G. Plaut. 1985. Humoral aspects of mucosal immunity. In *Advances in Host Defense Mechanisms,* ed. J.I. Gallin and A.S. Fauci, 31-61. New York: Raven Press.

Tomasi, T.B., F.B. Trudeau, D. Czerwinski, and S. Erredge. 1982. Immune parameters in athletes before and after strenuous exercise. *Journal of Clinical Immunology* 2: 173-178.

Trinchieri, G. 1989. Biology of natural killer cells. *Advances in Immunology* 47: 187-376.

Tsuji, Y., and F.M. Torti. 1992. Tumor necrosis factor: Structure and function. In *Cytokines in Health and Disease,* ed. S.L. Kunkel and D.G. Remick, 131-150. New York: Dekker.

Tvede, N., C. Heilmann, J. Halkjaer-Kristensen, and B. Pedersen. 1989a. Mechanisms of B-lymphocyte suppression induced by acute physical exercise. *Journal of Clinical and Laboratory Immunology* 30: 169-173.

Tvede, N., M. Kappel, J. Halkjaer-Kristensen, H. Galbo, and B.K. Pedersen. 1993. The effect of light, moderate and severe bicycle exercise on lymphocyte subsets, natural and lymphokine activated killer cells, lymphocyte proliferative response and interleukin 2 production. *International Journal of Sports Medicine* 14: 275-282.

Tvede, N., M. Kappel, K. Klarlund, S. Duhn, J. Halkjaer-Kristensen, M. Kjaer, H. Galbo, and B.K. Pedersen. 1994. Evidence that the effect of bicycle exercise on blood mononuclear cell proliferative responses and subsets is mediated by epinephrine. *International Journal of Sports Medicine* 15: 100-104.

Tvede, N., B.K. Pedersen, F.R. Hansen, T. Bendix, L.D. Christensen, H. Galbo, and J. Halkjaer-Kristensen. 1989b. Effect of physical exercise on blood mononuclear cell subpopulations and in vitro proliferative responses. *Scandinavian Journal of Immunology* 29: 383-389.

Tvede, N., J. Steensberg, J. Baslund, J. Halkjaer-Kristensen, and B.K. Pedersen. 1991. Cellular immunity in highly trained elite racing cyclists during periods of training with high and low intensity. *Scandinavian Journal of Science and Medicine in Sport* 3: 163-166.

Ullum, H., P.C. Gotzsche, J. Victor, E. Dickmeiss, P. Skinhoj, and B.K. Pedersen. 1995. Defective natural immunity: An early manifestation of human immunodeficiency virus infection. *Journal of Experimental Medicine* 182: 789-799.

Ullum, H., P. Martin, M. Diamant, J. Palmo, J. Halkjaer-Kristensen, and B.K. Pedersen. 1994a. Bicycle exercise enhances plasma IL-6 but does not change IL-1α, IL-1β, or TNF-α pre-mRNA in BMNC. *Journal of Applied Physiology* 77: 93-97.

Ullum, H.P., J. Palmo, J. Halkjaer-Kristensen, M. Diamant, M. Klokker, A. Kruuse, A. LaPerriere, and B.K. Pedersen. 1994b. The effect of acute exercise on lymphocyte subsets, natural killer cells, proliferative responses, and cytokines in HIV-seropositive persons. *Journal of Acquired Immune Deficiency Syndrome* 7: 1122-1133.

Vena, J.E., S. Graham, M. Zielezny, J. Brasure, and M.K. Swanson. 1987. Occupational exercise and risk of cancer. *American Journal of Clinical Nutrition* 45: 318-327.

Vena, J.E., S. Graham, M. Zielezny, M.K. Swanson, R.E. Barnes, and J. Nolan. 1985. Lifetime occupational exercise and colon cancer. *American Journal of Epidemiology* 122: 357-365.

Verde, T.J., S.G. Thomas, P.N. Shek, and R.J. Shephard. 1993. The effects of heavy training on two in vitro assessments of cell-mediated immunity in conditioned athletes. *Clinical Journal of Sports Medicine* 3: 211-216.

Verde, T.J., S.G. Thomas, and R.J. Shephard. 1992. Potential markers of heavy training in highly trained distance runners. *British Journal of Sports Medicine* 26: 167-175.

Viti, A., M. Muscettola, L. Paulesu, V. Bocci, and A. Almi. 1985. Effect of exercise on plasma interferon levels. *Journal of Applied Physiology* 59: 426-428.

Watson, R.R., S. Moriguchi, J.C. Jackson, L. Werner, J.H. Wilmore, and B.J. Freund. 1986. Modification of cellular immune function in humans by endurance exercise training during beta-adrenergic blockade with atenolol or propranolol. *Medicine and Science in Sports and Exercise* 18: 95-100.

Weidner, T.G. 1994a. Literature review: Upper respiratory illness and sport and exercise. *International Journal of Sports Medicine* 15: 1-9.

Weidner, T.G. 1994b. Reporting behaviors and activity levels of intercollegiate athletes with an URI. *Medicine and Science in Sports and Exercise* 26: 22-26.

Weight, L.M., D. Alexander, and P. Jacobs. 1991. Strenuous exercise: Analogous to the acute-phase response? *Clinical Science* 81: 677-683.

Welliver, R.C., and P.L. Ogra. 1988. Immunology of respiratory viral infections. *Annual Review of Medicine* 39: 147-62.

Wells, C.L., J.R. Stern, and L.H. Hecht. 1982. Hematological changes following a marathon race in male and female runners. *European Journal of Applied Physiology* 48: 41-49.

Wewers, M.D. 1992. Cytokines and macrophages. In *Cytokines in Health and Disease,* ed. S.L. Kunkel and D.G. Remick, 327-351. New York: Dekker.

Wilder, R.L. 1995. Neuroendocrine-immune system interactions and autoimmunity. *Annual Review of Immunology* 13: 307-338.

Wilmore, J.H., and D.L. Costill. 1994. *Physiology of Sport and Exercise*. Champaign, IL: Human Kinetics.

Wit, B. 1984. Immunological response of regularly trained athletes. *Biology of Sport* 1: 221-235.

Wong, D.W., H.L. Thompson, Y.H. Thong, and J.R. Thornton. 1990. Effect of strenuous exercise stress on chemiluminescence response of alveolar macrophages. *Equine Veterinary Journal* 22: 33-35.

Woodruff, J.F. 1980. Viral myocarditis: A review. *American Journal of Pathology* 101: 427-479.

Woods, J.A., and J.M. Davis. 1994. Exercise, monocyte/macrophage function, and cancer. *Medicine and Science in Sports and Exercise* 26: 147-157.

Woods, J.A., J.M. Davis, M.L. Kohut, A. Ghaffar, E.P. Mayer, and R.R. Pate. 1994a. Effects of exercise on the immune response to cancer. *Medicine and Science in Sports and Exercise* 26: 1109-1115.

Woods, J.A., J.M. Davis, E.P. Mayer, A. Ghaffar, and R.R. Pate. 1993. Exercise increases inflammatory macrophage antitumor cytotoxicity. *Journal of Applied Physiology* 75: 879-886.

Woods, J.A., J.M. Davis, E.P. Mayer, A. Ghaffar, and R.R. Pate. 1994b. Effects of exercise on macrophage activation for antitumor cytotoxicity. *Journal of Applied Physiology* 76: 2177-2185.

Index

Note: Page numbers followed by *t* or *f* refer to tables or figures, respectively. Authors indexed include only the main authors of "et al." citations.

D

E

F

G

J

K

L

About the Author

Dr. Laurel T. Mackinnon has conducted research on the immune response to exercise since 1985 and is internationally recognized for her work on overtraining and immune function in athletes. She is associate professor in the Department of Human Movement Studies, The University of Queensland, Brisbane, Australia.

In 1992, Dr. Mackinnon authored *Exercise and Immunology*, the first book to explore the intriguing relationship between exercise and the immune system. She serves as associate editor of exercise and immunology for *Medicine and Science in Sport and Exercise* and is on the editorial board of *Exercise Immunology Review*. She has received grant funding for projects related to overtraining and immune function in athletes.

Dr. Mackinnon is a fellow of Sports Medicine Australia and the American College of Sports Medicine. She is a former board member of the International Society of Exercise and Immunology (ISEI) and the Australian Association for Exercise and Sports Science. In 1997, she served as program chair for the international symposium of ISEI in Germany.

Dr. Mackinnon earned her PhD in exercise science at the University of Michigan. She previously held the position of research assistant professor at the University of New Mexico School of Medicine.

Related Journal from Human Kinetics

Exercise Immunology Review
Editor: Hinnak Northoff, PhD
Frequency: Annual
First Issues: March 1995
Call for current subscription rates
ISSN: 1077-5552 • Item: JEIR

Related Books from Human Kinetics

Exercise and Immunology
[Current Issues in Exercise Science, Monograph Number 2]
Laurel T. Mackinnon, PhD
1992 • Paper • 128 pp • Item BMAC0347 • ISBN 0-87322-347-0 • $22.00 ($32.95 Canadian)

Biochemistry of Exercise X
Mark Hargreaves, PhD, and Martin Thompson, PhD, Editors
1998 • Hardcover • Approx. 352 pp • Item BHAR0758 •ISBN 0-88011-758-3

Biochemistry of Exercise IX
Ronald J. Maughan, PhD, and Susan M. Shirreffs, BSc, Editors
1996 • Hardcover • 608 pp • Item BMAU0486 •ISBN 0-88011-486-X • $69.00 ($103.50 Canadian)

Exercise Metabolism
Mark Hargreaves, PhD, Editor
1995 • Hardcover • 272 pp • Item BHAR0453 •ISBN 0-87322-453-1 • $40.00 ($59.95 Canadian)

Antioxidants and Exercise
Jan Karlsson, PhD
1996 • Hardcover • 224 pp • Item BKAR04896 •ISBN 0-87322-896-0 • $39.00 ($58.50 Canadian)

Biochemistry Primer for Exercise Science
Michael E. Houston, PhD
1995 • Paper • 144 pp • Item BHOU0577 •ISBN 0-87322-577-5 • $23.00 ($34.50 Canadian)

To request more information or to order, U.S. customers call 1-800-747-4457, e-mail us at humank@hkusa.com or visit our Web site at www.humankinetics.com. Persons outside the U.S. can contact us via our Web site or use the appropriate telephone number, postal address, or e-mail address shown in the front of this book.

Human Kinetics
The Information Leader in Physical Activity